Detection and Treatment of Emerging Contaminants in Wastewater

Detection and Treatment of Emerging Contaminants in Wastewater

Edited by

Sartaj Ahmad Bhat, Vineet Kumar,
Fusheng Li and Pradeep Verma

IWA PUBLISHING

Published by **IWA Publishing**
Unit 104–105, Export Building
1 Clove Crescent
London E14 2BA, UK
Telephone: +44 (0)20 7654 5500
Fax: +44 (0)20 7654 5555
Email: publications@iwap.co.uk
Web: www.iwapublishing.com

First published 2024

British Library Cataloguing in Publication Data
A CIP catalogue record for this book is available from the British Library

ISBN: 9781789063745 (paperback)
ISBN: 9781789063752 (eBook)
ISBN: 9781789063769 (ePub)

Doi: 10.2166/9781789063752

This eBook was made Open Access in April 2024.

Contents

The Editors

Sartaj Ahmad Bhat works as a JSPS postdoctoral researcher at the River Basin Research Center, Gifu University, Japan. He received his PhD in environmental sciences from Guru Nanak Dev University, Amritsar, India in 2017. His research interests focus on the vermicomposting treatment of various solid wastes, especially for investigations on the fate and behavior of emerging pollutants during the biological treatment of organic wastes. He has published more than 65 papers in peer-reviewed journals and edited over 15 books published by Elsevier, Springer, CRC Press, IWA, and RSC. Dr Bhat serves as an associate/academic editor and editorial board member/advisory board member of more than 15 journals published by Frontiers, Springer, Elsevier, PLOS, Wiley, Hindawi, and De Gruyter. Dr Bhat is a recipient of several prestigious awards such as the JSPS Postdoctoral Fellowship to pursue research at River Basin Research Center, Gifu University, Japan, the Basic Scientific Research Fellowship (BSR JRF, SRF) by the University Grants Commission (UGC) India, the DST-SERB National Postdoctoral Fellowship at CSIR-NEERI, Nagpur, India, and Swachhta Saarthi Fellowship by the Government of India. He has also received the 2020 Outstanding Reviewer Award by the International Journal of Environmental Research and Public Health, MDPI, and Top Peer Reviewer 2019 award in Environment and Ecology by Web of Science and has more than 750 Verified Reviews and 70 Editor Records to his credit.

Vineet Kumar works as a national postdoctoral fellow in the Department of Microbiology, School of Life Sciences at the Central University of Rajasthan, Rajasthan, India. He received his MSc (2008) and MPhil (2012) in microbiology from the Department of Microbiology at Ch. Charan Singh University, Meerut, India. Subsequently, he earned his PhD (2018) in environmental microbiology from Babasaheb Bhimrao Ambedkar (A Central) University, Lucknow, India. Dr Kumar's research work mainly focuses on wastewater treatment and solid waste management. He has published more than 50 articles in peer-reviewed international journals of repute, 24 books, and 52 book chapters, on various aspects of science and engineering, with more than 2500 citations, and h-index 30. Dr Kumar has served as a guest editor and reviewer on more than 65 prestigious

International Journals, and on the editorial board of various reputed journals. He has presented several papers relevant to his research areas at national and international conferences. He is also a recipient of various prestigious fellowships and awards, such as the Young Scientist Award, the Rajiv Gandhi National Fellowship by University Grants Commission (UGC), and National Postdoctoral Fellowship by the Science and Engineering Research Board (SERB), Government of India. He is an active member of numerous scientific societies including the Microbiology Society (UK), the Indian Science Congress Association (India), the Association of Microbiologists of India (India), etc. He is the founder of the Society for Green Environment, India (website: www.sgeindia.org).

Fusheng Li is a professor in the Division of Water System Safety and Security Studies and the Graduate School of Engineering at Gifu University, Japan. He received his BS in environmental engineering from Lanzhou Jiaotong University of China in 1986, an MS from Kitami Institute of Technology of Japan in 1994, and a PhD from the Gifu University of Japan in 1998. Dr Li is directing the Division of Water Quality Studies that covers the fields from water quality to water and wastewater treatment, and recently to resource and energy recovery from organic waste. The ongoing research projects in his lab include adsorption; membrane filtration, enhanced coagulation, disinfection; biological water and wastewater treatment; vermicomposting treatment of vegetable waste and activated sludge; microbial fuel cell; physicochemical water quality assessment; and biological water quality assessment. He has over 350 scholarly publications, including more than 200 in peer-reviewed journal papers. As principal supervisor, he has already guided 50 masters and 21 doctorate graduate students to the completion of their degrees. Dr Li is the recipient of awards from several academic societies and associations for his research work on water treatment and water quality dynamics studies.

Pradeep Verma works as a professor in the Department of Microbiology, School of Life Sciences at Central University of Rajasthan, Rajasthan, India. He is a well-rounded researcher with more than 21 years of experience in leading, supervising, and undertaking research in the broader field of bioprocess and bioenergy production from lignocellulosic waste with a focus on waste management. He earned his PhD in microbiology from Sardar Patel University, Gujarat, India in 2002. His research area of expertise involves microbial diversity, bioremediation, bioprocess development, lignocellulosic, and algal biomass-based biorefinery. He has more than 74 research articles in peer-reviewed international journals and contributed to 46 book chapters in different edited books with citations of more than 6000, and an h-index of 40. He has also edited four books by international publishers such as Springer, Taylor and Francis CRC Press, and Elsevier. He also holds 12 International patents in the field of microwave-assisted biomass pretreatment and bio-butanol production. He is a guest editor to several journals such as Biomass Conversion and Biorefinery (Springer), Frontier in Nanotechnology (Frontiers), and International Journal of Environmental Research and Public Health (MDPI). He is also an editorial board member for the Journal of Current Nanomedicine (Bentham Sciences). He is acting as a reviewer for more than 40 journals in different publication houses such as Springer, Elsevier, RSC, ACS, Nature, Frontiers, MDPI, etc. He is also a recipient of various prestigious fellowships and awards, such as the JSPS Post-Doctoral Fellowship and the Ron Cockcroft Award by the Swedish Society, UNESCO Fellow ASCR Prague. He has been awarded with Fellow of Mycological Society of India (MSI-2020), Prof. P.C. Jain Memorial Award, Mycological Society of India 2020 and Fellow of Biotech Research Society, India (2021). He is a member of various national and international societies/academies.

doi: 10.2166/9781789063752_xv

Preface

It is with great pleasure and enthusiasm that we present this comprehensive book, *Detection and Treatment of Emerging Contaminants in Wastewater*, which delves into the critical challenges posed by emerging contaminants in wastewater and explores innovative detection and treatment methods.

Wastewater management has always been a matter of paramount importance for the preservation of our green environment and the well-being of society. However, with the rampant advancements in industrial and technological sectors, new contaminants have emerged, presenting unprecedented challenges for conventional wastewater treatment processes. Emerging contaminants, such as micro- and macro plastics, pharmaceuticals, personal care products, pesticides, and industrial chemicals, possess the potential to adversely impact on aquatic ecosystems and human health.

Sustainable domestic and industrial wastewater treatment with emerging contaminants is very challenging for several reasons, including recyclability and scalability issues. Considering the need for sustainable wastewater treatment, this book will be a timely contribution that will be extremely useful in identifying and comprehensively addressing the assessment, mitigation, and treatment of emerging contaminants in wastewater and/or sludges.

This book aims to address the urgent need for effective detection and treatment strategies to mitigate the risks associated with emerging contaminants in wastewater. It brings together a multidisciplinary approach, combining the expertise of researchers and practitioners from various fields, including environmental engineering, chemistry, toxicology, and public health. Their collective knowledge and experiences have been harnessed to create a comprehensive compilation of the latest research findings, methodologies, and technological advancements in the field.

This book presents an up-to-date and comprehensive collection of chapters contributed by prominent experts in the field of wastewater working in the top institutions globally. It promotes the development of green and eco-friendly technologies for removing emerging contaminants in wastewater treatment plants. It also highlights the need for collaboration among researchers, industry stakeholders, and policymakers to develop robust regulations and guidelines that foster sustainable and efficient wastewater management practices.

The 12 chapters cover the different aspects of the detection and treatment of emerging contaminants, such as microplastics, antibiotics and antibiotic resistance genes, and pharmaceuticals and personal care products in water and wastewater. The initial chapters provide an overview of emerging contaminants, their sources, fate in the environment, and associated risks. Subsequent sections delve into the various analytical techniques employed for detection and monitoring, including chromatographic and spectroscopic methods, biosensors, and molecular techniques. Furthermore, the book explores advanced treatment processes, such as membrane filtration, advanced oxidation processes, and biological treatment, specifically tailored to address the challenges posed by emerging contaminants.

Chapter 1 discusses the behavior of microplastics in different wastewater treatment plant units, as well as their accumulation in the sludge. The chapter also examines the impact of microplastics on fauna when they enter the environment. Chapter 2 discusses the occurrence of pharmaceutically active compounds in wastewater, their potential environmental impacts, and the necessary procedures for accurately quantifying these compounds. The chapter also addresses the possibilities of using constructed wetlands and bioelectrochemical systems as sustainable methods for eliminating pharmaceutically active compounds from wastewater. Chapter 3 focuses on the main occurrence of emerging contaminants in municipal wastewater and sludge. The chapter also discusses various treatment technologies, including anaerobic digestion, aerobic composting, and advanced oxidation, for dealing with different types of emerging contaminants. Chapter 4 discusses recent technological advances in the removal of microplastics from wastewater. Chapter 5 reviews the global distribution of antibiotics in the aquatic environment, their effects on the microbial community, and the assessment of antibiotic risks. Chapter 6 covers various types of nanomaterials and nanotechnologies that are useful in wastewater treatment for remediating emerging pollutants in the environment. Chapter 7 discusses the biological and physicochemical processes used to remove emerging contaminants from wastewater and sludge. Chapter 8 covers the mechanisms by which fungi remove emerging pollutants from sludge, including biosorption, biodegradation, and enzyme production. Chapter 9 provides a brief overview of emerging contaminants, their main categories, occurrences, points of discharge, and toxicity in natural and engineered systems. Chapter 10 focuses on the sources, types, effects, monitoring, and suitable removal techniques for different pharmaceuticals and personal care products in wastewater treatment systems. Chapter 11 summarizes emerging contaminants in wastewater, their occurrence, detection, and removal efficiency using advanced analytical techniques. Finally, Chapter 12 covers new trends in the development of greener nanomaterials and evaluates their performance for the abatement of pharmaceutical compounds from wastewater.

We express our heartfelt gratitude to all the authors who have contributed their expertise and valuable insights to this book. Their dedication and commitment have made this endeavor possible. We also extend our appreciation to the International Water Association (IWA) Publishing, United Kingdom for their support and belief in the significance of this book.

Finally, we hope that this book serves as a comprehensive reference for researchers, professionals, and students who are passionate about advancing the field of wastewater management and ensuring a sustainable future. Together, let us embark on a journey to detect, understand, and effectively treat emerging contaminants in wastewater, thereby safeguarding our precious water resources and promoting a healthier environment for generations to come.

doi: 10.2166/9781789063752_0001

Chapter 1

Fate and behavior of microplastics in wastewater, accumulation in organisms and effects

Agata Egea-Corbacho[1,2], Ana Amelia Franco[1], Ana Pilar Martín-García[1], José María Quiroga[1] and María Dolores Coello[1]*

[1]Department of Environmental Technologies, Faculty of Marine and Environmental Sciences, INMAR-Marine Research Institute, CEIMAR International Campus of Excellence of the Sea, University of Cadiz, Campus Universitario de Puerto Real, 11510 Cádiz, Spain

[2]Materials and Sustainability Group, Department of Engineering, Universidad Loyola Andalucía, Avda, de las Universidades s/n, 41704 Dos Hermanas, Seville, Spain

**Corresponding author: agata.egea@uca.es*

ABSTRACT

Studies on how microplastics (MPs) behave in wastewater treatment plants (WWTPs) are increasing day by day. Although conventional WWTPs can efficiently remove MPs (64–99%), when considering the daily discharge rate, this percentage would not be sufficient. The total amount of MPs would still be discharged daily into the environment; therefore, the final effluent can act as one of the main routes of entry of MPs into aquatic environments. This chapter reviews the behavior of MPs in the different WWTP units, as well as their accumulation in the sludge. Subsequently, a discussion on how the MPs from the WWTP can reach the receiving media, such as aquatic or terrestrial media (water line and sludge line), to finally discuss how the fauna is affected by the entry of the MPs into the environment. These MPs can be ingested by aquatic life forms, leading to their bioaccumulation and biomagnification along the food chain, and causing negative effects on tissues, organs, and metabolism. MPs can also act as transport vehicles for other emerging pollutants such as pharmaceuticals and pesticides, increasing their hazardousness.

Keywords: microplastics, wastewater, sludge, biota, bioaccumulation, biotoxicity

1.1 INTRODUCTION

Plastics are synthetic polymeric materials widely used in our daily life, and due to their main characteristics, low weight, flexibility, low cost, high plasticity, and above all durability, the global consumption of plastics has increased especially in recent decades (Andrady, 2011). The use, management, and disposal of plastics has become one of the issues of greatest concern worldwide in recent years. The extensive use of this material for a multitude of applications such as construction (16%), the textile industry (9%), the manufacture of consumer household products (10%), or packaging

(31%), among others, has generated a high demand and production of plastics (Organisation for Economic Co-operation and Development [OECD], 2022). The characteristics of many of the products currently manufactured mean that they are used only once or for a very short time (single-use plastics), generating large amounts of waste that, in many cases, end up in landfills due to the impossibility of reusing or recycling them.

The OECD estimates that by 2060 the use of plastics such as high-density polyethylene (HDPE) and low-density polyethylene (LDPE), polyethylene terephthalate (PET), polypropylene (PP), or polystyrene (PS) will be more than double from 246 to 616 Mt. These polymers alone already account for 50% of all plastics used worldwide. According to a new OECD report (OECD, 2022), the world generates twice as much plastic waste as it did two decades ago, with most of it ending up in landfills, incineration, or leaking into the environment, with only 9% being successfully recycled. Accumulation of plastic waste in aquatic ecosystems is a well-known problem. The United Nations Environment Programme (UNEP) estimates that the world's oceans by 2025 will have accumulated between 100 and 250 million tons of plastic debris (Alimba & Faggio, 2019; UNEP, 2014).

By 2060, in regions such as the USA and Europe, both the use and consumption of plastics and the amount of plastic waste generated will have doubled. Currently, it is estimated that in these regions of the planet, some 305 Mt of plastics end up in natural aquatic ecosystems, with the consequent threat that this implies for life (OECD, 2022). These wastes reach the environment, where they can remain for several decades to hundreds of years because they are very stable and resistant materials. Larger plastics can fragment due to the action of natural factors such as wind, solar radiation, or tidal movement, which produce mechanical degradation and generate smaller plastic particles (Martín-García *et al.*, 2023). Particles ranging in size from 1 μm to 5 mm are known as microplastics (MPs) and have attracted the attention of the scientific community in recent years due to their high persistence in the environment, the difficulty of elimination, and the effects they can have on living organisms, which can even enter the trophic chain.

According to Waldschläger *et al.* (2020), MP enters the environment through various pathways, such as surface runoff, wind and rain, or effluent from wastewater treatment plants (WWTP). Sources of these micropollutants are plastic production activities, the construction industry, sports fields, landfills, car tires, garbage, cosmetics, washing wastewater, or fishing gear losses. As mentioned above, wastewater treatment plants are the route through which MP enters the environment, it is known that one of the largest sources of microplastics entering the environment comes from wastewater treatment plants (WWTPs) (Pittura *et al.*, 2021; Turan *et al.*, 2021). These systems are capable of removing organic matter and large plastic particles. However, they cannot remove particles smaller than 100 μm, and their influent and effluent tend to contain similar amounts of these smaller particles (Freeman *et al.*, 2020).

Although WWTPs are not designed to treat and remove MP, several authors have reported that conventional wastewater treatment methods can remove 79–99% of the MP present in the water line (Nandakumar *et al.*, 2022). However, the reality is that this fraction is not removed as such but is removed from the liquid phase but accumulates in the solid fraction of the wastewater, the sludge (Franco *et al.*, 2023). The current approach is a circular economy based on the reuse of maximum products without the need to extract new raw materials. WWTPs play a very important role in this way, as both water and sludge are reused. This reduction of waste and pollution is also carried out to achieve sustainable development goal (SDG) 6, which focuses on ensuring sustainable water management and sanitation for all, and SDG 12, which is dedicated to sustainable consumption and production, including improved waste reduction and recycling. However, it should be noted that both liquid and solid fractions of WWTPs may contain pollutants such as MPs, which are not removed from WWTPs by current water purification mechanisms.

It should be noted that the Directive on urban wastewater treatment currently in force is more than 30 years old and although the quality of European rivers, lakes, and seas has improved considerably, certain types of pollution are not covered by the current regulations, a situation that must be corrected to achieve a pollution-free environment by 2050. At present, micropollutants, such as residues of pharmaceutical and cosmetic products and MPs, are also not included (Directive 91/271/EEC). In addition, Regulation (EU) 2020/741 of the European Parliament and of the Council of 25 May 2020 concerning minimum requirements for water reuse of recent implementation also does not take into account MPs.

In this chapter, the focus will be on the problem of MPs in WWTPs, the general characteristics of these pollutants in wastewater and their distribution and fate along the different treatments commonly found in conventional WWTPs around the world. It present a review of the behavior of MPs in the different WWTP units, as well as their accumulation in the sludge. Subsequently, a discussion on how the MPs from the WWTP can reach the receiving media, such as aquatic or terrestrial media (water line and sludge line), to finally discuss how the fauna is affected by the entry of the MPs into the environment.

1.2 MICROPLASTICS IN WASTEWATER TREATMENT PLANTS

The presence of MPs in wastewater was denoted relatively a few years ago (Browne *et al.*, 2011; Zhang *et al.*, 2015), however, numerous recent studies have attempted to address the problematic of these contaminants to explain their presence, distribution, fate, removal, and characteristics in conventional wastewater and WWTPs (Bayo *et al.*, 2020; Lares *et al.*, 2018; Lee *et al.*, 2023; Monira *et al.*, 2023; Sol *et al.*, 2020; Talvitie *et al.*, 2017; Ziajahromi *et al.*, 2017).

1.2.1 Arrival of MPs at WWTPs: sources

The MPs (1 μm–5 mm) that can be found in wastewater come from a multitude of anthropogenic activities, both domestic and industrial. MPs can be further classified into primary and secondary MPs. Primary MPs are those that are industrially produced with such size for various applications: cosmetics (exfoliants, toothpastes, facial cleansing gels, shampoos and shower gels, make-up (Nawalage & Bellanthudawa, 2022), cleaning products (Anik *et al.*, 2021), abrasive materials, and so on. Although some regions of the world such as Europe have promoted regulations banning the use of MPs in cosmetics and personal care products, in many other places these materials are still used. Sun (2020) reported that up to more than 2000 MPs/g could be found in personal care products, which end up irretrievably reaching urban WWTPs.

Secondary MPs are those that come from the fragmentation and deterioration of other larger plastic materials. In both industrial and domestic environments, plastics are often exposed to aggressive agents and through tearing, friction, or other mechanical means, small fragments break up from the original material and can end up in wastewater flows. A clear example of this is textile fibers, one of the main MPs commonly found in wastewater, which come from domestic and industrial washing of clothes that are woven with synthetic plastic materials such as polyester (PES) or nylon (PA). Several studies claim that these synthetic fabrics can release from 800 to 1.3×10^7 microfibers in a single wash (Sillanpää & Sainio, 2017; Yang *et al.*, 2019), depending on the washing conditions and fabric type. Other authors have demonstrated the persistence of these microfibers when they reach natural water bodies and the negative effects, they can have on aquatic living things (Kim *et al.*, 2021, 2023; Mishra *et al.*, 2019).

Whatever their origin, a large amount of MPs end up in both industrial and urban wastewater and reach WWTPs, where they are partially removed from the water stream and accumulated in the sludge generated.

1.2.2 Presence and removal of MPs in wastewater treatment units

The goal of wastewater treatment is to effectively remove or reduce contaminants in water that represent a hazard to people and the environment if discharged into surface water and/or groundwater without appropriate treatment (Jasim & Aziz, 2020).

Whatever their origin, a large amount of MPs end up in both industrial and urban wastewater and reach WWTPs. Conventional wastewater treatment involves a combination of both physical and biological processes to remove inorganic solids, organic matter, and nutrients from wastewater. The general terms used to describe the different degrees of treatment, in increasing order, are preliminary, primary, secondary, and tertiary or advanced wastewater treatment (Janssen *et al.*, 2002).

The presence of plastics and microplastics in water bodies has become a major environmental challenge of increasing importance. A key challenge is that the available analytical techniques are relatively inadequate and prevent a thorough understanding of the fate of microplastics in water and the difficulty of comparing results between authors (Enfrin *et al.*, 2019). The occurrence of microplastics in wastewater treatment plants raises concern about the quality of treated water and the reception of treated water into the environment.

One of the main problems in comparing the effectiveness of WWTPs in removing MPs is that not only are the processes installed in each of the WWTPs different, but there is no standardized method for analyzing MPs. Some examples of treatments used for the extraction of MPs are shown in Table 1.1, these treatments range from only a filtration, an enzymatic treatment, and alkaline digestion (KOH 2 M), an advanced oxidation only (H_2O_2 30%; Wet peroxide oxidation (WPO)), advanced oxidation with a density separation (Fenton oxidation, density separation, Wet peroxide oxidation (WPO), density separation) being this density separation $ZnCl_2$ solution (1.8 g/L), $ZnCl_2$ solution (1.5 g/cm^3), NaCl 5 M and LMT solution (1.62 g/mL). Even some authors like Dronjak *et al.* (2023) used a combination of all the treatments, Fenton oxidation, alkaline digestion (KOH 2 M), enzymatic digestion (2–3 days) + density separation with $ZnCl_2$ solution (1.8 g/L). These differences in treatment for MP extraction make it difficult to compare the effectiveness and quality of WWTPs. Therefore, a common method for MP analysis should be standardized.

Another difficulty encountered in the comparison of MPs in WWTPs lies in the size of MPs under study, Table 1.1 shows the papers that study MPs up to 10 μm (Liu *et al.*, 2020; Mintening *et al.*, 2017) to up to 200 μm (Le *et al.*, 2023).

Table 1.1 shows the MPs removal efficiency of the 14 documents under study. These documents showed a removal efficiency ranging from 68.8 to 99.9%. The WWTP that showed the worst, and the best removal of MPs was Southern and Central Vietnam, with a treatment process costing Coarse screen, grit chamber, AS/sequencing batch reactor with UV/trickling filters, and aerated lagoons (Le *et al.*, 2023). The extraction of microplastics used by the authors was enzymatic treatment. The concentration of MPs at the inlet of WWTPs varies from 0.92 MPs/L in Hvidovre, Denmark (Liu *et al.*, 2020) to 1058 in Barcelona, Spain (Dronjak *et al.*, 2023). In effluent, the variations between different studies range from 0.01 MPs/L in Oldenburg, Germany (Mintening *et al.*, 2017) to 73.25 MPs/L in Turkey (Akdemir & Gedik, 2023). According to the polymer types identified, PP is the only polymer reported in all studies, followed by PE which is reported in all the documents except in Vancouver and Canada (Gies *et al.*, 2018). The remaining polymers identified in most studies are PS, PVC, PET, PA, and PES.

1.2.3 Presence and accumulation of MPs in sewage sludge

As discussed previously, the presence of MPs in wastewater has been widely analyzed, reporting that these pollutants are highly removed during the treatment on the water line, although there are no specific treatments for the removal of MPs in these facilities, the MPs are eliminated from the water line, being retained, and accumulated in the sewage sludge line (Gies *et al.*, 2018). The sewage sludge is the main by-product generated at WWTPs, it can be classified into primary sludge, secondary sludge, and mixed sludge, depending on the operational step of the WWTP in which they are produced (Casella *et al.*, 2023). The sewage sludge contains solids from the mechanical treatment primary setting tank, extracellular polymeric substances (lipids, nucleic acids, proteins, polysaccharides, bacteria, or microorganisms) from the biological treatment or secondary tank settling tank and water (Melo *et al.*, 2022); however, the characteristics of each sludge depend on several variables such as; the source,

Table 1.1 WWTP location, treatment process, microplastics extraction, influent and effluent concentration, size, removal efficiency, polymer identified, and reference.

WWTP Location	Treatment Process	Microplastic Extraction	Influent Concentration (MPs/L)	Effluent Concentration (MPs/L)	MPs size (μm)	MPs removal Efficiency (%)	Polymers Identified	References
Gumi, South Korea	Grit removal, A^2O, two-stage sedimentation, rapid sand filtration, UV treatment	Fenton oxidation, density separation with LMT solution (1.62 g/mL)	102–266	0.05–0.56	Up to 20	>99	PP, PE, acrlylic, PES/PET PS, PA, SBR, PVC, PU	Kim *et al.* (2022)
Beijing, China	Aerated grit chamber, primary sedimentation, A^2O, secondary sedimentation, denitrification, ultra-filtration, ozonation, UV	Wet peroxide oxidation (WPO) + density separation with $ZnCl_2$ solution (1.5 g/cm^3)	12.03	0.59	Up to 50	95.16	PET, PES, PP, PE	Yang *et al.* (2019)
Cádiz, Spain	Coarse and fine screening, degritting and degreasing, EA biological reactor, secondary sedimentation, intermittent sand filtration	Wet peroxide oxidation (WPO) + density separation with NaCl 5 M	185.4–897.6	0.3–2.4	Up to 100	99.7	PE, PES, PET, PP	Martín-García *et al.* (2022)
Vancouver, Canada	Vertical screening bars, primary clarification, trickling filters, solids contact tank, secondary clarification, chlorination	H_2O_2 30% (7 days)	31.1	0.5	Up to 63	99	PS, PES, PA, PP	Gies *et al.* (2018)
Glasgow, Scotland	Coarse and fine screening, grit and grease removal, primary settling, aeration, secondary clarification	Filtration	15.7	0.25	Up to 65	98.4	Acrylic, PET, PA, PES, PE, PS, PU, PVC, PP	Murphy *et al.* (2016)

(Continued)

Table 1.1 WWTP location, treatment process, microplastics extraction, influent and effluent concentration, size, removal efficiency, polymer identified, and reference (*Continued*).

WWTP Location	Treatment Process	Microplastic Extraction	Influent Concentration (MPs/L)	Effluent Concentration (MPs/L)	MPs size (μm)	MPs removal Efficiency (%)	Polymers Identified	References
Karmiel, Israel	Bar screens, grit removal, primary clarification, biological nutrient removal tank, activated sludge, secondary clarification, filtration, chlorination	Fenton oxidation + density separation with NaCl 5 M	65–130	1.97	Up to 0.45	97	PE, PVC, PP, PC, PTFE, PO, PS, PU, PA	Ben-David *et al.* (2021)
Mikkeli, Finland	Screening, grit separation, primary clarification, activated sludge, secondary clarification, disinfection, MBR pilot plant	Wet peroxide oxidation (WPO)	57.6	0.4–1	Up to 250	98.3–99.4	PES, PA, PE, PP	Lares *et al.* (2018)
Hvidovre, Denmark	Biological nutrient removal, secondary clarification, biofilter (pilot-scale)	H_2O_2 50% (2 days), enzyme treatment (6 days), fenton oxidation + density separation with $ZnCl_2$ solution (1.8 g/cm^3)	0.92	0.1	Up to 10	87.9–95.6	PE, PP, PVC, PES, PS, acrylic, PA	Liu *et al.* (2020)
Barcelona, Spain	Sieving, grit remover, degreaser, primary clarification, biological reactor, secondary clarification	Fenton oxidation, alkaline digestion (KOH 2 M), enzymatic digestion (2–3 days) + density separation with $ZnCl_2$ solution (1.8 g/L)	369–1058	13–26	Up to 20	96–98	PE, PP, PVC, PE-PP, PA, PAN, PES	Dronjak *et al.* (2023)

(*Continued*)

Viikinmäki, Finland	Primary clarification, CAS, denitrification, discfilter (pilot-scale)/RSF/DAF/MBR	–	–	0.005–0.3	Up to 20	95 (DAF) – 99.9 (MBR)	PES, PE, PP, PS, PU, PVC, PA, acrylamide, polyacrylate, EVA, PPO	Talvitie *et al.* (2017)
Oldenburg, Germany	Primary treatment (skimming tank), secondary treatment (nutrient removal), tertiary treatment (maturation ponds)	Enzymatic treatment	–	0.01–9	Up to 10	93–98	PP, PE, PA, PVC, PS, PU, EVA, ABS, PLA, PEST	Mintening *et al.* (2017)
Southern and central Vietnam	Coarse screen, grit chamber, AS/sequencing batch reactor with UV/trickling filters and aerated lagoons	Enzymatic treatment	1.86–125	0.14–0.81	Up to 200	68.8–99.9	PE, PP, PES, acrylic	Le *et al.* (2023)
USA	Grit removal, primary clarification, AS/trickling filters, chlorination/sand filtration/anMBR (pilot-scale)	Sieving	133	5.9	Up to 20	93.8–99.4	–	Michielssen *et al.* (2016)
Turkey	Primary + secondary treatment (not specified)	H_2O_2 30% (3–4 days)	–	20.5–73.25	Up to 45	–	PET, PP, PA, PE, PS, PVC	Akdemir and Gedik (2023)

the volume to treat, the generation of waste by the population equivalent, and the type of wastewater treated (van Haandel & van der Lubbe, 2019).

Sewage sludge has become a valuable resource for the energy and agriculture sector, as it can be used for bioenergy production and/or as fertilizer for soil amendment following the circular economy principles of decreasing the generation of waste and encouraging the reuse of a product previously considered a waste. According to the Sewage Sludge Directive 86/278/EEC of 12 June 1986, the treated sludge must go through biological, chemical, heat, or any other process to obtain a final product that does not present any threat to the environment to be reused in agriculture (Helmecke *et al.*, 2020). The sludge treatment processes include thickening, stabilization, dewatering, and thermal drying (Mahon *et al.*, 2017).

The presence of MPs in sludge has been addressed worldwide, these pollutants have been encountered in sludge during all the processes taking place in the sewage sludge (Hassan *et al.*, 2023), therefore a review of the presence, accumulation, effects, and fate of the MPs along the sewage sludge line is covered below.

Table 1.2 shows the documents submitted for study on WWTP location, sampling collection, the type of pretreatment for the analysis of microplastics, the type of sludge, the concentration of MPs, the shape, size, and types of polymers identified. As with the water line, the lack of a standardized method for the extraction of MPs as well as different types of WWTP treatments makes it difficult to compare the different studies. Studies show that the predominant form is fibers, present in all studies where they have been considered, followed by fragments or particles. Regarding the size, the particles analyzed range from up to 20 μm (Liu *et al.*, 2020; Menendez-Manjon *et al.*, 2022; Mintening *et al.*, 2017) to 5 mm (Gies *et al.*, 2018).

Considering the type of sludge, it can be observed that primary sludge in most of the articles studied presents a higher concentration of MPs (0.23–24.6 MPs/g DW), compared to secondary sludge (0.05–23 MPs/g DW). The dewatered presents a higher amount of MPs, reaching 240.3 MPs/g DW (Liu *et al.*, 2020). The polymers present in most of the samples are as in the water line PP and PE, as well as PA.

1.3 CIRCULAR ECONOMY, REGENERATED WATER, AND SLUDGE AS SOIL AMENDMENT: ENVIRONMENTAL ISSUES

Nowadays, the main challenge for both industrial and municipal utilities is related to the increase of environmental protection standards and recommendations, which have been included in the circular economy package (EC).

The European Union Regulation (EU) 2020/741 concerning minimum requirements for water reuse states in its Annex II, point A) Main elements of risk management, paragraph 6, that consideration should be given to water quality requirements and their control that are additional to or more stringent than those specified in Annex I, Section 2, or both, where necessary and appropriate to ensure adequate protection of the environment and human and animal health, especially if there is scientific evidence that the risk is from reclaimed water and not from other sources, in particular paragraph e takes into account other substances of emerging concern, such as micropollutants and microplastics.

In the case of a CE 'Wastewater Treatment Plant of the Future', the recovery of water, energy, and raw materials from available waste streams is strongly recommended. The implementation of CE solutions in the analyzed facilities is incorporated into many strategies and policy frameworks, such as national and international (including European) documents (Smol, 2023). Despite this point, it does not strictly state that MPs must be removed from water and in what percentage. WWTPs are a pathway for the entry of MPs into both terrestrial and marine environments, so that terrestrial and aquatic organisms may be exposed. Water reuse, whether from urban, agricultural, industrial, or even environmental uses, can be an important source of MP input to the environment that must be taken into account. Many actions are currently being taken to upgrade and build WWTPs that can respond

Table 1.2 WWTP location, sampling collection, microplastics extraction, type of sludge, MPs concentration, shape, size, MPs identified, and reference.

WWTP Location	Sampling Collection	Microplastic Extraction	Type of Sludge	MP Concentration (MPs/g DW)	Shape	Size (μm)	MPs Identified	References
Cádiz, Spain	Glass jars and screening though metal sieve	WPO ($H_2O_2 + FeSO_4$) digestion, UTS treatment, and density separation (NaCl)	Primary	6. 58–20.40	Fibers and fragments (87%)	Up to 100	Acrylate, PMMA, PE, PP	Franco *et al.* (2023)
			Secondary	0.98–1.89				
			Digested–anaerobic digestion	0.02*–57.18				
Australia	Glass Jars	PLE extraction, and Pyr-GC/MS	Digested–composted	0.4–23.5 mg MPs/g DW	n.d.	n.d.	PE, PVC, PE, PP	Okoffo *et al.* (2020)
Chengdu, China	Stainless-steel shovel and stored aluminum bag	WPO ($H_2O_2 + FeSO_4$) digestion, density separation (NaCl)	Primary	0.23–0.75				
			Secondary	0.05–0.06	Particles, debris, and fibers	Up to 500	PP, PE, PVC, and PS	Wei *et al.* (2022)
			Digested–composting	0.04–0.11				
Falconara Marittina, Italy	Steel sieves	H_2O_2 digestion, density separation (NaBr)	Primary	1.67	Fragments (>70%)	Up to 63	PE, PP	Pittura *et al.* (2021)
			Secondary–WAS	5.3				
			Digested – dewatered	4.74				
Vancouver, Canada	Glass jar	Oil extraction protocol + H_2O digestion	Primary	14.9	Fibers	64–5000	PS, Modified cellulose	Gies *et al.* (2018)
			Secondary	4.4				
Beijing, China	Glass jars	Density separation ($ZnCl_2$)	Digested–dehydrated	2.93–5.33	Pellets and microbeads	Up to 50	PBA, PE, PA, PP, rayon, and PET	Xu *et al.* (2020a)
Newport United Kingdom	Glass bottles	WPO ($H_2O_2 + FeSO_4$) digestion and density separation ($ZnCl_2$)	Primary	24.6	–	Up to 1000	–	Lofty *et al.* (2022)

(*Continued*)

Table 1.2 WWTP location, sampling collection, microplastics extraction, type of sludge, MPs concentration, shape, size, MPs identified, and reference (*Continued*).

WWTP Location	Sampling Collection	Microplastic Extraction	Type of Sludge	MP Concentration (MPs/g DW)	Shape	Size (μm)	MPs Identified	References
Speyer, Germany	Aluminum tins	H_2O digestion and density separation (SPT)	Digested sludge–compost	97.66	Fibers	Up to 100	PES, PO	Tagg *et al.* (2022)
Mikkeli, Finland	Glass flask	WPO (H_2O_2+$FeSO_4$) digestion and density separation ($ZnCl_2$)	Secondary – activated sludge	23.0	Fibers	Up to 63	PES, PE, PA	Lares *et al.* (2018)
			Digested	170.9				
			MBR	27.3				
Leganés, Spain	Steel mesh	H2O digestion + density separation (NaCl)	Digested sludge–anaerobic	133	Fibers and fragments	36–4720	PE, PP, Acrylic	Edo *et al.* (2020)
			Dewatering–soil amendment	101				
Wuhan, China	Glass beaker	Density separation	Dewatered	240.3	Fibers and fragments	Up to 20	PA	Liu *et al.* (2020)
Murcia, Spain	Glass jars and screening though metal sieve	WPO (H_2O_2+ $FeSO_4$) digestion and density separation (ZnCl2)	Dewatered	12–39	Fragments and fibres	Up to 20	PET, PS, PA, PVC	Menendez-Manjon *et al.* (2022)
Olderburg, Germany	Stainless steel shovel and stored in PVC container	Alkaline treatment	Primary	1–24	Fragments* (fibers not analysed)	Up to 20	PE, PP, PA, PS	Mintening *et al.* (2017)
Devon, United Kingdom	Stainless-steel bucket	WPO (H_2O_2+ $FeSO_4$) digestion and density separation (ZnCl2)	Mixed sludge–primary and secondary	107.5	Fibers and particle	Up to 50	PES, PVA, PE.	Harley-Nyang *et al.* (2022)
			Digested–anaerobic digestion	97.2				
			Digested – lime stabilization	37.7				

to current and future challenges related to environmental protection and should consider the removal of MPs from both the water line and the sludge line.

Microplastic pollution emerged from the oceans, but it is estimated that soil receives 4–25 times more plastic debris annually compared to the marine environment. Lofty *et al.* (2022) indicated that European agricultural soils are contaminated with among 31 000 and 42 000 tons (considering MPs 1000–5000 μm in size) or 8.6×10^{13}–7.1×10^{14} MPs particles (considering MPs 25–5000 μm in size). The origin of these pollutants is fertilizers reused from wastewater, a practice aimed precisely at saving raw materials and favoring the circular economy.

The application of sewage sludge on agricultural land has been an acceptable practice until the increasing presence of MPs appeared as a new environmental threat to terrestrial ecosystems through the deposition of MPs on agricultural soils. In Canada, concentrations of up to 541 particles MPs/kg of soil were found in agricultural soils where sewage sludge had been applied, compared to 4 particles MPs/kg in control soils where no sewage sludge had been applied, meaning that land application of sewage sludge is contaminating agricultural soil with MPs (Crossman *et al.*, 2020).

According to these results, it is estimated that wastewater treatment plants in countries such as Denmark, Sweden, or Norway apply between 63 000 and 430 000 tons MPs/year to agricultural soils (Nizzetto *et al.*, 2016).

Based on average application rates of sewage sludge on agricultural soils, it is estimated that the mass of MPs reaches values between 31 000 and 42 000 tons/year for microplastics size between 1000 and 5000 μm. To face this problem, it is necessary to understand the transport of MPs in the environment, the processes of removing MP in sewage treatment plants, the concentration of MPs in the sewage sludge and to provide more information on the MP balance and the contamination of agricultural soils. The removal of MPs from the sludge line of a WWTP is paramount to the use of soil amendment or fertilizers from sewage sludge, as we would be adding a MPs contamination problem to the terrestrial environment.

1.4 ACCUMULATION OF MICROPLASTICS IN ORGANISMS AND EFFECTS

WWTPs are a pathway for the entry of MPs into both terrestrial and marine environments, so terrestrial and aquatic organisms may be exposed. Ingestion is the main interaction between organisms and microplastics, possibly due to confusion with food, although adsorption processes may also occur (Dovidat *et al.*, 2020; Ribeiro *et al.*, 2019). The potential of microplastics to cause harm to marine organisms has been widely documented and leads to the following adverse effects: reduced feeding rate, reduced predatory performance, physical damage, induction of oxidative stress, effects on reproduction, decreased neurofunctional activity, oxidative damage, development of pathologies, mortality, among others (de Sá *et al.*, 2018).

In conjunction with this, the large specific surface area of microplastics facilitates their adsorption of organic contaminants, as demonstrated for polybrominated diphenyl ethers (PBDEs), polycyclic aromatic hydrocarbons (PAHs), polychlorinated biphenyls (PCBs), endocrine disruptors, and heavy metals. This ability of microplastics to serve as vectors may result in bioaccumulation of these and potentially other contaminants (Xu *et al.*, 2020b).

Table 1.3 shows the accumulation of microplastics in organisms and effects: Specie analyzed, type of polymer, characteristics of the MPs, effects, scale study, and reference. The difficulty of analyzing MPs in the field, in uncontaminated individuals and with specific concentrations or types of polymers, means that to better understand the toxic effects, this must be done in a controlled laboratory environment. PE, PS, PET, and PVC are the most common polymers used for the studies because of their composition or abundance. Some of the effects caused by MPs are delayed germination due to their accumulation in the seed case in vegetables, growth, and reproductive stunting or even inhibition (Schöpfer *et al.*, 2020; Zhang *et al.*, 2023a; Zhong *et al.*, 2021), increased bioaccumulation of ATZ in species, especially in aged species (Song *et al.*, 2023), and transport of MPs through the vascular pool to the vapor and leaves (Li *et al.*, 2020).

Table 1.3 Accumulation of microplastics in organisms and effects. Specie analyzed, type of polymer, characteristics of the MPs, effects, scale study and reference.

Specie Analyzed	Polymer	Characteristics of the MPs	Effects	Scale Study	References
Eisenia Fetida *Metaphire Guillelmi*	PE PE-aged	Shape: Spheres: Fragments Size: - MPs concentration: 0.2 g	– Increase the bioaccumulation of ATZ in both species, especially in aged-PE	Lab	Song *et al.* (2023)
Bufo gargarizans (Tadpoles)	PS	Shape: spheres Size: 1 and 10 μm MPs concentration: 2.5% (w/v)	– Growth and development of tadpoles delayed – 10 μm MPs bioaccumulated in the digestive tract: altered gut biota changing the homeostasis. – 1 μm MPs affected host tissues: altered the cellular response and neural functions, upregulated protein synthesis and mitochondrial energy	Lab	Zhang *et al.* (2023b)
Eisenia Fetida	PE	Shape: - Size: 100–200 μm MPs concentration: 2000, 50 000 and 200 000 particles/kg sludge	– MPs size decrease in earthworm casts – Oxidative stress and neurotoxicity damage at high concentrations of MPs	Lab	Zhong *et al.* (2021)
Folsomia candida	PVC	Shape: - Size: 80–250 μm MPs concentration: 1 g/kg soil	– Reduction in growth and reproduction – Increased bacterial diversity and altered gut microbiota – Enhanced isotopic composition ($\delta^{15}N$ and $\delta^{13}C$) values of collembolan tissues	Lab	Zhu *et al.* (2018)
Achatina fulica	PET	Shape: Fibers Size: 76.3–1257 μm MPs concentration: 0.014, 0.14, 0.71 g/kg soil	– Reduced food intake and excretion – Villi damage in the gastrointestinal walls – Oxidative stress in the individuals of the 0.71 g/kg soil	Lab	Song *et al.* (2019)
Eisenia Fetida	PE PLA	Shape: - Size: 70–250 μm MPs concentration: 0.5%, 1%, 2%, 5%, 7%, 14% (w/w)	– SOD, CAT, POD, GST, and AchE activities showed an inhibition-stimulation pattern. – Indicating neurotoxicity and oxidative stress – Gut microbiota variation decreased – No difference between PLA and PE on day 28	Lab	Yu *et al.* (2022)
Caenorhabditis elegans	LDPE PLA PBAT	Shape: fragments Size: LDPE 57 ± 40 μm PLA/PBAT 40 ± 31 μm MPs concentration: 1, 10, 100 mg/L	– Reproduction inhibition	Lab	Schöpfer *et al.* (2020)

(*Continued*)

Triticum aestivum L. *Lactuca sativa L.*	PS PMMA	Shape: spheres Size: PS 0.21 ± 0.05 μm–1.93 μm ± 0.09 μm PMMA: 0.18 ± 0.05 μm–2 ± 0.1 μm MPs concentration: 0.5, 5 and 50 mg/L	– Transportation of MPs through vascular assembly to steam and leaves	Lab	Li *et al.* (2020)
Daucus carota	PES, PA, PP, LDPE, PET, PU, PS, PC	Shape: Fibers, films, foams, fragments Size: Fibers up to 5 mm Films, foams and fragments up to 5 mm²	– Shoot and root mass increased in presence of MPs – MPs negatively influenced soil aggregation and microbial activity	Lab	Lozano *et al.* (2021)
Spirodela polyrhiza	PS	Shape: Spheres Size: 50–500 nm MPs concentration: 10^2 to 10^6 particles/mL	– External adsorption of MPs to the roots – No significant effects on growth or chlorophyll concentrations	Lab	Dovidat *et al.* (2020)
Glycine max L. Merrill	PS	Shape: - Size: 100 nm–100 μm MPs concentration: 10 mg/kg	– MPs damaged the root and inhibited its activity, decreasing the abundance of microbial in rhizosphere – Genotoxicity in presence of MPs was detected – MPs enhanced the toxic effects of polycyclic aromatic hydrocarbons such as phenanthrene	Lab	Xu *et al.* (2021)
Oryza sativa	PS	Shape: - Size: 20 nm MPs concentration: 10–100 mg/L	– MPs decreased root length and weight – PS induced oxidative stress and damage in rice roots	Lab	Zhou *et al.* (2021)
Cucumis sativus L.	PE, PA, PLA	Shape: - Size: 13–500 μm MPs concentration: 40–1000 mg/L	– PE MPs cases higher Cr (VI) accumulation and phytotoxicity than PA and PLA – MPs type affect negatively plant growth and chlorophyll content	Lab	Zhang *et al.* (2023a, 2023b)
Lepidum sativum	–	Shape: - Size: 50–4800 nm MPs concentration: 10^3–10 particles/mL	– MPs cause late germination due to accumulation on seed case – MPs cause significant impact on root growth	Lab	Bosker *et al.* (2019)

REFERENCES

Akdemir T. and Gedik K. (2023). Microplastic emission trends in Turkish primary and secondary municipal wastewater treatment plant effluents discharged into the Sea of Marmara and Black Sea. *Environmental Research*, **231**(Part 2), 116188, https://doi.org/10.1016/j.envres.2023.116188

Alimba G. G. and Faggio C. (2019). Microplastics in the marine environment: current trends in environmental pollution and mechanisms of toxicological profile. *Environmental Toxicology and Pharmacology*, **68**, 61–74, https://doi.org/10.1016/j.etap.2019.03.001

Andrady A. L. (2011). Microplastics in the marine environment. *Marine Pollution Bulletin*, **62**(8), 1596–1605, https://doi.org/10.1016/j.marpolbul.2011.05.030

Anik A. H., Hossain S., Alam M., Sultan M. B., Hasnine M. D. T. and Rahman M. M. (2021). Microplastics pollution: a comprehensive review on the sources, fates, effects, and potential remediation. *Environmental Nanotechnology, Monitoring & Management*, **16**, 100530, https://doi.org/10.1016/j.enmm.2021.100530

Bayo J., López-Castellanos J. and Olmos S. (2020). Membrane bioreactor and rapid sand filtration for the removal of microplastics in an urban wastewater treatment plant. *Marine Pollution Bulletin*, **156**, 111211, https://doi.org/10.1016/j.marpolbul.2020.111211

Ben-David E. A., Habibi M., Haddad E., Hasanin M., Angel D. L., Booth A. M. and Sabbah I. (2021). Microplastic distribution in a domestic wastewater treatment plant: removal efficiency, seasonal variation and influence on sampling technique. *Science of the Total Environment*, **752**, 141880, https://doi.org/10.1016/j.scitotenv.2020.141880

Bosker T., Bouwman L. J., Brun N. R., Behrens P. and Vijver M. G. (2019). Microplastics accumulate on pores in seed capsule and delay germination and root growth of the terrestrial vascular plant *Lepidium sativum*. *Chemosphere*, **226**, 774–781, https://doi.org/10.1016/j.chemosphere.2019.03.163

Browne M. A., Crump P., Niven S. J., Teuten E., Tonkin A., Galloway T. and Thompson R. (2011). Accumulation of microplastic on shorelines worldwide: sources and sinks. *Environmental Science & Technology*, **45**, 9175–9179, https://doi.org/10.1021/es201811s

Casella C., Sol D., Laca A. and Díaz M. (2023). Microplastics in sewage sludge: a review. *Environmental Science and Pollution Research*, **30**, 63382–63415, https://doi.org/10.1007/s11356-023-27151-6

Council Directive. Council Directive 91/271/EEC of 21 May 1991 concerning urban waste-water treatment.

Council Directive. Council Directive 86/278/EEC of 12 June 1986 on the protection of the environment, and in particular of the soil, when sewage sludge is used in agriculture.

Crossman J., Hurley R. R., Futter M. and Nizzetto L. (2020). Transfer and transport of microplastics from biosolids to agricultural soils and the wider environment. *Science of the Total Environment*, **724**, 138334, https://doi.org/10.1016/j.scitotenv.2020.138334

de Sá L. C., Oliveira M., Ribeiro F., Lopes Rocha T. and Futter T. N. (2018). Studies of the effects of microplastics on aquatic organisms: what do we know and where should we focus our efforts in the future?. *Science of the Total Environment*, **645**, 1029–1039, https://doi.org/10.1016/j.scitotenv.2018.07.207

Dovidat L. C., Brinkmann B. W., Vijver M. G. and Bosker T. (2020). Plastic particles adsorb to the roots of freshwater vascular plant *Spirodela polyrhiza* but do not impair growth. *Limnology and Oceanography*, **5**(1), 37–45, https://doi.org/10.1002/lol2.10118

Dronjak L., Exposito N., Serra J., Schuhmacher M., Florencio K., Corzo B. and Rovira J. (2023). Tracing the fate of microplastics in wastewater treatment plant: a multi-stage analysis of treatment units and sludge. *Environmental Pollution*, **333**, 122072, https://doi.org/10.1016/j.envpol.2023.122072

Edo C., González-Pleiter M., Leganés F., Fernández-Piñas F. and Rosal R. (2020). Fate of microplastics in wastewater treatment plants and their environmental dispersion with effluent and sludge. *Environmental Pollution*, **259**, 113837, https://doi.org/10.1016/j.envpol.2019.113837

Enfrin M., Dumée L. F. and Lee J. (2019). Nano/microplastics in water and wastewater treatment processes – origin, impact and potential solutions. *Water Research*, **161**, 621–638, https://doi.org/10.1016/j.watres.2019.06.049

Franco A. A., Martín-García A. P., Egea-Corbacho A., Arellano J. M., Albendín G., Rodríguez-Barroso R., Quiroga J. M. and Coello M. D. (2023). Assessment and accumulation of microplastics in sewage sludge at wastewater treatment plants located in Cádiz, Spain. *Environmental Pollution*, **317**, 120689, https://doi.org/10.1016/j.envpol.2022.120689

Freeman S., Booth A. M., Sabbah I., Tiller R., Dierking J., Klun K., Rotter A., Ben-David E., Javidpour J. and Angel D. L. (2020). Between source and sea: the role of wastewater treatment in reducing marine microplastics. *Journal of Environmental Management*, **266**, 110642, https://doi.org/10.1016/j.jenvman.2020.110642

Gies E. A., LeNoble J. L., Noël M., Etemadifar A., Bishay F., Hall E. R. and Ross P. S. (2018). Retention of microplastics in a major secondary wastewater treatment plant in Vancouver, Canada. *Marine Pollution Bulletin*, **133**, 553–561, https://doi.org/10.1016/j.marpolbul.2018.06.006

Harley-Nyang Memon F. A., Jones N. and Galloway T. (2022). Investigation and analysis of microplastics in sewage sludge and biosolids: a case study from one wastewater treatment works in the UK. *The Science of the Total Environment*, **823**, 153735–153735, https://doi.org/10.1016/j.scitotenv.2022.153735

Hassan F., Prasetya K. D., Hanun J. N., Bui HM. ., Rajendran S., Kataria N., Khoo S. S., Wang Y. F., You S. J. and Jiang J. J. (2023). Microplastic contamination in sewage sludge: abundance, characteristics, and impacts on the environment and human health. *Environmental Technology & Innovation*, **31**, 103176, https://doi.org/10.1016/j.eti.2023.103176

Helmecke M., Fries E. and Schulte C. (2020). Regulating water reuse for agricultural irrigation: risks related to organic micro-contaminants. *Environmental Sciences Europe*, **32**, 4, https://doi.org/10.1186/s12302-019-0283-0

Janssen P. M. J., Meinema K. and van der Roest H. F. (eds) (2002). Biological Phosphorus Removal: Design and Operation Manual. STOWA, London, IWA.

Jasim N. A. and Aziz H. A. (eds) (2020). The design for wastewater treatment plant (WWTP) with GPS X modelling. *Cogent Engineering*, **7**, 1, https://doi.org/10.1080/23311916.2020.1723782

Kim D., Kim H. and An Y. J. (2021). Effects of synthetic and natural microfibers on *Daphnia magna* – are they dependent on microfiber type? *Aquatic Toxicology*, **240**, 105968, https://doi.org/10.1016/j.aquatox.2021.105968

Kim M.-J., Na S.-H., Batool R., Byun I.-S. and Kim E.-J. (2022). Seasonal variation and spatial distribution of microplastics in tertiary wastewater treatment plant in South Korea. *Journal of Hazardous Materials*, **438**, 129474, https://doi.org/10.1016/j.jhazmat.2022.129474

Kim M. J., Kim J. A., Song J. A., Kho K. E. and Choi C. Y. (2023). Synthetic microfiber exposure negatively affects reproductive parameters in male medaka (*Oryzias latipes*). *General and Comparative Endocrinology*, **334**, 114216, https://doi.org/10.1016/j.ygcen.2023.114216

Lares M., Ncibi M. C., Sillanpää M. and Sillanpää M. (2018). Occurrence, identification and removal of microplastic particles and fibers in conventional activated sludge process and advanced MBR technology. *Water Research*, **133**, 236–246, https://doi.org/10.1016/j.watres.2018.01.049

Lee J. H., Kim M. J., Kim C. S., Cheon S. J., Choi K. I., Kim J., Jung J., Yoon J. K., Lee S. H. and Jeong D. H. (2023). Detection of microplastic traces in four different types of municipal wastewater treatment plants through FT-IR and TED-GC-MS. *Environmental Pollution*, **333**, 122017, https://doi.org/10.1016/j.envpol.2023.122017

Le T.-M.-T., Truong T.-N.-S., Nguyen P.-D., Le Q.-D.-T., Tran Q.-V., Le T.-T., Nguyen Q.-H., Kieu-Le T.-C. and Strady E. (2023). Evaluation of microplastics removal efficiency of wastewater-treatment plants in developing country, Vietnam. *Environmental Technology & Innovation*, **29**, 102994, https://doi.org/10.1016/j.eti.2022.102994

Li L., Luo Y., Li R., Zhou Q., Peijnenburg W. J. G. M., Yin N., Yang J., Tu C. and Li Y. Z. (2020). Effective uptake of submicrometre plastics by crop plants via a crack-entry mode. *Nature Sustainability*, **3**, 929–937, https://doi.org/10.1038/s41893-020-0567-9

Liu F., Nord N. B., Bester K. and Vollertsen J. (2020). Microplastics removal from treated wastewater by a biofilter. *Water*, **12**, 1085, https://doi.org/10.3390/w12041085

Lofty J., Muhawenimana V., Wilson C. A. M. E. and Ouro P. (2022). Microplastics removal from a primary settler tank in a wastewater treatment plant and estimations of contamination onto European agricultural land via sewage sludge recycling. *Environmental Pollution*, **304**, 119198, https://doi.org/10.1016/j.envpol.2022.119198

Lozano Y. M., Lehnert T., Linck L. T., Lehmann A. and Rillig M. C. (2021). Microplastic shape, polymer type, and concentration affect soil properties and plant biomass. *Frontiers in Plant Science*, **12**, 616645, https://www.frontiersin.org/articles/10.3389/fpls.2021.616645

Mahon A. M., O'Connell B., Healy M. G., O'Connor I., Officer R., Nash R. and Morrison L. (2017). Microplastics in sewage sludge: effects of treatment. *Environmental Science & Technology*, **51**, 810, https://doi.org/10.1016/B978-0-12-812271-6.00181-2

Martín-García A. P., Egea-Corbacho A., Franco A. A., Albendín G., Arellano J. M., Rodríguez-Barroso R., Coello M. D. and Quiroga J. M. (2022). Application of intermittent sand and coke filters for the removal of microplastics in wastewater. *Journal of Cleaner Production*, **380**(Part 1), 134844, https://doi.org/10.1016/j.jclepro.2022.134844

Martín-García A. P., Egea-Corbacho A., Franco A. A., Rodríguez-Barroso R., Coello D., Quiroga J. M. (2023). Grab and composite samples: Variations in the analysis of microplastics in a real wastewater treatment plant in the South of Spain. *Journal of Environmental Chemical Engineering*, **11**(2), 109486, https://doi.org/10.1016/j.jece.2023.109486

Menendez-Manjon A., Martinez-Diez R., Sol D., Laca A., Laca A., Rancano A. and Diaz M. (2022). Long-term occurrence and fate of microplastics in WWTPs: a case study in southwest Europe. *Applied Science*, **12**(4), 2133, https://doi.org/10.3390/app12 042133

Melo A., Quintelas C., Ferreira E. C. and Mesquita D. P. (2022). The role of extracellular polymeric substances in micropollutant removal. *Frontiers in Chemical Engineering*, **4**, 778469, https://doi.org/10.3389/fceng.2022.778469

Michielssen M. R., Michielssen E. R., Ni J. and Duhaime M. B. (2016). Fate of microplastics and other small anthropogenic litter (SAL) in wastewater treatment plants depends on unit processes employed. *Environmental Science: Water Research & Technology*, **2**, 1064–1073, https://doi.org/10.1039/C6EW00207B

Mintening S. M., Int-Veen I., Löder M. G. J., Primpke S. and Gerdts G. (2017). Identification of microplastics in effluents of waste water treatment plants using focal plane array-based micro-Fourier-transform infrared imaging. *Water Research*, **108**, 365–372, https://doi.org/10.1016/j.watres.2016.11.015

Mishra S., Rath C. C. and Das A. P. (2019). Marine microfiber pollution: a review on present status and future challenges. *Marine Pollution Bulletin*, **140**, 188–197, https://doi.org/10.1016/j.marpolbul.2019.01.039

Monira S., Roychand R., Hai F. I., Bhuiyan M., Dhar B. R. and Pramanik B. K. (2023). Nano and microplastics occurrence in wastewater treatment plants: a comprehensive understanding of microplastics fragmentation and their removal. *Chemosphere*, **334**, 139011, https://doi.org/10.1016/j.chemosphere.2023.139011

Murphy F., Ewins C., Carbonnier F. and Quinn B. (2016). Wastewater Treatment Works (WwTW) as a source of microplastics in the aquatic environment. *Environmental Science and Technology*, **50**, 5800–5808, https://doi.org/10.1021/acs.est.5b05416

Nandakumar V. K., Palani S. G. and Raja Raja Varma M. (2022). Interactions between microplastics and unit processes of wastewater treatment plants: a critical review. *Water Science and Technology*, **85**, 496–514, https://doi.org/10.2166/wst.2021.502

Nawalage N. S K. and Bellanthudawa B. K. A. (2022). Synthetic polymers in personal care and cosmetics products (PCCPs) as a source of microplastic (MP) pollution. *Marine Pollution Bulletin*, **182**, 113927, https://doi.org/10.1016/j.marpolbul.2022.113927

Nizzetto N., Martyn Futter M. and Sindre Langaas S. (2016). Are agricultural soils dumps for microplastics of urban origin? *Environmental Science & Technology*, **50**(20), 10777–10779, https://doi.org/10.1021/acs.est.6b04140

Pittura L., Foglia A., Akyol Ç., Cipolletta G., Benedetti M., Regoli F., Eusebi A. L., Sabbatini S., Tseng L. Y., Katsou E., Gorbi S. and Fatone F. (2021). Microplastics in real wastewater treatment schemes: comparative assessment and relevant inhibition effects on anaerobic processes. *Chemosphere*, **262**, 128415, https://doi.org/10.1016/j.chemosphere.2020.128415

OECD. (2022). Global Plastics Outlook: Economic Drivers, Environmental Impacts and Policy Options. OECD Publishing, Paris, https://doi.org/10.1787/de747aef-en

Okoffo Tscharke B. J., O'Brien J. W., O'Brien S., Ribeiro F., Burrows S. D., Choi P. M., Wang X., Mueller J. F. and Thomas K. V. (2020). Release of plastics to Australian land from biosolids end-use. *Environmental Science & Technology*, **54**(23), 15132–15141, https://doi.org/10.1021/acs.est.0c05867

Regulation (EU). Regulation (EU) 2020/741 of the European Parliament and of the Council of 25 May 2020.

Regulation (EU). Regulation (EU) 2020/741 on minimum requirements for water reuse.

Ribeiro F., O'Brien J. W., Galloway T. and Thomas K. V. 2019. Accumulation and fate of nano- and micro-plastics and associated contaminants in organisms. *TrAC Trends in Analytical Chemistry*, **111**, 139–147, https://doi.org/10.1016/j.trac.2018.12.010

Schöpfer L., Menzel R., Schnepf U., Ruess L., Marhan S., Brummer F., Pagel H. and Kandeler E. (2020). Microplastics effects on reproduction and body length of the soil-dwelling nematode *Caenorhabditis elegans*. *Frontiers Environmental Science*, **8**, 41, https://doi.org/10.3389/fenvs.2020.00041

Sillanpää M. and Sainio P. (2017). Release of polyester and cotton fibers from textiles in machine washings. *Environmental Science and Pollution Research*, **24**, 19313–19321, https://doi.org/10.1007/s11356-017-9621-1

Smol M. (2023). Economía circular en plantas de tratamiento de aguas residuales: recuperación de agua, energía y materias primas. *Energías*, **16**(9), 3911, https://doi.org/10.3390/en16093911

Sol D., Laca A., Laca A. and Díaz M. (2020). Approaching the environmental problem of microplastics: importance of WWTP treatments. *Science of the Total Environment*, **740**, 140016, https://doi.org/10.1016/j.scitotenv.2020.140016

Song Y., Cao C., Qiu R., Hu J., Liu M., Lu S., Shi H., Raley-Susman K. M. and He D. (2019). Uptake and adverse effects of polyethylene terephthalate microplastics fibers on terrestrial snails (*Achatina fulica*) after soil exposure. *Environmental Pollution*, **250**, 447–455, https://doi.org/10.1016/j.envpol.2019.04.066

Song W., Du Y., Li D., Xiao Z., Li B., Wei J., Huang X., Zheng C., Wang J., Wang J. and Zhu L. (2023). Polyethylene mulch film-derived microplastics enhance the bioaccumulation of atrazine in two earthworm species (*Eisenia fetida* and *Metaphire guillelmi*) via carrier effects. *Journal of Hazardous Materials*, **455**, 131603–131603, https://doi.org/10.1016/j.jhazmat.2023.131603

Sun Q., Ren S. Y., and Ni H. G. (2020). Incidence of microplastics in personal care products: An appreciable part of plastic pollution. *Science of The Total Environment*, **742**, 140218, https://doi.org/10.1016/j.scitotenv.2020.140218

Talvitie J., Mikola A., Koistinen A. and Setälä O. (2017). Solutions to microplastic pollution – removal of microplastics from wastewater effluent with advanced wastewater treatment technologies. *Water Research*, **123**, 401–407, https://doi.org/10.1016/j.watres.2017.07.005

Tagg Brandes E., Fischer F., Fischer D., Brandt J. and Labrenz M. (2022). Agricultural application of microplastic-rich sewage sludge leads to further uncontrolled contamination. *The Science of the Total Environment*, **806**(Pt 4), 150611–150611, https://doi.org/10.1016/j.scitotenv.2021.150611

Turan N. B., Erkan H. S. and Engin G. O. (2021). Microplastics in wastewater treatment plants: occurrence, fate and identification. *Process Safety and Environmental Protection*, **146**, 77–84, https://doi.org/10.1016/j.psep.2020.08.039

United Nation Environmental Programme (UNEP). (2014). Emerging Issues Update. United Nations Environment Programme, Nairobi, Kenya.

van Haandel A. and van der Lubbe J. (2019). Anaerobic Sewage Treatment: Optimization of Process and Physical Design of Anaerobic and Complementary Processes. IWA Publishing, London, https://doi.org/10.2166/9781780409627

Waldschläger K., Lechthaler S., Stauch G. and Schüttrumpf H. (2020). The way of microplastic through the environment – application of the source-pathway-receptor model (review). *Science of the Total Environment*, **713**, 136584, https://doi.org/10.1016/j.scitotenv.2020.136584

Wei Xu C., Chen C., Wang Y., Lan Y., Long L., Xu M., Wu J., Shen F., Zhang Y., Xiao Y. and Yang G. (2022). Distribution of microplastics in the sludge of wastewater treatment plants in Chengdu, China. *Chemosphere*, **287**, 132357–132357, https://doi.org/10.1016/j.chemosphere.2021.132357

Xu G. Y., Xu L., Shi W., Wang F., LeBlanc G. A., Cui S., An L. and Lei K. (2020a). Investigation of the microplastics profile in sludge from China's largest water reclamation plant using a feasible isolation device. *Journal of Hazardous Materials*, **388**, 122067–122067, https://doi.org/10.1016/j.jhazmat.2020.122067

Xu S., Ma J., Ji R., Pan K. and Miao A. J. (2020b). Microplastics in aquatic environments: occurrence, accumulation, and biological effects. *Science of the Total Environment*, **703**, 134699, https://doi.org/10.1016/j.scitotenv.2019.134699

Xu G., Liu Y. and Yu Y. (2021). Effects of polystyrene microplastics on uptake and toxicity of phenanthrene in soybean. *Science of the Total Environment*, **783**, 147016, https://doi.org/10.1016/j.scitotenv.2021.147016

Yang L., Li K., Cui S., Kang Y., An L. and Lei K. (2019). Removal of microplastics in municipal sewage from China's largest water reclamation plant. *Water Research*, **155**, 175–181, https://doi.org/10.1016/j.watres.2019.02.046

Yu H., Shi L., Fan P., Xi B. and Tan W. (2022). Effects of conventional versus biodegradable microplastic exposure on oxidative stress and gut microorganisms in earthworms: a comparison with two different soils. *Chemosphere*, **307**, 135940–135940, https://doi.org/10.1016/j.chemosphere.2022.135940

Zhang K., Gong W., Lv J., Xiong X. and Wu C. (2015). Accumulation of floating microplastics behind the Three Gorges Dam. *Environmental Pollution*, **204**, 117–123, https://doi.org/10.1016/j.envpol.2015.04.023

Zhang Q., Gong K., Shao X., Liang W., Zhang W. and Peng C. (2023a). Effect of polyethylene, polyamide, and polylactic acid microplastics on Cr accumulation and toxicity to cucumber (*Cucumis sativus* L.) in hydroponics. *Journal of Hazardous Materials*, **450**, 131022, https://doi.org/10.1016/j.jhazmat.2023.131022

Zhang Q., Lv Y., Liu J., Chang L., Chen Q., Zhu L., Wang B., Jiang J. and Zhu W. (2023b). Size matters either way: differently-sized microplastics affect amphibian host and symbiotic microbiota discriminately. *Environmental Pollution*, **328**, 121634–121634, https://doi.org/10.1016/j.envpol.2023.121634

Zhong H., Yang S., Zhu L., Liu C., Zhang Y. and Zhang Y. (2021). Effect of microplastics in sludge impacts on the vermicomposting. *Bioresource Technology*, **326**, 124777–124777, https://doi.org/10.1016/j.biortech.2021.124777

Zhu D., Ke X., Christie P. and Zhu Y. (2018). Exposure of soil collembolans to microplastics perturbs their gut microbiota and alters their isotopic composition. *Soil Biology & Biochemistry*, **124**, 275–276, https://doi.org/10.1016/j.soilbio.2017.10.027

Zhou C. Q., Lu C. H., Mai L., Bao L. J., Liu L. Y. and Zeng E. Y. (2021). Response of rice (*Oryza sativa* L.) roots to nanoplastic treatment at seedling stage. *Journal of Hazardous Materials*, **401**, 123412, https://doi.org/10.1016/j.jhazmat.2020.123412

Ziajahromi S., Neale P. A., Rintoul L. and Leusch F. D. L. (2017). Wastewater treatment plants as a pathway for microplastics: development of a new approach to sample wastewater-based microplastics. *Water Research*, **112**, 93–99, https://doi.org/10.1016/j.watres.2017.01.042

doi: 10.2166/9781789063752_0019

Chapter 2

Occurrence and detection of pharmaceuticals in wastewater and its subsequent treatment using constructed wetlands, bioelectrochemical systems and their combination

*Mahak Jain[1], Abhradeep Majumder[2], Pubali Mandal[3], Shalini Singh[4], Partha Sarathi Ghosal[1] and Manoj Kumar Yadav[4]**

[1]School of Water Resources, Indian Institute of Technology Kharagpur, Kharagpur 721302, India
[2]School of Environmental Science and Engineering, Indian Institute of Technology Kharagpur, Kharagpur 721302, India
[3]Department of Civil Engineering, Birla Institute of Technology Pilani, Pilani 333031, India
[4]Department of Civil and Environmental Engineering, Indian Institute of Technology Patna, Patna 801106, India
*Corresponding author: mkyadav@iitp.ac.in

ABSTRACT

Pharmaceutically active compounds (PhAC) are pervasive in aqueous environments, and their presence poses an ever-increasing threat to aquatic creatures and all associated living forms. Most PhACs are extremely hydrophilic and have a complicated molecular structure, preventing them from being destroyed by traditional wastewater treatment methods. In addition, these contaminants are present at such a low concentration that their detection poses a significant challenge. Researchers have utilized advanced oxidation processes to degrade these chemicals over time. However, most studies have been conducted on the lab scale and do not function well for real wastewater since many interfering substances are present. In addition, these techniques are expensive and result in the production of harmful byproducts. To combat the PhACs, it is vital to develop a sustainable economic strategy. This book chapter discusses the occurrence of PhACs in wastewater, their potential environmental impacts, and the necessary procedures for accurately quantifying these compounds. The book addresses the possibilities of biological systems, such as constructed wetlands (CW) and bioelectrochemical systems (BES), in the hunt for a sustainable method of eliminating PhACs. CWs have been selected because they are robust systems with several simultaneous removal mechanisms. BES have also demonstrated considerable potential for treating these substances in wastewater and producing bioelectricity. In addition, the chapter discusses an emerging technology, that is, hybrid CW–BES systems, which utilize the benefits of both CW and BES and may prove to be an efficient approach to treating wastewater, removing PhACs, and generating electricity simultaneously.

2.1 INTRODUCTION

The onset of the 21st century is marked by the rapid detection of pharmaceutically active compounds (PhACs) in various aqueous environments. The improvement in healthcare facilities and medicines has led to increased consumption of pharmaceuticals. Most unmetabolized fractions of pharmaceuticals are excreted as feces or urine by human bodies. As a result, PhACs and metabolites of pharmaceuticals are frequently detected in hospital and municipal wastewater. The increase in medicine consumption has resulted in the shift in research toward pharmaceutical removal and treatment of hospital wastewater in the last 10 years or so (Majumder *et al.*, 2021; Parida *et al.*, 2022). The country/region-wise studies over the last 20 years pertaining to the occurrence of PhACs in aqueous environments are depicted in Figure 2.1. Articles that come under the topic containing the words, 'occurrence of pharmaceuticals' AND 'water', were considered for the study. The search was restricted to document type 'articles' only. It was observed from Figure 2.1 that research on PhACs occurrence in the early part of the 21st century (2001–2005) was only restricted to developed countries, such as the USA, Germany, Spain, Italy, and France. After 2006, Asian countries, such as South Korea, Japan, and China, started contributing to this field, which may be due to the requirement of high-end analytical instruments, such as high-performance liquid chromatography (HPLC), gas chromatography (GC), liquid chromatography coupled with mass spectrometry (LC–MS), gas chromatography coupled with mass spectrometry (GC–MS), and others. Furthermore, the presence of other non-target organic compounds and interfering agents often make the detection and quantification of the PhACs, a challenging task (Boulard *et al.*, 2020; Stamatis & Konstantinou, 2013). However, with the passage of time and the advent of new technologies, research in this field got a significant boost. The number of research articles published between 2011 and 2020 was found to be almost five times higher than that between 2001 and 2010. Furthermore, the articles were reported from countries all over the globe, indicating the presence of PhACs in aqueous matrices is a global problem.

Although the PhACs are found in almost all aqueous environments, their concentration is very low, that is, in the range of μg/L to ng/L. However, the PhACs have the potential to cause significant harm to the environment even at low concentrations (Majumder *et al.*, 2019). As a result, it is important to detect the presence of these contaminants. In water and wastewater, there are innumerable PhACs and their metabolites detected in water. Each of these compounds has a different effect on non-target

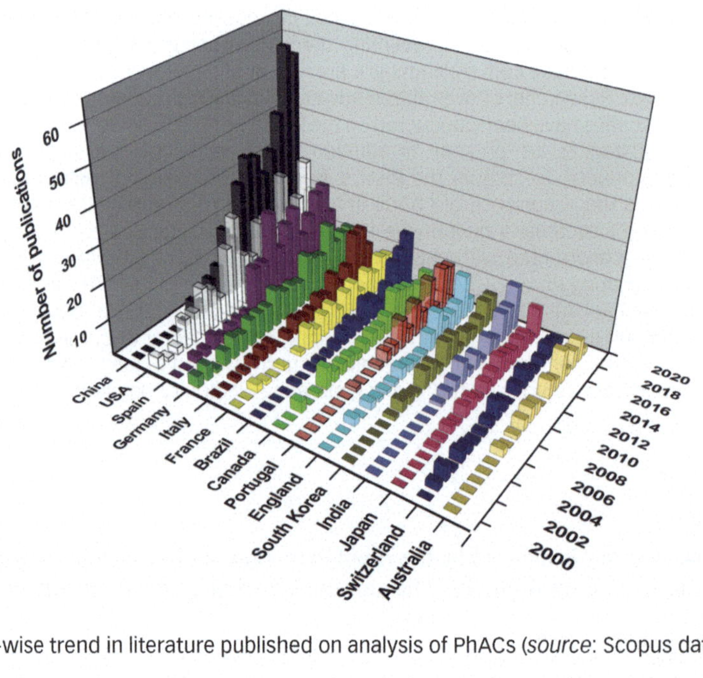

Figure 2.1 Country-wise trend in literature published on analysis of PhACs (*source*: Scopus database).

species. As a result, collectively, all these compounds can significantly disrupt ecosystems. Hence, it has become mandatory to address the removal of these compounds (Majumder *et al.*, 2019). PhACs are highly mobile in the aqueous phase and have complex molecular structures. Furthermore, many of the PhACs are hydrophilic in nature (Majumder *et al.*, 2019). Hence, it is difficult to remove the PhACs by conventional treatment methods. When conventional treatment methods are coupled with advanced treatment processes, the overall cost of the process increases. Hence, to achieve a sustainable treatment of PhACs from wastewater, it is necessary to simultaneously remove the contaminants and recover resources in some form. Constructed wetlands (CWs) have proved to be effective in removing a wide range of contaminants due to the multiple removal mechanisms taking place simultaneously. However, CWs need an extensive amount of aeration for their proper functioning. On the other hand, microbial fuel cells (MFC) have been known to convert wastewater to electrical energy. In this context, constructed wetland-microbial fuel cell (CW–MFC) systems have been developed that combine the benefits of CW and MFC (Fang *et al.*, 2015; Liu *et al.*, 2022; Lutterbeck *et al.*, 2022). While the CWs are responsible for bringing down the organic loading and degrading the PhACs, the energy recovery in the form of electricity may lower the costs of operation and maintenance of the treatment unit. CW–MFC has been used to treat sewage from homes, treat wastewater from factories, and control pollution from non-point sources in cities (Fang *et al.*, 2015; Liu *et al.*, 2022; Lutterbeck *et al.*, 2022).

In this chapter, the different types of PhACs detected in aqueous environments have been discussed, along with the concentration of few of the most commonly detected PhACs. The environmental impact

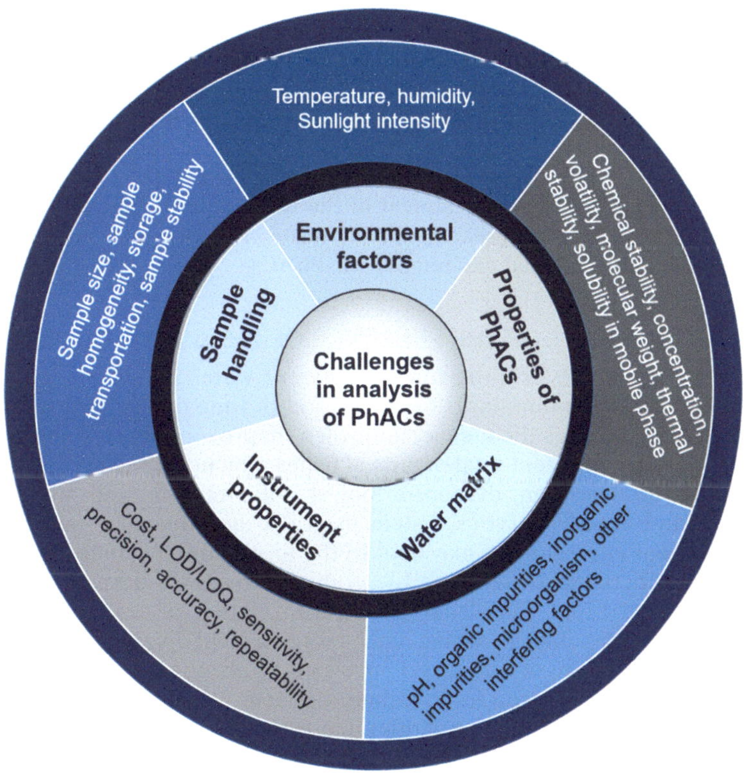

Figure 2.2 Challenges in the analysis of PhACs.

of the PhACs and the challenges in detecting these PhACs will also be discussed. The difficulties in removing the PhACs using conventional treatment processes and the applicability of CW and BES in removing PhACs from wastewater have been addressed. Lastly, the performance of CW–BES-based systems to simultaneously treat wastewater, remove PhACs and generate electricity.

2.2 TYPES OF PHACS DETECTED IN WASTEWATER AND THEIR PHYSICOCHEMICAL PROPERTIES

Researchers have documented the presence of a large range of different PhACs in variable concentrations in different aqueous environments (Majumder *et al.*, 2019). These compounds are difficult to remove due to their high mobility and hydrophilicity, and they remain in the aqueous environment for a prolonged duration (Majumder *et al.*, 2019). Therefore, research into the treatment of PhACs has been given a higher priority.

The most common types of PhACs detected in aqueous environments are analgesics, antibiotics, β-blockers, and endocrine-disrupting compounds (EDC). Apart from these, antiepileptics and stimulants, such as carbamazepine and caffeine, are also frequently detected in various aqueous environments (Majumder *et al.*, 2019). In Table 2.1, the different classes of pharmaceuticals, their physicochemical properties, and their concentration in surface water, municipal wastewater, and hospital wastewater have been depicted. In this chapter, the most commonly occurring PhACs in different aqueous environments have been considered (Majumder *et al.*, 2019; Parida *et al.*, 2022; Pubchem, 2023; Saidulu *et al.*, 2021). The PhACs which are associated with a low octanol–water partition coefficient ($\log k_{ow}$) do not get absorbed easily. Furthermore, the dissociation constants (pK_a) of the PhACs indicate the charge of these compounds in water (Majumder *et al.*, 2019; Tran *et al.*, 2018). If the particles are charged, then it is difficult to remove by conventional sedimentation or adsorption processes. Furthermore, the toxicity and complex structure of these compounds prevent biodegradation. Although these PhACs are removed by AOPs, they significantly increase the cost of treatment. Furthermore, the AOPs have only proved to be effective when the water or wastewater matrix is free from suspended organic or inorganic matter and interfering agents. Hence, AOPs are not standalone processes, and they need to be implemented after undergoing pre-treatment of the wastewater (Majumder *et al.*, 2022; Santos *et al.*, 2009; Tran *et al.*, 2018). Hence, researchers have shifted their focus to developing cost-effective, sustainable treatment technologies that can remove the PhACs.

2.3 ENVIRONMENTAL IMPACT OF THE PRESENCE OF PHACS IN WASTEWATER

The concentration of the PhACs is quite low in different water matrices (Table 2.1). However, prolonged exposure to these contaminants over a period of time can significantly affect the non-target species. Analgesics, such as diclofenac, ibuprofen, naproxen, paracetamol, and others, may cause cytological changes in different vital organs of fishes. The non-target species, upon exposure, may also develop gastric ulceration, dyspepsia, bowel inflammation, and mucosal damage, and their cardiovascular and central nervous system may also be affected (Majumder *et al.*, 2019). Antibiotics in the aqueous environment may lead to the formation of resistant genes and bacteria. The extended-spectrum beta-lactamase (ESBL) producing bacteria develop a resistance to the exposed antibiotics and multiply, thus leading to the formation of antibiotic-resistant bacteria. These resistant bacteria are capable of causing much more virulent diseases (Majumder *et al.*, 2021). Furthermore, the immune system of non-target species may also be get affected (Majumder *et al.*, 2019). β-blockers may hinder the growth of embryonic stem cells and cause cardiovascular and neural problems among non-target species. Antiepileptics and stimulants may hamper the growth of embryonic cells, cause panic disorders, and increase plasma epinephrine levels among non-target species. EDCs can lower sperm count among male fishes and lead to abnormal sexual development (Majumder *et al.*, 2019).

Table 2.1 Physicochemical properties of PhACs and their concentrations in different aqueous environments.

PhACs	Class	Molecular Weight	pK_a	Log k_{ow}	Surface Water (μg/L)	Municipal Wastewater (μg/L)	Hospital Wastewater (μg/L)
Codeine	Analgesics of non-steroidal inflammatory drug (NSAID)	299	10.6	1.19	0.31 ± 0.41	8.11 ± 13.70	
Diclofenac		296	4.15	4.52	1.22 ± 2.64	3.52 ± 6.59	1.50 ± 1.50
Ibuprofen		207	4.9	3.97	9.65 ± 24.12	21.19 ± 18.41	16.61 ± 15.11
Naproxen		229	4.15	3.18	1.67 ± 2.98	4.12 ± 3.70	13.82 ± 20.41
Acetaminophen		152	9.4	0.46	18.04 ± 39.45	222.71 ± 235.35	109.40 ± 131.86
Salicylic acid		137	3.49	1.19	5.35 ± 7.81	64.41 ± 97.79	
Ciprofloxacin	Antibiotics	332	6.25	0.28	1302.78 ± 2905.34	7.30 ± 9.89	53.21 ± 112.72
Sulfamethoxazole		254	1.6, 5.7	0.89	4.45 ± 11.65	13.42 ± 27.59	6.09 ± 7.86
Trimethoprim		291	7.12	0.91	1.78 ± 2.78	9.96 ± 23.68	6.23 ± 11.61
Erythromycin		734	8.8, 8.9	3.06	1.15 ± 2.38	3.75 ± 4.38	0.67 ± 0.51
Azithromycin		749	8.74	4.02	0.02	300.00±	7.25 ± 11.50
Norfloxacin		320	6.34, 8.75	0.46	260.07 ± 367.60	1.10 ± 1.40	4.57 ± 4.74
Levofloxacin		362	6.24, 8.74	−0.39	0.04	38.57 ± 74.29	8.80
Ofloxacin		362	5.97, 9.28	−0.39	27.52 ± 38.86	2.38 ± 3.20	6.99 ± 7.97
Atenolol	β-blockers	267	9.6	0.16	6.56 ± 11.04	38.99 ± 98.43	4.65 ± 5.10
Metoprolol		268	9.6	1.88	3.11 ± 4.02	190.29 ± 424.69	3.24 ± 2.49
Carbamazepine	Antiepileptics	237	13.9	2.45	0.54 ± 0.96	2.52 ± 5.39	2.00 ± 2.37
Caffeine	Stimulants	194	10.4	−0.07	35.57 ± 62.38	61.68 ± 60.72	124.50 ± 133.77
Estriol	EDCs	287	10.54	2.45	0.13 ± 0.22	0.28 ± 0.46	0.25 ± 0.18
17β estradiol		271	10.46	4.01	0.04 ± 0.04	0.07 ± 0.11	0.43 ± 0.44

Source: Majumder *et al.* (2019), Parida *et al.* (2022), Pubchem (2023) and Saidulu *et al.* (2021).

2.4 CHALLENGES IN DETECTING PHACS IN WASTEWATER AND STRATEGIES FOR THEIR EFFECTIVE ANALYSIS

One of the biggest challenges in the detection of PhACs is their low concentrations in the aqueous environment. Most of the instruments, that is, HPLC and GC have a limit of quantification of around 1 μg/L, whereas many of the PhACs in surface water or groundwater are present in the range of ng/L to pg/L (Bexfield *et al.*, 2019). Furthermore, since these are organic compounds, they may undergo transformations from the time of sampling to the time of analysis. The PhACs may undergo bacterial degradation or photolysis, thereby lowering their initial concentration. Furthermore, PhACs may get adsorbed onto the walls of the container or suspended solids present in the container, thereby leading to erroneous analysis. Additionally, since real wastewater samples contain many other organic substances and interfering agents apart from the target PhACs, their quantification may be affected (Mompelat *et al.*, 2013). The different issues and challenges in the detection of PhACs have been illustrated in Figure 2.1.

There are certain protocols that, if followed, may lead to accurate analysis and thereby help in overcoming the challenges pertaining to the detection and quantification of PhACs. Firstly, thorough blanking should be carried out. Blanking of samples should include field blank, instrument blank, equipment blank, method blank, and trip blank (USEPA, 2000). The sample containers should be properly washed and made sure that there is no prior contamination. To prevent microbial degradation, the samples can be preserved by adding sodium azide, formaldehyde, or methanol, which are toxic to microorganisms (Guzel *et al.*, 2019; Havens *et al.*, 2010; Mompelat *et al.*, 2013; Vanderford *et al.*, 2003, 2011). Lowering the pH of the sample by adding hydrochloric acid and nitric acid also prevents bacterial growth (Guzel *et al.*, 2019; Havens *et al.*, 2010; Mompelat *et al.*, 2013; Vanderford *et al.*, 2003, 2011). To prevent the degradation of the PhACs by photolysis, the samples should be stored in amber bottles. Amber bottles prevent the entry of light inside the storage containers. Also, storing the samples at less than 4°C prevents microbial activity (Mompelat *et al.*, 2013). These are few of the steps that should be carried out to maintain the concentration of the PhACs from the time of sampling to the time of analysis.

As mentioned earlier, a major problem associated with the quantification of PhACs is the concentration of the compounds in the aqueous environment (Bexfield *et al.*, 2019; Majumder *et al.*, 2019). Hence, it is required to increase the concentration of these compounds in the solution for their detection. However, increasing evaporation may lead to the breakdown of the organics. Even concentrating the solutions by nitrogen purging may be a time-consuming process. In this context, extraction is carried out using different methods, such as solid-phase extraction (SPE), solid-phase micro-extraction (SPME), liquid-phase extraction (LPE), accelerated solvent extraction (ASE), microwave-assisted extraction (MAE), ultrasonic-assisted extraction (UAE), Soxhlet extraction (SE), membrane extraction, lyophilization, and others (Fatta-Kassinos *et al.*, 2019; Kostopoulou & Nikolaou, 2008; Pavlović *et al.*, 2007). Among these processes, the SPE is the most commonly used process.

In the SPE, along with the concentration of the target analytes, the interfering agents can also be filtered out. The SPE technique is based on the basic principle of transferring target analytes from a liquid phase to a solid phase (cartridges containing a sorbent), which can retain the target analytes and can subsequently be stripped by an appropriate solvent. This transfer of target analytes from a liquid phase to a solid phase is accomplished by the SPE technique. A typical liquid sample that needs to be examined contains a number of components that cause interference in addition to the analyte that is of interest. The SPE process typically consists of four phases, which are carried out in the sequence listed in order to successfully isolate the target analyte from the interfering components (Andrade-Eiroa *et al.*, 2016; Meng *et al.*, 2021).

In the initial stage, cartridge columns are put through a first pass with either an organic solvent or water. This process is referred to as conditioning, and its primary purpose is to enhance the effective surface area of the sorbent material contained within the cartridge while simultaneously removing any interferences. After allowing the sorbent to dry out, the next step involves loading the cartridges with a water sample that contains PhAC. The interference analytes are allowed to pass through with the sample solution, while the target analytes are taken up by the sorbent and stored there. The subsequent stage is washing, which involves removing interferences and components from the sorbent that are not the target analytes. This phase comes after the process in which the target analytes are determined. The final stage is known as elution, and it is the process in which organic solvents are circulated through the cartridges. During this step, desorption takes place, which moves the target analytes from the solid phase (sorbent) to the liquid phase (organic solvent) (Andrade-Eiroa *et al.*, 2016; Meng *et al.*, 2021).

The whole concentration of the analytes that are present in the liquid sample is absorbed into the cartridges through the use of SPE, which is one of the primary reasons why this technique is so important to the quantification of the analytes. If the concentration of the analytes in the sample is low, a large volume of the sample can be run through the cartridges to maintain a sufficient quantity of the analyte in the SPE cartridges. This is possible since the concentration of the analyte in the sample is low. When these target analytes are eluted in a relatively lesser amount of organic solvent, the concentration of the analytes in the solvent gets enhanced by the sample-to-solvent volume ratio, which subsequently facilitates better detection and quantification. Before analyzing the sample, it is possible to further concentrate it by subjecting the elute containing the target analytes to a mild stream of nitrogen gas (Afsa *et al.*, 2020; Andrade-Eiroa *et al.*, 2016; Biel-Maeso *et al.*, 2018; Meng *et al.*, 2021). Only after this can real water samples containing PhACs be accurately quantified using HPLC or GC.

2.5 CHALLENGES IN REMOVING PHACS FROM WASTEWATER

Most of the PhACs are characterized by high molecular weight, complex molecular structure, high hydrophilicity, and pose toxicity to microorganisms. The complex molecular structure prevents these molecules from being easily degraded. Conventional wastewater treatment plants relying on activated sludge processes have not proved to be effective in removing the PhACs. This is mainly because the primary removal mechanism of the PhACs in these systems is biodegradation, and many of the PhACs are resistant to biodegradation due to their toxic nature. Another removal mechanism involved in the removal of PhACs is the adsorption in the suspended matter. However, many PhACs have low log k_{ow} values, which makes them hydrophilic. Hence, they do not get easily removed by adsorption. In the case of water treatment plants, the primary sedimentation tanks, clariflocculator, and sand filtration are not designed to remove the PhACs. Furthermore, it has been observed that when chlorination is carried out, the residual chlorine reacts with metabolites of the PhACs present in the effluent to produce toxic disinfection byproducts (Qadafi *et al.*, 2023).

Among membrane-based treatment processes, a high degree of PhAC removal has been observed only when nanofiltration or reverse osmosis is used. However, the drawback of these processes is that they require a high operating cost and the membranes are subjected to fouling (Perreault *et al.*, 2016; Prado *et al.*, 2017). Other advanced treatment processes, such as adsorption, have proved to be effective. However, adsorption produces a significant amount of sludge, which needs to be disposed of. Also, adsorption can be used only as a tertiary treatment process and cannot be used to treat raw wastewater (Bizi, 2020). Similarly, in AOPs, such as photocatalysis, Fenton process, electro-oxidation, and others, which rely on the generation of oxidizing radicals for the degradation of PhACs, a pre-treatment of the wastewater is necessary (Majumder *et al.*, 2022). Often due to the presence of suspended solids, organic matter, and other interfering agents, the oxidizing radicals get scavenged (Majumder *et al.*, 2022). This significantly affects the removal efficiency of the AOPs. Hence, AOPs should also follow a pre-treatment process where the majority of suspended matter and organics are removed.

Another downside to the degradation of PhACs is the formation of degradation byproducts or transformation products (TPs). Often during biological degradation or oxidative degradation of the PhACs, complete degradation is not attained. As a result, TPs are formed, which may at times, have toxicity more than the parent compound (Gogoi *et al.*, 2018; Li *et al.*, 2023; Villarín & Merel, 2020). The toxicity of the formed TPs may further affect the aquatic ecosystem. Hence, the treatment provided should be such that relatively low toxic products are formed, and even if they are formed, they are at concentrations low enough to not significantly harm the environment.

The removal of PhACs by conventional treatment processes has not proved to be effective while incorporating advanced oxidation processes (AOP)s with the conventional treatment processes increases the overall cost of the system. In this context, it has become necessary to come up with a low-cost sustainable treatment system that can address all the concerns stated above. In this context, the role of constructed wetlands, bioelectrochemical systems, and their combination in terms of the removal of PhACs have been discussed in the following sections.

2.6 PERFORMANCE OF CW IN REMOVING PHACS

The CWs have shown a high degree of PhAC removal as compared to other conventional biological processes because of the numerous removal mechanisms involved in their degradation (Jain *et al.*, 2023). However, the primary mechanisms involved in the removal of PhACs in CWs are the microbial degradation, plant uptake, and adsorption by the media (Jain *et al.*, 2023). CWs have been even more efficient in treating real wastewater on numerous occasions because it help removing soluble and insoluble organic matter. The substrate present in the CWs helps remove a major portion of the suspended and dissolved organic and inorganic matter, which allows the microbial degradation of the PhACs to take place. However, hydraulic retention time (HRT) is an important factor in CWs. Sufficient time should be provided to allow microbial degradation or plant uptake of the PhACs to take place (Kamilya *et al.*, 2023; Yates *et al.*, 2016).

The removal of PhACs using CW-based systems has been shown in Figure 2.3a. It has been observed that the removal efficiency of PhACs has varied quite a lot in the systems. Analgesics, such as naproxen and salicylic acid, exhibited excellent removal. On the other hand, carbamazepine and a few antibiotics showed very low removal. This may be because of the low biodegradability of these compounds (Saidulu *et al.*, 2021). Furthermore, many of the PhACs, such as antibiotics, are toxic to bacteria, thereby preventing microbial degradation. The other driving factor responsible for the removal of PhACs in the CWs is aeration. Often aeration facilitates the biodegradation process of the PhACs (Auvinen *et al.*, 2017; Sochacki *et al.*, 2018).

Apart from the removal of organic matter, suspended matter, and recalcitrant organics, PhACs can also be used to recover nutrients, such as ammonia and phosphorous. Heavy metal recovery is another advantage of CWs. These systems can significantly contribute to the circular economy by recovering valuable resources from wastewater (Guo *et al.*, 2020; Kamilya *et al.*, 2022). Therefore, the CWs have been a robust system that can tackle high-strength wastewater, remove PhACs, contribute to the circular economy, and also improve the aesthetics of the place. Furthermore, due to the ability of the CWs to tackle fluctuations in organic and hydraulic loading, any pre-treatment is not mandatory (Jain *et al.*, 2023). However, the drawbacks of this process involve high HRT and large land requirements (Jain *et al.*, 2020, 2023). The schematic of a typical CW and its applicability in various aspects of sustainable wastewater management options has been depicted in Figure 2.4.

2.7 PERFORMANCE OF BES IN REMOVING PHACS

Bio-electrochemical systems (BES) have the potential to convert the chemical energy present in wastewater and lignocellulose biomass into various forms of energy, including electricity, hydrogen, and other chemical compounds. In this process, the organic molecules are degraded by the bacterial

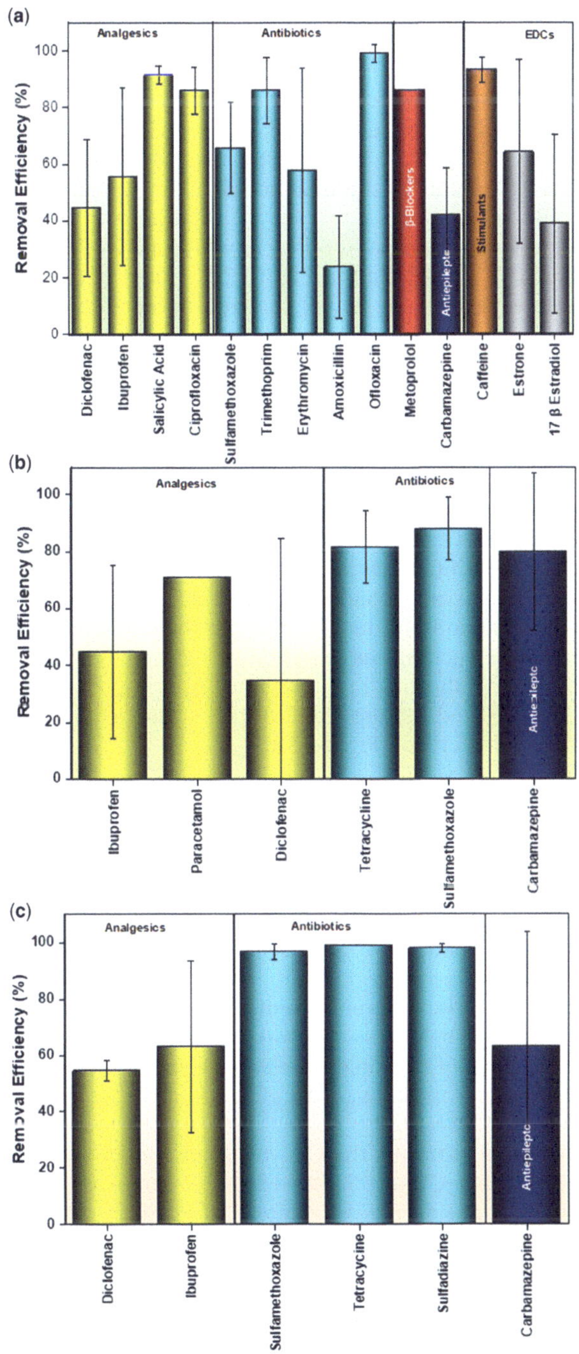

Figure 2.3 Removal of different PhACs using (a) CW, (b) BES, and (c) CW–BES based systems (source: Ahmad *et al.*, 2022; Hu *et al.*, 2021; Huang *et al.*, 2021; Kamilya *et al.*, 2023; Li *et al.*, 2019; Luo *et al.*, 2023; Pun *et al.*, 2019; Thapa *et al.*, 2022; Wang *et al.*, 2015; Xu *et al.*, 2022; Yan *et al.*, 2019).

population in the wastewater. Electrons generated through the process of oxidation can be harnessed to produce energy that can subsequently be utilized for the creation of practical applications. The prevalent types of bioelectrochemical systems (BES) are categorized based on the biocatalysts utilized or their mode of application. These include microbial fuel cells, plant microbial fuel cells, microbial electrolysis cells (MEC), enzymatic fuel cells (EFC), microbial solar cells, and others (Kelly & He, 2014; Pant *et al.*, 2012; Wang & Ren, 2014). The effectiveness of BES systems in removing PhACs has been shown in Figure 2.3b. The anode in the BES hosts a large consortium of microorganisms that help in the production of electricity and also in the degradation of PhACs. In the BES, the PhACs can get degraded at the cathodic and the anodic chambers via different redox processes (Thapa *et al.*, 2022; Zhang *et al.*, 2015). As depicted in Figure 2.3b, most of the systems have shown a very good removal efficiency. However, quite a few of the studies have been carried out using synthetic wastewater.

Apart from removing PhACs, BES are also efficient in removing organic matter. However, the removal of insoluble inorganic fractions or suspended solids from the wastewater using a stand-alone BES is not substantial (Kim *et al.*, 2016). As a result, for enhanced performance of the BES, employing a pre-treatment process to reduce the load of suspended solids is important. BES-based systems have been known to recover electricity and contribute toward the sustainability of the system and circular economy. Apart from electricity, modifications to BES can also bring about nutrient recovery in the form of nitrogen and phosphorous (Kelly & He, 2014; Pant *et al.*, 2012; Wang & Ren, 2014).

However, the operation and maintenance of the BES are not as convenient as that of the CWs. In BES, it is required to maintain anaerobic conditions in the anodic chamber and aerobic conditions in the cathodic chamber. Failing to maintain this will hamper the performance of the BES. As a result of this, the BES cannot also handle fluctuations in hydraulic loading. It is a much more sensitive system as compared to CWs and requires thorough maintenance to bring out the best performance of BES. The schematic of a typical BES and its applicability in various aspects of sustainable wastewater management options has been depicted in Figure 2.4.

2.8 PERFORMANCE OF HYBRID CW–BES SYSTEM IN REMOVING PHACS

In CW–BES systems, the dissolved oxygen of the wastewater decreases with the increase in substrate depth, and anaerobic conditions start to prevail. The cathode is usually kept at the top of the substrate, where it is exposed to air (air cathode) or water (in the case of floating wetlands) (Lutterbeck *et al.*, 2022; Mu *et al.*, 2020). CW–BES has the advantages of CWs, such as the ability to handle fluctuations in organic and hydraulic loading, resource recovery, multiple removal mechanisms for the removal of PhACs, and others. Also, due to the difference in dissolved oxygen in the cathodic chamber (above the substrate) and anodic chamber (below the substrate), there is a growth of electrogens at the anode, which uses wastewater as a source of carbon and releases electrons, thereby initiating the transfer of electrons from anode to cathode. This leads to the generation of electricity (Lutterbeck *et al.*, 2022; Mu *et al.*, 2020).

Apart from the generation of electricity, the CW–BES has shown the potential to degrade PhACs (Figure 2.3c). It was observed that CW–BES showed better removal of antibiotics as compared to other PhACs. However, studies considering the removal of PhACs from real wastewater are limited, and hence, it is too early to remark anything regarding the performance of the CW–BES in terms of the removal of particular PhACs. However, it can be estimated that it will perform at parity with CWs and BES since microbial degradation, plant uptake, and substrate adsorption will all take part in the removal of PhACs (Hartl *et al.*, 2021; Luo *et al.*, 2023). Furthermore, it has significant advantages over the other two systems because of its ability to produce electricity, recover nutrients, and handle fluctuations in organic and hydraulic loading. Additionally, the operation and maintenance of the CW–BES are not complicated. The schematic of a typical CW–BES and its applicability in various aspects of sustainable wastewater management options has been depicted in Figure 2.4.

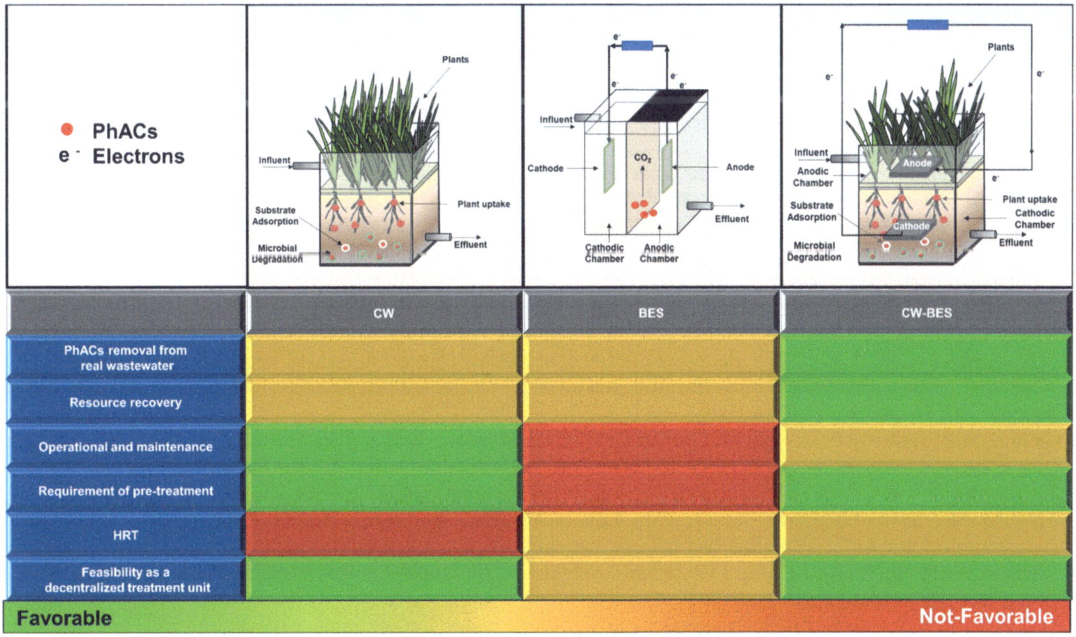

Figure 2.4 Schematic of CW, BES, and CW–BES and the favorability of the systems in terms of various aspects of sustainable wastewater management option.

2.9 SUMMARY

The PhACs are found in concentrations ranging from µg/L to ng/L in various aquatic environments, such as wastewater, surface water, and even groundwater. Even at such low concentrations, PhACs can significantly affect entire ecosystems. However, no legislation or standards have been made yet due to the lack of availability of sufficient data for PhACs. The reason behind the lack of data is the detection of such PhACs. PhACs require the use of high-end instruments, such as LC, GC, LC–MS, GC-MS, MALDI-ToF, and others, for their detection. Furthermore, the samples need to be properly collected, transported, stored, and analyzed for their precise measurement. An important step in measuring PhACs having concentrations lower than the detection limit of instruments is the use of SPE. The challenges in dealing with PhACs do not end there. Most of the PhACs have a complex molecular structure, high hydrophilicity, unfavorable dissociation constants, and other physicochemical properties that prevent them from being removed using conventional treatment methods. Employing advanced treatment methods, such as AOPs, membrane filtration, or adsorption, only increases the cost of treatment. Hence, to remove these PhACs in a sustainable, cost-effective manner, the use of CW, BES, and CW–BES has been explored, and it was found that CW–BES and CW have the potential to treat real wastewater contaminated with such PhACs. Due to their ability to recover resources in the form of nutrients and electricity, ability to remove soluble and insoluble organics, non-requirement of mandatory pre-treatment, and efficiency in removing PhACs, CW–BES may be a probable solution to tackle complex wastewater comprising of PhACs and other such recalcitrant organic compounds. However, further studies on the up-scaling of these systems and modifications to existing systems should be carried out to increase electricity generation, reduce HRT, and provide higher treatment efficiency.

REFERENCES

Afsa S., Hamden K., Lara Martin P. A. and Mansour H. B. (2020). Occurrence of 40 pharmaceutically active compounds in hospital and urban wastewaters and their contribution to Mahdia coastal seawater contamination. *Environmental Science and Pollution Research*, **27**(2), 1941–1955, https://doi.org/10.1007/s11356-019-06866-5

Ahmad A., Priyadarshani M., Das S. and Ghangrekar M. M. (2022). Role of bioelectrochemical systems for the remediation of emerging contaminants from wastewater: a review. *Journal of Basic Microbiology*, **62**(3–4), 201–222, https://doi.org/10.1002/jobm.202100368

Andrade-Eiroa A., Canle M., Leroy-Cancellieri V. and Cerdà V. (2016). Solid-phase extraction of organic compounds: a critical review (part I). *Trends in Analytical Chemistry*. Elsevier B.V., **80**, pp. 641–654, https://doi.org/10.1016/j.trac.2015.08.015

Auvinen H., Havran I., Hubau L., Vanseveren L., Gebhardt W., Linnemann V., Van Oirschot D., Du Laing G. and Rousseau D. P. L. (2017). Removal of pharmaceuticals by a pilot aerated sub-surface flow constructed wetland treating municipal and hospital wastewater. *Ecological Engineering*, **100**, 157–164, https://doi.org/10.1016/j.ecoleng.2016.12.031

Bexfield L. M., Toccalino P. L., Belitz K., Foreman W. T. and Furlong E. T. (2019). Hormones and pharmaceuticals in groundwater used as a source of drinking water across the United States. *Environmental Science and Technology*, **53**(6), 2950–2960, https://doi.org/10.1021/acs.est.8b05592

Biel-Maeso M., Baena-Nogueras R. M., Corada-Fernández C. and Lara-Martín P. A. (2018). Occurrence, distribution and environmental risk of pharmaceutically active compounds (PhACs) in coastal and ocean waters from the Gulf of Cadiz (SW Spain). *Science of the Total Environment*, **612**, 649–659, https://doi.org/10.1016/j.scitotenv.2017.08.279

Bizi M. (2020). Sulfamethoxazole removal from drinking water by activated carbon: kinetics and diffusion process. *Molecules (Basel, Switzerland)*, **25**(20), 1–18, https://doi.org/10.3390/molecules25204656

Boulard L., Dierkes G., Schlüsener M. P., Wick A., Koschorreck J. and Ternes T. A. (2020). Spatial distribution and temporal trends of pharmaceuticals sorbed to suspended particulate matter of German rivers. *Water Research*, **171**, 115366, https://doi.org/10.1016/j.watres.2019.115366

Fang Z., Song H. L., Cang N. and Li X. N. (2015). Electricity production from Azo dye wastewater using a microbial fuel cell coupled constructed wetland operating under different operating conditions. *Biosensors and Bioelectronics*, **68**, 135–141, https://doi.org/10.1016/J.BIOS.2014.12.047

Fatta-Kassinos D., Nikolaou A. and Ioannou-Ttofa L. (2019). Advances in analytical methods for the determination of pharmaceutical residues in waters and wastewaters. In: Encyclopedia of Environmental Health, J. Nriagu (ed.) Elsevier, pp. 1–12, https://doi.org/10.1016/B978-0-12-409548-9.11247-3

Gogoi A., Mazumder P., Tyagi V. K., Tushara Chaminda G. G., An A. K. and Kumar M. (2018). Occurrence and fate of emerging contaminants in water environment: a review. *Groundwater for Sustainable Development*, **6**, 169–180, https://doi.org/10.1016/j.gsd.2017.12.009

Guo X., Cui X. and Li H. (2020). Effects of fillers combined with biosorbents on nutrient and heavy metal removal from biogas slurry in constructed wetlands. *Science of the Total Environment*, **703**, 134788, https://doi.org/10.1016/j.scitotenv.2019.134788

Guzel E. Y., Cevik F. and Daglioglu N. (2019). Determination of pharmaceutical active compounds in Ceyhan River, Turkey: seasonal, spatial variations and environmental risk assessment. *Human and Ecological Risk Assessment*, **25**(8), 1980–1995, https://doi.org/10.1080/10807039.2018.1479631

Hartl M., García-Galán M. J., Matamoros V., Fernández-Gatell M., Rousseau D. P. L., Du Laing G., Garfí M. and Puigagut J. (2021). Constructed wetlands operated as bioelectrochemical systems for the removal of organic micropollutants. *Chemosphere*, **271**, 129593, https://doi.org/10.1016/j.chemosphere.2021.129593

Havens S. M., Hedman C. J., Hemming J. D. C., Mieritz M. G., Shafer M. M. and Schauer J. J. (2010). Stability, preservation, and quantification of hormones and estrogenic and androgenic activities in surface water runoff. *Environmental Toxicology and Chemistry*, **29**(11), 2481–2490, https://doi.org/10.1002/and so on.307

Hu X., Xie H., Zhuang L., Zhang J., Hu Z., Liang S. and Feng K. (2021). A review on the role of plant in pharmaceuticals and personal care products (PPCPs) removal in constructed wetlands. *Science of the Total Environment*, **780**, 146637, https://doi.org/10.1016/j.scitotenv.2021.146637

Huang X., Duan C., Duan W., Sun F., Cui H., Zhang S. and Chen X. (2021). Role of electrode materials on performance and microbial characteristics in the constructed wetland coupled microbial fuel cell (CW-MFC): a review. *Journal of Cleaner Production*, **301**, 126951, https://doi.org/10.1016/j.jclepro.2021.126951

Jain M., Majumder A., Ghosal P. S. and Gupta A. K. (2020). A review on treatment of petroleum refinery and petrochemical plant wastewater: a special emphasis on constructed wetlands. *Journal of Environmental Management*, **272**, 111057, https://doi.org/10.1016/j.jenvman.2020.111057

Jain M., Majumder A., Gupta A. K. and Ghosal P. S. (2023). Application of a new baffled horizontal flow constructed wetland-filter unit (BHFCW-FU) for treatment and reuse of petrochemical industry wastewater. *Journal of Environmental Management*, **325**, 116443, https://doi.org/10.1016/J.JENVMAN.2022.116443

Kamilya T., Majumder A., Yadav M. K., Ayoob S., Tripathy S. and Gupta A. K. (2022). Nutrient pollution and its remediation using constructed wetlands: insights into removal and recovery mechanisms, modifications and sustainable aspects. *Journal of Environmental Chemical Engineering*, **10**(3), 107444, https://doi.org/10.1016/J.JECE.2022.107444

Kamilya T., Yadav M. K., Ayoob S., Tripathy S., Bhatnagar A. and Gupta A. K. (2023). Emerging impacts of steroids and antibiotics on the environment and their remediation using constructed wetlands: a critical review. *Chemical Engineering Journal*, **451**, 138759, https://doi.org/10.1016/J.CEJ.2022.138759

Kelly P. T. and He Z. (2014). Nutrients removal and recovery in bioelectrochemical systems: a review. *Bioresource Technology*, **153**, 351–360, https://doi.org/10.1016/j.biortech.2013.12.046

Kim K.-Y., Yang W., Evans P. J. and Logan B. E. (2016). Continuous treatment of high strength wastewaters using air-cathode microbial fuel cells. *Bioresource Technology*, **221**, 96–101, https://doi.org/10.1016/j.biortech.2016.09.031

Kostopoulou M. and Nikolaou A. (2008). Analytical problems and the need for sample preparation in the determination of pharmaceuticals and their metabolites in aqueous environmental matrices. *TrAC – Trends in Analytical Chemistry*, **27**(11), 1023–1035, https://doi.org/10.1016/j.trac.2008.09.011

Li H., Zhang S., Yang X.-L., Yang Y.-L., Xu H., Li X.-N. and Song H.-L. (2019). Enhanced degradation of bisphenol A and ibuprofen by an up-flow microbial fuel cell-coupled constructed wetland and analysis of bacterial community structure. *Chemosphere*, **217**, 599–608, https://doi.org/10.1016/j.chemosphere.2018.11.022

Li S., Wang C., Liu Y., Liu Y., Cai M., Zhao W. and Duan X. (2023). S-scheme MIL-101 (Fe) octahedrons modified Bi2WO6 microspheres for photocatalytic decontamination of Cr (VI) and tetracycline hydrochloride: synergistic insights, reaction pathways, and toxicity analysis. *Chemical Engineering Journal*, **455**, 140943.

Liu S., Lu F., Qiu D. and Feng X. (2022). Wetland plants selection and electrode optimization for constructed wetland-microbial fuel cell treatment of Cr(VI)-containing wastewater. *Journal of Water Process Engineering*, **49**, 103040, https://doi.org/10.1016/J.JWPE.2022.103040

Luo S., Zhao Z. Y., Liu Y., Liu R., Liu W. Z., Feng X. C., Wang A. J. and Wang H. C. (2023). Recent advancements in antibiotics containing wastewater treatment by integrated bio-electrochemical-constructed wetland systems (BES-CWs). *Chemical Engineering Journal*, **457**, 141133, https://doi.org/10.1016/j.cej.2022.141133

Lutterbeck C. A., Colares G. S., Oliveira G. A., Mohr G., Beckenkamp F., Rieger A., Lobo E. A., Rodrigues L. H. R. and MacHado Ê. L. (2022). Microbial fuel cells and constructed wetlands as a sustainable alternative for the treatment of hospital laundry wastewaters: assessment of load parameters and genotoxicity. *Journal of Environmental Chemical Engineering*, **10**(3), 108105, https://doi.org/10.1016/J.JECE.2022.108105

Majumder A., Gupta B. and Gupta A. K. (2019). Pharmaceutically active compounds in aqueous environment: a status, toxicity and insights of remediation. *Environmental Research*, **176**, 108542, https://doi.org/10.1016/j.envres.2019.108542

Majumder A., Gupta A. K., Ghosal P. S. and Varma M. (2021). A review on hospital wastewater treatment: a special emphasis on occurrence and removal of pharmaceutically active compounds, resistant microorganisms, and SARS-CoV-2. *Journal of Environmental Chemical Engineering*, **9**(2), 104812, https://doi.org/10.1016/j.jece.2020.104812

Majumder A., Gupta A. K. and Sillanpää M. (2022). Insights into kinetics of photocatalytic degradation of neurotoxic carbamazepine using magnetically separable mesoporous Fe_3O_4 modified Al-doped ZnO: delineating the degradation pathway, toxicity analysis and application in real hospital wastewater. *Colloids and Surfaces A: Physicochemical and Engineering Aspects*, **648**, 129250, https://doi.org/10.1016/j.colsurfa.2022.129250

Meng Y., Liu W., Liu X., Zhang J., Peng M. and Zhang T. (2021). A review on analytical methods for pharmaceutical and personal care products and their transformation products. *Journal of Environmental Sciences (China)*, **101**, 260–281. Chinese Academy of Sciences, https://doi.org/10.1016/j.jes.2020.08.025

Mompelat S., Jaffrezic A., Jardé E. and LeBot B. (2013). Storage of natural water samples and preservation techniques for pharmaceutical quantification. *Talanta*. Elsevier B.V., **109**, pp. 31–45, https://doi.org/10.1016/j.talanta.2013.01.042

Mu C., Wang L. and Wang L. (2020). Performance of lab-scale microbial fuel cell coupled with unplanted constructed wetland for hexavalent chromium removal and electricity production. *Environmental Science and Pollution Research*, **27**(20), 25140–25148, https://doi.org/10.1007/s11356-020-08982-z

Pant D., Singh A., Van Bogaert G., Irving Olsen S., Singh Nigam P., Diels L. and Vanbroekhoven K. (2012). Bioelectrochemical systems (BES) for sustainable energy production and product recovery from organic wastes and industrial wastewaters. *RSC Advances*, **2**(4), 1248–1263, https://doi.org/10.1039/c1ra00839k

Parida V. K., Sikarwar D., Majumder A. and Gupta A. K. (2022). An assessment of hospital wastewater and biomedical waste generation, existing legislations, risk assessment, treatment processes, and scenario during COVID-19. *Journal of Environmental Management*, **308**, 114609, https://doi.org/10.1016/J.JENVMAN.2022.114609

Pavlović D. M., Babić S., Horvat A. J. M. and Kaštelan-Macan M. (2007). Sample preparation in analysis of pharmaceuticals. *TrAC – Trends in Analytical Chemistry*, **26**(11), 1062–1075, https://doi.org/10.1016/j.trac.2007.09.010

Perreault F., Jaramillo H., Xie M., Ude M., Nghiem L. D. and Elimelech M. (2016). Biofouling mitigation in forward osmosis using graphene oxide functionalized thin-film composite membranes. *Environmental Science and Technology*, **50**(11), 5840–5848, https://doi.org/10.1021/acs.est.5b06364

Prado M., Borea L., Cesaro A., Liu H., Naddeo V., Belgiorno V. and Ballesteros F. (2017). Removal of emerging contaminant and fouling control in membrane bioreactors by combined ozonation and sonolysis. *International Biodeterioration and Biodegradation*, **119**, 577–586, https://doi.org/10.1016/j.ibiod.2016.10.044

Pubchem. (2023). PubChem Compound – NCBI. National Center for Biotechnology Information, USA, https://www.ncbi.nlm.nih.gov/pccompound

Pun Á., Boltes K., Letón P. and Esteve-Nuñez A. (2019). Detoxification of wastewater containing pharmaceuticals using horizontal flow bioelectrochemical filter. *Bioresource Technology Reports*, **7**, 100296, https://doi.org/10.1016/j.biteb.2019.100296

Qadafi M., Rosmalina R. T., Pitoi M. M. and Wulan D. R. (2023). Chlorination disinfection by-products in Southeast Asia: a review on potential precursor, formation, toxicity assessment, and removal technologies. *Chemosphere*, **316**, 137817, https://doi.org/10.1016/j.chemosphere.2023.137817

Saidulu D., Gupta B., Gupta A. K. and Ghosal P. S. (2021). A review on occurrences, eco-toxic effects, and remediation of emerging contaminants from wastewater: special emphasis on biological treatment based hybrid systems. *Journal of Environmental Chemical Engineering*, **9**(4), 105282, https://doi.org/10.1016/j.jece.2021.105282

Santos J. L., Aparicio I., Callejón M. and Alonso E. (2009). Occurrence of pharmaceutically active compounds during 1-year period in wastewaters from four wastewater treatment plants in Seville (Spain). *Journal of Hazardous Materials*, **164**(2–3), 1509–1516, https://doi.org/10.1016/j.jhazmat.2008.09.073

Sochacki A., Felis E., Bajkacz S., Nowrotek M. and Miksch K. (2018). Removal and transformations of diclofenac and sulfamethoxazole in a two-stage constructed wetland system. *Ecological Engineering*, **122**, 159–168, https://doi.org/10.1016/j.ecoleng.2018.07.039

Stamatis N. K. and Konstantinou I. K. (2013). Occurrence and removal of emerging pharmaceutical, personal care compounds and caffeine tracer in municipal sewage treatment plant in Western Greece. *Journal of Environmental Science and Health – Part B Pesticides, Food Contaminants, and Agricultural Wastes*, **48**(9), 800–813, https://doi.org/10.1080/03601234.2013.781359

Thapa B., Sen Pandit S., Patwardhan S. B., Tripathi S., Mathuriya A. S., Gupta P. K., Lal R. B. and Tusher T. R. (2022). Application of microbial fuel cell (MFC) for pharmaceutical wastewater treatment: an overview and future perspectives. *Sustainability*, **14**, 8379, https://doi.org/10.3390/SU14148379

Tran N. H., Reinhard M. and Gin K. Y. H. (2018). Occurrence and fate of emerging contaminants in municipal wastewater treatment plants from different geographical regions–a review. *Water Research*, **133**, 182–207, https://doi.org/10.1016/j.watres.2017.12.029

USEPA. (2000). 7 Sampling and preparation for. In: Multi-Agency Radiation Survey and Site Investigation Manual, USEPA, United States, **7**, 1–28.

Vanderford B. J., Pearson R. A., Rexing D. J. and Snyder S. A. (2003). Analysis of endocrine disruptors, pharmaceuticals, and personal care products in water using liquid chromatography/tandem mass spectrometry. *Analytical Chemistry*, **75**(22), 6265–6274, https://doi.org/10.1021/ac034210g

Vanderford B. J., Mawhinney D. B., Trenholm R. A., Zeigler-Holady J. C. and Snyder S. A. (2011). Assessment of sample preservation techniques for pharmaceuticals, personal care products, and steroids in surface and drinking water. *Analytical and Bioanalytical Chemistry*, **399**(6), 2227–2234, https://doi.org/10.1007/s00216-010-4608-5

Villarín M. C. and Merel S. (2020). Paradigm shifts and current challenges in wastewater management. *Journal of Hazardous Materials*, **390**, 122139, https://doi.org/10.1016/j.jhazmat.2020.122139

Wang H. and Ren Z. J. (2014). Bioelectrochemical metal recovery from wastewater: a review. *Water Research*, **66**, 219–232, https://doi.org/10.1016/J.WATRES.2014.08.013

Wang H., Heil D., Ren Z. J. and Xu P. (2015). Removal and fate of trace organic compounds in microbial fuel cells. *Chemosphere*, **125**, 94–101, https://doi.org/10.1016/j.chemosphere.2014.11.048

Xu W., Zou R., Jin B., Zhang G., Su Y. and Zhang Y. (2022). The ins and outs of pharmaceutical wastewater treatment by microbial electrochemical technologies. *Sustainable Horizons*, **1**, 100003, https://doi.org/10.1016/J.HORIZ.2021.100003

Yan W., Xiao Y., Yan W., Ding R., Wang S. and Zhao F. (2019). The effect of bioelectrochemical systems on antibiotics removal and antibiotic resistance genes: a review. *Chemical Engineering Journal*, **358**, 1421–1437, https://doi.org/10.1016/j.cej.2018.10.128

Yates C. N., Varickanickal J., Cousins S. and Wootton B. (2016). Testing the ability to enhance nitrogen removal at cold temperatures with *C. aquatilis* in a horizontal subsurface flow wetland system. *Ecological Engineering*, **94**, 344–351, https://doi.org/10.1016/j.ecoleng.2016.05.064

Zhang L., Yin X. and Li S. F. Y. (2015). Bio-electrochemical degradation of paracetamol in a microbial fuel cell-Fenton system. *Chemical Engineering Journal*, **276**, 185–192, https://doi.org/10.1016/J.CEJ.2015.04.065

doi: 10.2166/9781789063752_0035

Chapter 3

Emerging contaminants in municipal sewage/sludge: occurrence, risk assessment, and treatment technologies

Bing Wang[1,2], *Tao Jiang*[2], *Nana Wang*[2] *and Qianqian Zou*[1]

[1]*College of Resources and Environmental Engineering, Guizhou University, Guiyang, Guizhou 550025, China*
[2]*Key Laboratory of Karst Georesources and Environment (Guizhou University), Ministry of Education, Guiyang, Guizhou 550025, China*

ABSTRACT

Emerging contaminants (ECs) have received widespread attention globally due to their potential ecological and human health impacts. Sludge, a byproduct of the sewage treatment processes, accumulates amounts of ECs or toxic byproducts that have not been fully degraded. Without proper treatment, it may affect the resource utilization of sludge and damage the ecological environment. Therefore, it is significant to investigate the generation mechanism of ECs in municipal sewage/sludge and explore effective treatment technologies to remove them. This chapter first reviews the main occurrence of ECs in municipal sewage/sludge. Secondly, the potential environmental risks of ECs are evaluated, and the treatment technologies of different ECs are discussed in detail, such as anaerobic consumption, aerobic composting, and advanced oxidation. Finally, suggestions and prospects are put forward for the removal of ECs from municipal sewage/sludge.

Keywords: municipal sludge, emerging pollutants, treatment technologies, risk assessment, resource utilization

3.1 INTRODUCTION

With the progress of industry, agriculture, and urbanization, various emerging contaminants (ECs), traditional organic pollutants, and heavy metals have also emerged in the environment, posing a serious threat. Among them, ECs persist in the environment at low concentrations. Common ECs include endocrine disruptors (EDCs), pharmaceuticals and personal care products (PPCPs), perfluorinated compounds (PFCs), and disinfection by-products (DBPs) of drinking water and microplastics (Cheng *et al.*, 2021). ECs have attracted global attention due to their concealment, persistence, and complexity in environmental governance. ECs can be directly discharged through various channels, such as aquaculture sewage, domestic sewage, and industrial sewage, or degraded and treated by wastewater treatment plants (WWTPs) before being discharged into the environment (Dubey *et al.*, 2021). As one of the essential places for treating ECs, WWTPs can still detect low concentrations of ECs in the effluent after treatment. Therefore, discharging water containing trace amounts of ECs into the environment may cause secondary pollution of the aquatic environment. In addition, sludge is a byproduct generated during the sewage treatment process, accumulating incompletely degraded ECs or toxic byproducts. Suppose untreated or improperly treated sludge is directly applied as fertilizer to soil or prepared into value-added products. In that case, it may indirectly cause harm to human health through the food chain and limit the resource utilization of sludge. Therefore, how to efficiently degrade ECs in sewage/sludge is of great significance for reducing ecological risks. This chapter systematically summarizes the sources of ECs in municipal sewage/sludge, and evaluates the environmental risks brought by ECs. At the same time, a detailed discussion is conducted on the treatment technology and degradation mechanisms of ECs in existing sewage/sludge. Finally, suggestions and prospects for removing ECs from sewage/sludge are proposed.

3.2 OCCURRENCE OF ECS IN MUNICIPAL SEWAGE/SLUDGE

The ECs originating from sewage are collected through the municipal system and flow to WWTPs. Urban sewage is a type of mixed wastewater, including domestic sewage, industrial wastewater, and surface runoff rainwater. The research results indicate that the removal efficiency of ECs from sewage by WWTPs is only 30%, and most of the ECs were transferred to municipal sludge through adsorption. Therefore, the concentration of ECs in municipal sludge may be higher after enrichment than in inflow sewage. The sludge produced by WWTPs may cause serious environmental risks in the subsequent resource utilization process (Barret *et al.*, 2012). Municipal sludge can be used for biogas production and as a raw material for biofuels, and it can also be digested and then used for soil improvement in agriculture. However, the ecotoxicological effects of residual ECs in municipal sludge on microorganisms, animals, and plants are currently a concern in this field. Hence, this section aims to summarize the occurrence of ECs in municipal sewage/sludge and trace their possible sources. It is helpful to well design the removal technologies of ECs in WWTPs. The primary sources of ECs in WWTPs are summarized in Figure 3.1.

Antibiotics are widely used in multiple industries, such as healthcare, veterinarians, and agriculture, and are also one of the most common ECs in sewage. However, most antibiotics are not utilized by organisms and are released into the environment as parent compounds or metabolites. The main sources of antibiotics in WWTPs are medical wastewater, domestic wastewater, and aquaculture wastewater. The types of antibiotics used in the aquaculture industry mainly include tetracyclines, sulfonamides, and quinolones. The amount of antibiotics discharged by this industry far exceeds the total human use. For example, Zhi *et al.* (2020) investigated the pollution status of veterinary antibiotics on household farms in the Erhai region of China. The results showed that antibiotics were found in soil, wastewater, and feces, the content of antibiotics in feces was the highest. Tetracyclines (mainly chlortetracycline) had the highest concentration (404.95 mg/kg) compared with other antibiotics. The main types of antibiotics on chicken farms were quinolones and macrolides. In addition to antibiotics, high concentrations of pharmaceuticals generated during production can remain in pharmaceutical wastewater. However, pharmaceutical companies have limited wastewater treatment capacity, resulting in many pharmaceuticals entering WWTPs. Furthermore, approximately 60–85% of pharmaceuticals are excreted from the body as raw compounds or metabolites through feces or urine after ingestion and then enter the WWTPs.

Antibiotic resistance genes (ARGs) are typical ECs that have attracted attention due to their adverse effects on treating pathogenic infections during antibiotic therapy. As the previous text shows, widely used antibiotics cannot be completely metabolized by humans or animals, and residuals would be discharged into the WWTPs. If low concentrations of antibiotics existed in the environment for a long time, it could increase the probability of producing ARGs and promote their proliferation. Similarly, the sources of ARGs in WWTPs are extensive, mainly from aquaculture, pharmaceuticals, hospitals, household wastewater, and initial rainwater runoff (Qin *et al.*, 2020). The urban drainage system enters WWTPs by collecting the initial rainwater runoff, and ARGs in the septic tank of the breeding farm and

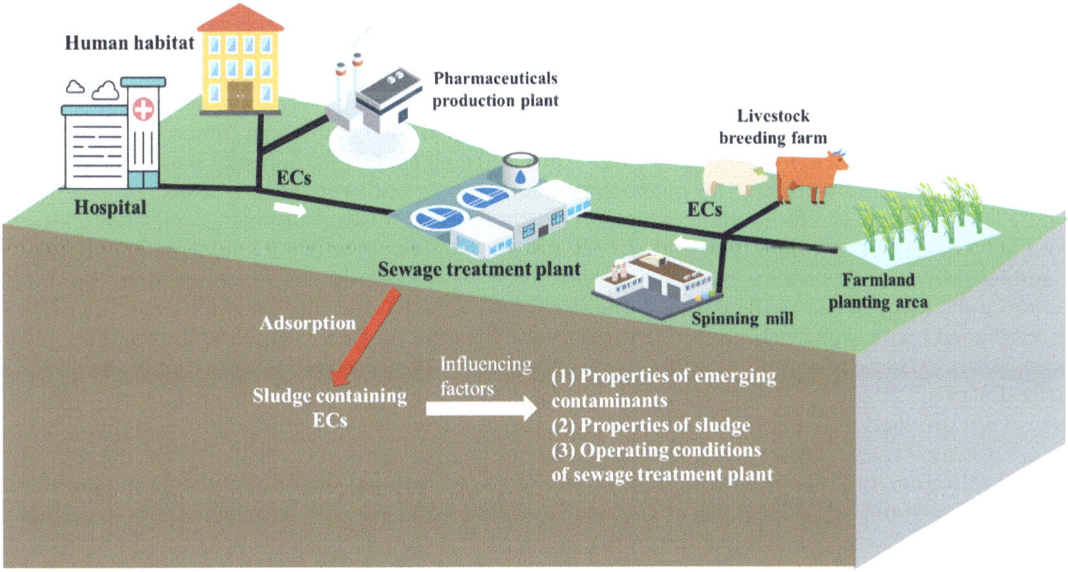

Figure 3.1 Primary sources of ECs in WWTPs.

field irrigation enter WWTPs along with the rainwater runoff. The increased hydraulic load of WWTPs may reduce the removal efficiency of ARGs. Garner *et al.* (2017) detected as many as 121 kinds of ARGs in a single urban river sample, and five kinds of ARGs *sul*1, *sul*2, *tet* (O), *tet* (W), and *erm* (F) in urban inland rivers were increased after a rainstorm under the same background value. However, the complex composition of rainwater may promote the variation and proliferation of ARGs, while limited research systematically elucidates the interaction between ARGs and rainwater components.

Natural and synthetic steroid hormones belong to a category of endocrine disruptors, causing serious harm to the environment and humans, even in trace amounts. Natural steroid hormones, such as progestogen, mineralocorticoid, androgen, and estrogen, are secreted by the human and animal ovaries, testes, placentas, and adrenal cortex. The concentration of steroid hormones in sewage is closely related to the composition and distribution of the population. Steroid estrogens in WWTPs exist in inactive glucuronic acid glycosides, and sulfate complexes, or free forms. The source of steroid estrogen in the pharmaceutical industry is by-products produced by preparing oral contraceptives, which inflow WWTPs along with the sewage. According to a current study, the urine excreted by males every day contained estrone (3.9 μg), 17β-estradiol (1.6 μg), and estriol (1.5 μg), while pregnant women had a higher excretion quantity, reaching 600 μg estrogen, 259 μg 17β-estradiol, and 6000 μg estriol (Johnson *et al.*, 2000). In another study, Garner *et al.* (2017) detected the daily emissions of 12 natural estrogens in pig and cow urine. The results showed that the daily urinary excretion rates of 12 natural estrogens in boars and sows ranged from 322.5 to 575.4 μg/d and 330.3 to 1100.7 μg/d, respectively. The urinary excretion quantity of 12 natural estrogens in non-pregnant beef cattle ranged from 338.2 to 2093.2 μg/d, the urine excretion quantity of pregnant beef cattle was 4974.9 μg/d, and the total estrogen excreted by a male beef cow was 1201.1 μg/d. The natural estrogen in livestock urine is the main source of estrogen in livestock wastewater, and the ecological risks of its discharge into the environment should also be of concern.

Microplastics are a kind of EC inevitably produced in daily life. Their particle and fiber sizes are less than 5 nm, characterized by their small volume and strong adsorption capacity. In the WWTPs, the types of microplastics are mainly microfiber, microplastic fragments, and plastic particles. Currently, the main sources of microplastics in the WWTPs include personal care products, cosmetics, and laundry textile fibers in domestic sewage, the breaking of large pieces of plastic in industrial wastewater, and the plastic brought by rainwater runoff. Praveena *et al.* (2018) found the existence of microplastics in all facial cleanser/scrub samples, mainly low-density polyethene and polypropylene. In another study, Chang (2015) found that the bestselling facial cleanser in the United States was characterized by white, opaque balls with particle sizes ranging from 60 to 800 μm. These microplastics were similar to the color of plankton, and were easy to mistakenly eaten by fish, and were then transferred to a higher trophic level through the food chain. This noteworthy finding raises significant ecological concerns about the potential bioaccumulation and biomagnification of microplastics within aquatic ecosystems. In addition, fiber microplastics mainly originate from washing daily clothes, rotor spinning, fabric friction, and cutting in the textile process. The studies showed that the content of fiber microplastics in the roughened textile was five times higher than that in the unprocessed textile, and the number of fiber microplastics extracted from the traditional cut textile was 3–31 times higher than that of the laser cut textile (Cai *et al.*, 2020). Pinlova *et al.* (2022) systematically studied the existence of fiber microplastics during yarn production. Fiber microplastics have always existed in yarn production, and the spinning process parameters significantly affected the production of fiber microplastics, including ring spinning, compact spinning, rotor spinning, and air jet spinning. A study found that the inflowing wastewater of the WWTPs contained 15.70 (±5.23) microplastics/L, and 0.25 (±0.04) microplastics/L were detected during emission. Even though a small number of microplastics were released per liter of water, many microplastics would inflow into the environment due to the enormous amount of water treated.

Rainwater and sewage confluence exist due to unreasonable urban planning, and plastic wastes may be carried in the surface runoff, forming microplastics through friction decomposition in the flow process. Cheung *et al.* (2019) detected the microplastic abundance of Hong Kong rivers after rainfall and obtained that the abundance of microplastics in the river after rainfall was 7.428/m^3, almost twice the observed value of coastal sea level in the same area. The microplastics transported by discontinuous and explosive surface runoff would multiply during rainfall; thus, surface runoff is also one of the sources of microplastics in WWTPs that cannot be ignored.

The ECs in municipal sludge originate from sewage, their content is positively correlated with the concentration of ECs in the sewage, and adsorption is considered an essential mechanism for transferring ECs from sewage to sludge. The current concentration range of ECs in sludge ranges from ng/kg to mg/kg. The transfer of ECs from sewage to sludge depends on the physical and chemical properties of ECs (Dubey *et al.*, 2021). Li *et al.* (2013) detected the presence of 18 antibiotics in sewage/sludge samples collected from 45 WWTPs in 23 cities in China, including seven quinolones, six sulfonamides, and five macrolides. These sludge samples all showed similar antibiotic composition characteristics, and there was a significant correlation between total organic carbon and the total concentration of antibiotics. Besides, the types of ECs in sludge are closely related to the type of sludge. Martín *et al.* (2015) found that nonsteroidal anti-inflammatory pharmaceuticals, estrogen, and antiepileptic pharmaceuticals were more concentrated in the primary sludge, while antibiotics β-receptor blockers and lipid modulators had higher concentrations in the secondary sludge. Meanwhile, the concentration of ECs in sludge increased with the extension of sludge retention time. Positively charged ECs were more beneficial to adsorption onto sludge, such as positively charged amitriptyline, clozapine, verapamil, risperidone, and oxazine (Stevens-Garmon *et al.*, 2011), but neutral compounds only adsorb onto sludge by hydrophobic interactions. Differently, Yan *et al.* (2014) reported that high concentrations of quinolones in sludge were adsorbed onto sewage/sludge by chelating cations.

Various ECs have been found in sewage/sludge from global countries, and ECs in sludge are enriched into sludge by adsorption, influenced by multiple factors. However, current studies mainly focus on the interaction between ECs in wastewater and solid phases, and their relationships with colloids and suspended solids are unclear. Besides, although the analysis methods of ECs in sewage have been relatively mature, the analysis methods in sludge are still lacking. It is urgent to establish standard analysis methods to determine the abundance and types of ECs (such as microplastics and viruses) in sludge, analyze the components of ECs in sludge, and reduce the risks to the ecological environment.

3.3 RISK ASSESSMENT OF ECS IN MUNICIPAL SEWAGE/SLUDGE

3.3.1 Ecological risk assessment

ECs usually have characteristics such as biological toxicity, environmental persistence, and bioaccumulation, and their cumulative level and detection frequency in the environment are increasing day by day. Consequently, it is imperative to assess the ecological risk associated with ECs in the environment. The United Nations Environment Programme has explicitly stated that the impact of ECs such as antibiotics and pharmaceuticals on aquatic ecosystems must be considered. Ecological risk assessment is the process of predicting the likelihood of harmful effects of ECs on the ecosystem. It refers to the risks borne by the ecosystem and its components, or the threat posed by uncertain accidents or disasters within a certain area, such as an individual, population, community composition, or the entire ecosystem. Ecological risk assessment, as a systematic analysis method, provides forward-looking prevention strategies for protecting ecosystems. At present, the four-step method is commonly used for ecological risk assessment based on dose–effect assessment and exposure assessment. Figure 3.2 shows the basic steps of ecological risk assessment. Among them, hazard identification is to determine the potential risk of pollutants to the ecological environment based on field research, data collection, and risk source identification. Exposure assessment is based on hazard identification, accurately describing the exposure intensity, exposure pathway, and spatiotemporal range of ecological receptors. Dose–effect assessment is to analyze the possible ecological effects of ecological receptors exposed to a certain risk source. Risk characterization is the combination of both exposure assessment and dose–effect

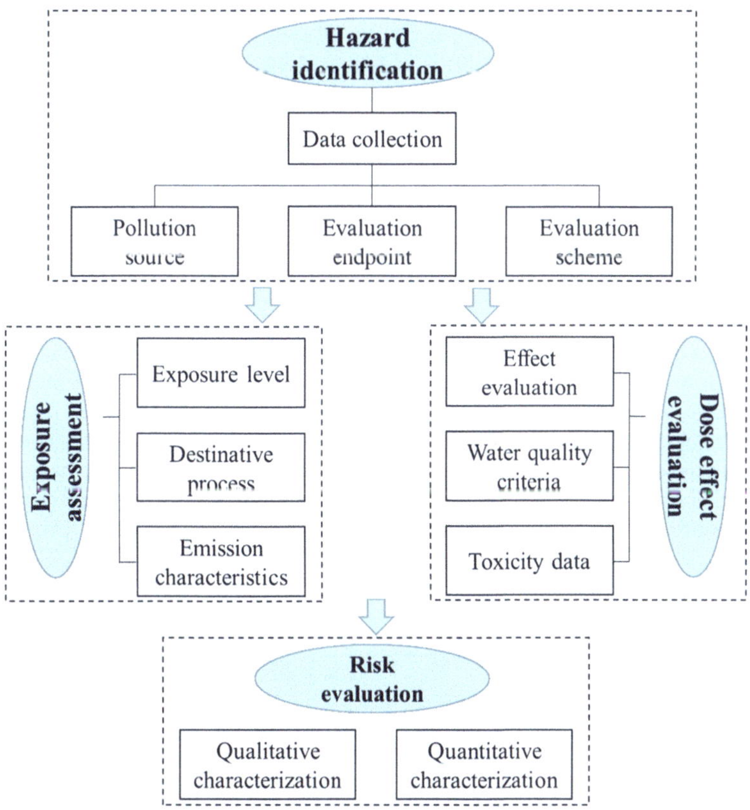

Figure 3.2 Basic steps of ecological risk assessment.

assessment to evaluate the risk and magnitude of stress factors on ecosystems and their components and to analyze the uncertainty of the evaluation results (Wu *et al.*, 2023; Yang *et al.*, 2022a).

The main purpose of ecological risk assessment is to integrate the relationship between exposure assessment and effect assessment and then characterize the risk. There are two commonly used expressions for risk characterization: qualitative and quantitative. When data and information resources are sufficient, quantitative evaluation is usually used. Quantitative risk assessment methods mainly include entropy method, safety threshold method, and probability method.

The entropy method is to compare the measured environmental concentration (MEC) or predicted environmental exposure concentration (PEC) of pollutants in the water environment with the maximum concentration, representing that the aquatic ecosystem is not endangered, and obtain the risk entropy (RQ). That is RQ = PEC/PNEC. The PNEC value is usually the ratio of acute and chronic toxicity data (lethal concentration half (LC_{50}), effect concentration half (EC_{50}), and maximum no effect concentration (CNOE) to the assessment factor (AF). The corresponding risk levels were classified (Ren *et al.*, 2023) as shown in Table 3.1. Wu *et al.* (2023) used the entropy method to calculate 22 types of PPCPs found in WWTPs and surface water and quantitatively allocate risks from specific sources. The findings indicated that there were significant differences in the ecological risks of PPCPs from six sources (medical sewage, farmland drainage, aquaculture, WWTPs, domestic sewage, and livestock discharge), but none of them reached high risks. Among them, although domestic sewage poses the most significant threat to the aquatic ecosystem, its contribution in terms of source proportion is lower compared to that of medical sewage. The incidence of potential risks resulting from urban domestic sewage ($RQ > 0.01$) is higher at 88.9% compared to rural domestic sewage at 75.9%. Camotti Bastos *et al.* (2020) evaluated the potential ecological toxicity of pharmaceutical compounds in sewage/sludge to the environment using the entropy method. The results showed that the RQ values of trimethoprim (25.20), ciprofloxacin (8.98), and norfloxacin (7.55) against bacteria in lime sludge were higher than 1. Additionally, the RQ values for sulfamethoxazole against invertebrates and algae in digested sludge were 32.47 and 46.70, respectively, indicating extremely high ecological risks. The entropy method is suitable for evaluating the toxic effects of individual compounds and is relatively simple to apply. However, it is only a rough estimation of ecological risks, which leads to uncertainty in its calculation results. For example, the measured total amount is related to the actual intake of the organism, individual exposure differences within the population, and sensitive differences (Thomaidi *et al.*, 2016).

The safety threshold method characterizes the ecological risk of ECs by comparing the safety threshold of biological communities with the exposure concentration of pollutants. The ratio of the concentration at 10% on the cumulative distribution curve of species sensitivity or toxicity data (SSD_{10}) to the concentration at 90% on the cumulative distribution curve of environmental exposure concentration (EXD_{90}) is used to analyze the degree of overlap and characterize the risk, that is, $MOS_{10} = SSD_{10}/EXD_{90}$. The safety threshold method utilizes both the distribution curve of pollutant toxicity effects and the curve of pollutant environmental exposure concentration, which is an extension of the entropy method. When $MOS_{10} > 1$, it indicates that there is no risk; when $MOS_{10} \leq 1$, it signifies a significant overlap between the two distributions, indicating that the pollutant has the potential to pose a risk to the environment. Moreover, a smaller MOS_{10} value corresponds to

Table 3.1 Classification of ecological risk levels.

RQ Value	Ecological Risk Levels
$RQ < 0.01$	Mild risk
$0.01 \leq RQ < 0.1$	Low risk
$0.1 \leq RQ < 1$	Medium risk
$RQ \geq 1$	High risk

Note: The higher the RQ value, the higher the ecological risk that the pollutant poses to the environment.

a greater degree of overlap, indicating a higher ecological risk associated with the pollutant (He *et al.*, 2019). Liu *et al.* (2020) evaluated the ecological risk of tris (1,3-dichloro-2-propyl) phosphate (TDCPP) using the safety threshold method. The findings indicated that both the toxicity data and exposure data followed a normal distribution after logarithmic transformation. The resulting MOS_{10} value of 0.51 suggested that TDCPP has the potential to pose ecological risks due to its toxicity to growth and development. The safety threshold method considers the uncertainty between environmental exposure concentration and toxicity data and is a more reasonable risk assessment method. However, when using the safety threshold method, it is necessary to use the cumulative probability distribution map of logarithmic toxicity data. When a certain pollutant does not have sufficient toxicity data, this method cannot be used for ecological risk assessment.

The probability method involves comparing the distributions of exposure concentration and toxicity data concentration to determine the probability of the expected entropy value being greater or less than the threshold value for determination. Specifically, Monte Carlo models were used to fit the distribution of exposure concentration and toxicity data (Xu *et al.*, 2015), followed by random sampling of the two distributions (such as 10,000 samples) to determine the ratio of exposure concentration to toxicity data and to obtain the probability of exceeding a specific RQ. The calculation formula is: DBQ = EXD/SSD. The probability method lies in the determination of the actual exposure level or probability curve of exposure concentration and toxicity reference values or toxicity parameter probability curves. The actual exposure level or probability curve of exposure concentration is related to the concentration of pollutants and their proportion of bioavailability and effects. The probability curve of toxicity reference values or toxicity parameters is associated with the characteristics of pollutants and ecological receptors, as well as the mechanism of interaction between them (Kooistra *et al.*, 2005). The probability method requires the collection of a large amount of data and information, and the computational process is relatively complex. Its evaluation results have uncertainty, which limits its use to some extent (Ren *et al.*, 2021).

In summary, the entropy method can provide a relatively objective evaluation result, but it is sensitive to the weight of indicators, and careful consideration should be given to indicator selection and weight allocation when applying it. The safety threshold method can quickly evaluate specific pollution events, but setting the threshold requires considering multiple factors, including risk acceptance and uncertainty. The probability method requires high data requirements, reliable data support, and consideration of uncertain factors. Therefore, to efficiently determine the risk level of ECs in the ecological environment, multi-level risk assessment methods can be adopted in the future, such as combining the entropy method and the probability method, and using various methods and means to conduct ecological risk assessment from simple to complex, providing support for the treatment of ECs in sewage/sludge.

3.3.2 Health risk assessment

ECs have been proven to interfere with the human endocrine system, affect the normal metabolism and immune function of the human body, and then cause diseases. Environmental epigenome believes that, in addition to chemical structure and concentration, ECs can lead to epigenetic changes through DNA methylation and affect human health. To clarify the harm of ECs in sewage/sludge to human health, there is a necessity for an objective health risk assessment of ECs in sewage/sludge. Predict the estimated likelihood of adverse effects of ECs on the human body through health risk assessment, and quantify the adverse effects on human health . To ensure a comprehensive health risk assessment, it is essential to adhere to four key steps, which are hazard identification, exposure assessment, dose-effect assessment, and risk characterization. Exposure assessment is the main basis of health risk assessment, and the accuracy of exposure parameter data directly determines the reliability of health assessment results. Exposure pathways include respiratory, dietary, and skin contact pathways (Hu *et al.*, 2020).

Based on the carcinogenicity of pollutants, health risk assessment is categorized into two main branches: carcinogenic risk assessment and non-carcinogenic risk assessment. In the evaluation of

ECs in sewage and sludge, non-carcinogenic risk assessment is commonly employed for health risk assessment. The non-carcinogenic risk assessment is conducted by utilizing a risk index known as the Hazard Quotient (HQ). The HQ is determined by dividing the long-term daily intake dose (CDI) resulting from pollutant exposure by the reference dose (RfD). When HQ ≥ 1, it indicates that ECs pose a high risk to human health; when HQ < 1, it means that ECs have a low health risk (Bruce *et al.*, 2010). Margenat *et al.* (2020) found that the concentration range of lincomycin, ciprofloxacin, and azithromycin in lettuce leaves grown in sludge-improved soil was 0.7–4.2 ng/g, and its total risk index (THQ) was less than 1, indicating that the level of antibiotics in lettuce was not a threat to human health. Similarly, eating radishes would not result in adverse effects on human health, as indicated by HQ values below 1 for azithromycin and sulfamethoxazole (You *et al.*, 2020). Therefore, a human health risk assessment of EC in sewage/sludge is important for the safe disposal of sewage/sludge to clarify the extent of the risk to human health. Future health risk assessments should consider the comprehensive impact of multiple factors, including environmental factors, lifestyle, and genetic inheritance. Meanwhile, long-term follow-up research should also be implemented to better appreciate the long-term effects of specific factors on human health through long-term observation and data collection, which can help improve the scientific and practical nature of health risk assessment.

3.4 TREATMENT TECHNOLOGIES OF ECS IN MUNICIPAL SEWAGE/SLUDGE

3.4.1 Treatment technologies of ECs in municipal sewage

3.4.1.1 Adsorption

As a traditional sewage treatment technology, the adsorption method utilizes porous solids (adsorbents) to adsorb pollutants onto the surface of the adsorbent, thereby purifying sewage. The commonly used adsorbents are activated carbon, biochar, activated alumina, zeolite, and clay minerals. These adsorbents usually have suitable pore size and surface structure and do not undergo chemical reactions with the adsorbent or medium (Varsha *et al.*, 2022). Among them, activated carbon (AC), as one of the adsorbents, has been widely used due to its rich functional groups, strong stability, and good recovery performance. It has been successfully used for the adsorption of ECs, such as perfluorooctanoic acid, pharmaceuticals, antibiotics, and microplastics. Research reports that the different pore structures and morphological characteristics of AC can affect the adsorption efficiency of target pollutants. For example, AC with micropores (width <2 nm) can better adsorb small-molecule pollutants, while it is generally difficult to achieve ideal adsorption effects for large-molecule pollutants. Therefore, Bedia *et al.* (2018) increased the porous structure and surface area of AC by activating it with $FeCl_3$, thereby improving the removal of antipyrine. Xu *et al.* (2023) used S/Fe co-doping to activate AC, increasing its specific surface area to 1194.14 m^2g^{-1} and enriching its porous structure, the removal rate of triclosan also reached 91.5%. Vieira *et al.* (2021) also found that activating AC through K_2CO_3 significantly increased the porous quantity of AC and effectively removed atrazine. In addition, AC can be divided into granular activated carbon (GAC) and powdered activated carbon (PAC) based on its appearance and morphology. Generally, PAC has a shorter adsorption time and stronger adsorption capacity than GAC. However, PAC may remain largely in activated sludge after use, making it difficult to recover and separate, and it is usually used as a disposable adsorbent. Therefore, based on the characteristics of the adsorbent and its regeneration cost, AC is still not considered the optimal adsorbent material, and many pharmaceuticals cannot achieve good removal effects through AC.

The other adsorbent is clay minerals, a non-toxic, cheap, and natural mineral considered a good adsorbent material. Currently, clay minerals are mainly used as adsorbents to remove antibiotics from ECs. Their adsorption mechanisms mainly include ion exchange, electrostatic interaction, hydrophobic interaction, and so on. Meanwhile, it was found that after modification treatment (heat treatment, acid activation, chemical treatment), clay minerals can improve their specific surface area, biocompatibility, and regeneration, and as well as their adsorption capacity for target pollutants. In addition, biochar, a porous carbon-rich material prepared by pyrolysis under anaerobic conditions, has a longer life cycle than activated carbon. Compared with other adsorbents, biochar has a relatively rich source, from agricultural wastes, animal and plant residues, and urban waste, and is considered the most promising adsorbent material. The adsorption mechanism of biochar on ECs in water mainly includes electrostatic interactions, $\pi-\pi$ interactions, and pore filling. Finally, pore filling is a unique pollutant removal mechanism of biochar that is closely related to the microporous surface area of biochar. Besides, modification can further improve the physical and chemical properties of biochar to enhance its adsorption capacity.

The three materials mentioned above are currently commonly used or the most promising adsorbents, and adsorbent materials such as silica gel, alumina, and zeolite also have certain application value in practical environments. Although adsorption can easily and quickly remove pollutants, it cannot currently be used as a large-scale sewage treatment method due to limitations in adsorption capacity, cost, and regeneration performance. Therefore, the future research direction of adsorption methods should be to develop cheaper and more efficient adsorbents.

3.4.1.2 Biological treatment

According to the treatment environment of sewage, biological treatment can be divided into aerobic and anaerobic biological treatment. Among them, the activated sludge method in aerobic treatment is a commonly used biological treatment process in WWTPs (Mpongwana & Rathilal, 2022). ECs adsorbed into activated sludge are degraded or transformed through microbial metabolism. At the same time, some ECs can also serve as a source of nutrients for microorganisms, promoting their metabolic processes. The degradation processes of ECs through the activated sludge method mainly involve hydroxylation, carboxylation, oxidation, and ring opening, leading to their degradation into H_2O and CO_2. For example, tetracycline undergoes oxidation during biodegradation (Taşkan *et al.*, 2016). Gemcitabine, cephalosporins, and penicillin antibiotics may undergo carboxylation reactions during biodegradation (Kamal *et al.*, 2023; Kong *et al.*, 2019). Sulfonamide antibiotics undergo processes such as hydroxylation and acetylation (Wang *et al.*, 2023). However, considering that ECs mostly exist in trace concentrations in the environment, the energy generated during the degradation process may not be sufficient to sustain microbial life activities. This may require the addition of additional nutrients to maintain the growth and metabolism of microorganisms. In addition, the degradation efficiency of the activated sludge method is also affected by the characteristics of sludge, operating conditions, and the complex properties of the ECs themselves. When the sludge index is high, it may cause sludge bulking, reducing its ability to adsorb pollutants. The high or low pH of the solution and changes in water temperature can affect the activity of enzymes, thereby affecting the biodegradation effect.

Anaerobic biological treatment refers to the decomposition of ECs into CO_2 and CH_4 under anaerobic conditions and the combined action of multiple microorganisms. Compared with aerobic processes, anaerobic processes have lower sludge production and operating costs. In recent decades, many studies have been focused on the anaerobic depletion treatment of ECs. Some studies have found that anaerobic biodegradation can reduce nitro, demethylation, and hydrolysis reactions, breaking toxic ECs into simpler, less toxic small-molecule substances. However, some studies have shown that most perfluorinated compounds are not biodegradable. This is attributed to the high polarity of C–F bonds in perfluorinated pollutants, which increases the hydrophobicity and liposolubility of these ECs and hinders the biodegradation process. The degradation mechanisms of other ECs through anaerobic biological treatment mainly include adsorption, non-biological reactions, and biodegradation. Different enzymes produced during microbial metabolism may affect the degradation effect of ECs during the biodegradation process. For example, Carneiro *et al.* (2020) found that acidic enzymes may increase their affinity for organic groups, thereby improving the removal rate of some ECs. Perfluorooctanoic acid and perfluorooctane sulfonic acid achieved removal rates of 60% through acidic microorganisms (Huang & Jaffé, 2019). In addition, peroxidase, nitroreductase, and cellobiose dehydrogenase also affect the degradation effect. However, due to the high-cost limitations of biocatalysts in large-scale applications and the possibility of some efflux of chemicals damaging enzyme activity. Therefore, it is necessary to develop green, economical, and efficient remediation technologies in the future.

3.4.1.3 Advanced oxidation processes

Advanced oxidation processes (AOPs) change the structure of pollutants by generating strong oxidizing free radicals (OH) and chain reactions, degrading them into small-molecule substances, and then finally oxidizing them to CO_2 and H_2O (Jiang *et al.*, 2023). Commonly used AOPs include photocatalysis, Fenton oxidation, and ozone oxidation. These processes have been extensively applied to various ECs, such as antibiotics, PPCPs, and perfluorinated compounds. Photocatalytic oxidation mainly involves the degradation of pollutants through strong oxidizing free radicals generated by semiconductor catalysts under ultraviolet/visible light. Photocatalytic degradation mainly undergoes the following processes: (1) The pollutants are adsorbed on the surface of the catalyst. (2) The pollutants a bsorb suitable photons and generate holes in the valence band. (3) The pollutants are degraded through a redox reaction under the generated electron–hole pairs, and the H_2O and O_2 on the catalyst surface also generate hydroxyl and superoxide-free radicals through the reaction, which indirectly degrade the pollutants. Common photocatalysts include nanoprecious metals, metal oxides, and composite catalysts, and using different photocatalysts can produce different degradation effects. For example, TiO_2 has strong photocatalytic ability and a low price, but its utilization range for visible light is small. Similarly, as another type of metal oxide catalyst, ZnO also exhibits excellent catalytic performance only under UV irradiation. Nonmetallic photocatalysts such as graphene, organic semiconductors, and covalent organic frameworks have attracted extensive research in the field of photocatalysis due to their good stability, suitable bandgap width, and no metal leaching. The degradation pathways of ECs through photocatalytic oxidation can be divided into three types: (1) The pollutants undergo photoisomerization before degradation. (2) Under light, many reactive oxygen species are generated to oxidize and decompose pollutants. (3) Some organic pollutants absorb photons to produce living oxygen substances and then degrade their excited state. Table 3.2 lists the removal rates of ECs by some photocatalysts .

Ozone, as a strong oxidant, can directly or indirectly oxidize ECs. Direct oxidation is a selective reaction between ozone molecules and electron-rich sites in ECs. The indirect reaction mainly involves decomposing ozone into more reactive and less selective ·OH to oxidize ECs. Solid catalysts have been used for heterogeneous catalytic ozonation reactions to avoid toxic by-products (such as bromate) that may form during the ozone oxidation process. Solid catalysts have also been used to promote the decomposition of more ·OH by O_3. For example, $LaCoO_3$ as a catalyst for ozonation degradation of typical ECs, can improve the degradation performance of benzotriazole in individual ozonation and reduce the production of toxic by-products (Zhang *et al.*, 2019b). Cai *et al.* (2021) used $CoFe_2O_4$ as

Table 3.2 Removal rate of ECs by different photocatalysts.

Photocatalysts	Target Pollutants	Removal Rate (%)	Reaction Time (min)	References
C–N–S tridoped TiO_2	Tetracycline	99	180	Wang *et al.* (2011)
Multiwall carbon nanotubes/BiOI	Antipyrine	100	120	Gao *et al.* (2020)
TiO_2/g-C_3N_4	Diclofenac	93	90	González-González *et al.* (2022)
Bi_2WO_6	Tetracycline	97	120	Chu *et al.* (2016)
β-FeOOH@g-C_3N_4	Carbamazepine	92	30	Wang *et al.* (2020)
BiVO/CHCOO(BiO)	Bisphenol A	99	180	Zhang *et al.* (2019a)
InS/GdO	Oxytetracycline	80	50	Murugalakshmi *et al.* (2020)
ZnO-doped g-C_3N_4	Ciprofloxacin	91.2	150	Van Thuan *et al.* (2022)
BiOBr/BiVO	Organic dye	96	120	Liu *et al.* (2022)
SnO_2/CeO_2	Tetracycline	97	120	Mohammad *et al.* (2021)
Mn-CCMN	Crystal violet	97	120	Yang *et al.* (2022b)

a solid catalyst for catalytic ozonation, greatly improving the degradation rate of clofibric acid and increasing the mineralization rate from 29.3% in a single ozonation to 72.7%. Pokkiladathu *et al.* (2022) studied the AC/CeO_2/ZnO catalyzed ozonation to remove bisphenol A from water, resulting in a 25% increase in mineralization efficiency compared to non-catalytic ozonation. In summary, using solid catalysts to catalyze ozonation can accelerate the oxidation reaction and generate more unselective reactive oxygen species (·OH, $^1O_2$, $O_2^{\cdot-}$ and so on), thereby enhancing the degradation and mineralization of pollutants. In addition, the coupling process of photocatalytic ozonation has also been applied to the degradation of ECs, such as acetaminophen, antipyrine, and diclofenac. Due to O_3 being more electron-friendly than O_2, more ·OH is generated, and the photocatalytic ozonation significantly increases the oxidation rate of ECs.

Persulfate (PS) oxidation has been a hot research topic in recent years for AOPs. The PS process mainly degrades ECs through SO_4^-. Compared with other free radicals, SO_4^- has a higher oxidation-reduction potential and a wider pH tolerance range. SO_4^- also has non-selectivity and can directly react with various ECs, but the reaction rate is lower when PS directly reacts with ECs. Therefore, it is necessary to activate PS to improve its efficiency in removing ECs. Pirsaheb *et al.* (2020) compared the removal rate of amoxicillin using PS alone and ultraviolet (UV)-activated PS and proved that activated PS has a higher removal rate of pollutants. Mainly attributed to the UV activation of PS, which leads to O–O bond breakage and produces two SO_4^-, thereby improving the removal rate of ECs. In addition, ultrasonic activation of PS is also one of the commonly used methods, and ECs are mainly degraded through the following two pathways: (1) Ultrasound causes the rupture of cavitation bubbles, resulting in a high-temperature and high-pressure environment, which activates PS to degrade ECs. (2) The high-temperature and high-pressure environment generated by the rupture of cavitation bubbles decomposes water molecules into ·OH, thereby activating PS to degrade ECs. In addition, some transition metals and their metal oxides have high activation ability for PS and are currently one of the most commonly used methods for PS activation to produce SO_4^-. ECs can be degraded by metal ions in a single homogeneous phase.

3.4.1.4 Membrane treatment

Membrane treatment is the separation and purification of a mixture through potential, pressure, and concentration differences. According to the pore size of the membrane, it can be divided into four categories (microfiltration, ultrafiltration, nanofiltration, and permeation). Among them, ultrafiltration technology has removed various ECs with smaller pore sizes (pore sizes range from 0.001 to 0.02 μm). For example, Shakak *et al.* (2020) synthesized a nanocomposite ultrafiltration membrane (polysulfone/polyvinylpyrrolidone/SiO_2) to evaluate for removal of amoxicillin. Due to the presence of SiO_2, the hydrophilicity, porosity, and membrane flux of the composite membrane were increased. With increasing SiO_2 nanoparticles from 0 to 4 wt%, the amoxicillin separation performance increased from 66.52% to 89.81%. However, ultrafiltration technology may not be able to completely and effectively remove certain ECs (pore sizes 100–1000 times larger than ECs). Microfiltration technology is a membrane process driven by static pressure differences, utilizing the 'screening' effect of a mesh filter medium membrane for separation. Nanofiltration technology is a membrane separation method that utilizes a pressure gradient as the driving force. Due to the presence of charged groups on the surface of most nanofiltration membranes, the removal mechanisms for ECs mainly include the charge effect and the screening effect. Perfluorooctane sulfonic acid, as a persistent EC, is negatively charged in aqueous solutions and can be effectively removed through nanofiltration technology. In addition, infiltration technology is divided into two categories: forward osmosis (FO) and reverse osmosis (RO), and ECs can achieve good removal rates through infiltration technology. Guo *et al.* (2020) fabricated a thermoplastic polyurethane/polysulfone (TPU/PSF) composite membrane using electrospinning, and then loaded UiO-66-NH_2 particles onto the membrane. The results showed that the UiO/TPU/PSF forward osmosis membrane achieved a retention efficiency of 99.64% for tetracycline. Another permeation technology, RO, is a membrane separation technology that uses pressure difference as the driving force to separate solvents from solutions, with a membrane pore size between 0.5 and 10 nm. Alonso *et al.* (2018) found that RO can achieve a 99.96% removal effect of ciprofloxacin. The mechanisms by which ECs are removed through membrane technology mainly include electrostatic interactions, size exclusion, biodegradation, and hydrophobic interactions. Some non-polar ECs are mainly removed by size exclusion and adsorption onto the membrane surface, while polar ECs are mainly removed through biodegradation.

The urgent problem and challenge that membrane technology needs to solve is membrane pollution and blockage. Due to the potential for performance degradation and shortened membrane lifespan during long-term use, it may affect the degradation effect of ECs. Therefore, new membrane materials need to be developed to address the shortcomings of traditional membrane technologies. For example, Mendes *et al.* (2018) prepared the cellulose acetate silica-mixed ultrafiltration membrane with good permeability, hydrophilicity, and mechanical strength. Zhao *et al.* (2018) studied the thin-film nanocomposite forward osmosis membranes. The future membrane technology of ECs should develop composite membranes with strong stability, a large surface area, and efficient electron transfer. Meanwhile, the energy consumption generated by aeration should be reduced.

3.4.2 Treatment technologies of ECs in municipal sludge

It can be seen from Section 3.2 that ECs in sewage are enriched into sludge by adsorption, such as polychlorinated biphenyls, endocrine disruptors, and microplastics. Due to the low biodegradability and high persistence of ECs, they pose a greater threat than traditional organic pollutants. Effectively reducing the concentration of EC in sludge is of great significance to the environment and humanity (Golet *et al.*, 2002). The main treatment technologies and advantages of ECs in sludge are summarized in Figure 3.3.

3.4.2.1 *Aerobic composting*

Aerobic composting is a technology that utilizes aerobic microorganisms to decompose organic matter in sludge. It can achieve a harmless and stable treatment of sludge. Factors affecting the effectiveness

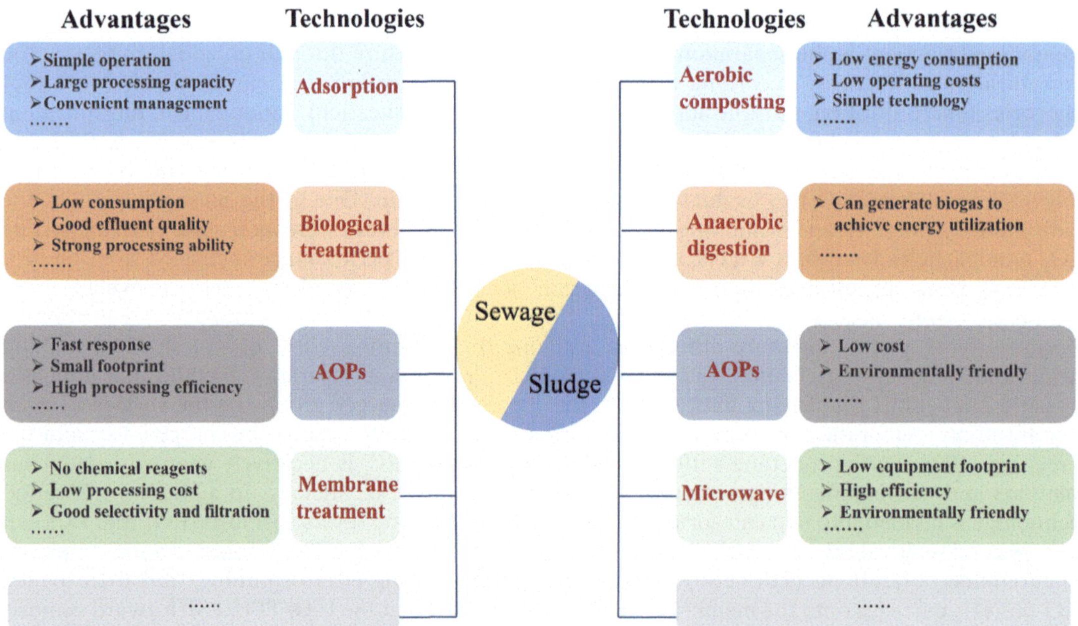

Figure 3.3 Treatment technologies and advantages of ECs in sludge.

of composting include composting fertilizer, composting conditions, microbial action, or the type and initial concentration of ECs. The most critical influencing factor is the microbial activity. In addition, when the concentration of ECs in sludge is high, the decomposition of organic matter can be correspondingly delayed, and the composting efficiency and quality can also change. Different types of ECs have different synthesis methods and physicochemical properties, which may lead to differences in degradation effects. Generally, the removal rate of tetracycline, sulfa, and macrolide antibiotics by compost is higher than that of quinolone antibiotics because most quinolone antibiotics are synthetic compounds with complex structures and are not easily affected by microorganisms. Khadra *et al.* (2018) used plant by-products and excess sludge for composting and found that ampicillin, clarithromycin, lincomycin, tetracycline, or trimethoprim could be eliminated, but ofloxacin and ciprofloxacin did not degrade well in this process. Furthermore, the study found that different types of sulfa antibiotics in sludge had different removal effects under aerobic composting, and the degradation order was sulfamethoxazole > sulfadimethoxine > sulfadimidine. Thermophilic conditions have a crucial impact on the composting process. High temperatures can promote tetracycline and sulfonamide antibiotics in sludge more effectively than low-temperature conditions. Finally, different composting fertilizers also affect the process. Lin *et al.* (2017) found that compost using pig manure as a substrate was more effective in removing sulfonamide antibiotics than compost using chicken manure. In summary, composting has good application prospects in sludge reduction treatment and removing ECs from sludge. However, the driving mechanisms of aerobic composting for the removal of ECs from sludge have not been systematically explored, and further exploration is needed.

3.4.2.2 Anaerobic digestion

Anaerobic digestion microorganisms decompose ECs in sludge in an anaerobic environment and convert sludge into high-value-added products. The process of anaerobic digestion primarily consists of four stages: hydrolysis, acidogenesis, acetogenesis, and methanogenesis. Among them, the hydrolysis stage is the most critical, where complex organic compounds in sludge are decomposed into glycerol, amino acids, and monosaccharide small-molecule compounds. Therefore, a part of the ECs is removed during hydrolysis, thus reducing the number of ECs in sludge. The factors affecting the anaerobic digestion process include temperature and sludge types. Temperature can affect the activity of microorganisms. There are certain differences in temperature between different seasons, resulting in changes in the concentration of ECs and the abundance of microbial communities in sludge. When granular sludge is subjected to crushing treatment, the distribution of microorganisms could be disrupted, causing direct contact between microbial communities and ECs to suffer toxic effects, thereby reducing the treatment efficiency of ECs. Importantly, when ECs exist in sludge, the anaerobic digestion process would be subject to many disturbances, such as inhibition of biogas production and imbalance of the microbial community, making ECs unable to be effectively degraded. For example, Liu *et al.* (2018) found that the removal of tetracycline in sludge through industrial-scale anaerobic digestion was challenging, with a total removal rate of less than 18%. Elevated concentrations of sulfamethoxazole exhibited significant toxicity towards microbial communities, leading to inhibition of substrate utilization and biogas production (Cetecioglu *et al.*, 2015). Based on this, additional measures can be taken to reduce the inhibition of ECs in sludge on microorganisms and improve anaerobic digestion performance.

Advanced anaerobic digestion is a commonly used method to improve anaerobic digestion performance, including ozone oxidation, ultrasonic combined ozone oxidation, and the addition of iron-based compounds. Ozone oxidation accelerates the decomposition of ECs and enhances the utilization of ECs by microorganisms. During the ultrasonic treatment of sludge, the rupture of bubbles can create high pressure, high temperature, and a strong shear force, which is very helpful for ECs reduction in sludge. The coupling of ultrasound and ozone oxidation aids in enhancing the generation of reactive oxygen species during ozone oxidation. This synergistic effect accelerates the degradation of ECs while reducing their inhibitory impact on microorganisms. It is worth noting that the addition of iron-based compounds affects the hydrolysis process of sludge, and iron is beneficial to the occurrence of enzyme catalysis. The research showed that the antibiotics, with the exception of ofloxacin, were effectively removed in the sewage sludge at a dosage of 1000 mg/L Fe^0, 20 d of solid retention time, and an antibiotic concentration of 20 μg/L, and *Erysipelotrichia*, *Verrucomicrobia*, *Clostridia*, *Caldiserica*, and *Alphaproteobacteria* of the class were dominated microorganisms in the anaerobic digestion (Zhou *et al.*, 2020).

The utilization of anaerobic digestion proves to be a successful and efficient means of treating ECs present in sludge. However, multiple ECs are present in sludge, and there is a lack of sufficient research on the synergistic and antagonistic effects of each EC. Although most of the ECs can inhibit the methanogenesis process, a few ECs (such as azithromycin and cefalexin) may promote methane formation under specific conditions, while the mechanisms of action are still unclear.

3.4.2.3 Advanced oxidation processes

AOPs are of interest for the removal of ECs from sludge due to their ability to produce highly oxidizing reactive oxygen species. Among them, ozone oxidation is a commonly used AOPs for removing ECs in sludge, promoting the desorption of ECs in sludge into the water, and reducing the difficulty of ECs removal. The removal of ECs from sludge by ozone oxidation mainly includes the following processes: (1) ECs are oxidized by ozone in wastewater. (2) ECs are desorbed and then oxidized by ozone. (3) ECs are directly oxidized by ozone and combined with sludge. Marce *et al.* (2017) found a variety of pharmaceuticals (ciprofloxacin, sulfamethoxazole, carbamazepine, diclofenac, and ibuprofen) that existed in sludge had been removed to a certain extent after the addition of ozone. However, the removal rate of pollutants depended on their transferred ozone dose ranges. In fact, when there are differences in the transferred ozone dose ranges of ECs, their removal rates may go to two extremes. For example, provided the transferred ozone dose range is 5 mg/g SS, the removal rate of ibuprofen is about 35–45%. Nevertheless, if the transferred ozone dose range is 10 mg/g SS, the removal rate would reach 99%.

In recent years, AOPs based on calcium peroxide (CaO_2) have been an emerging research direction in sludge pre-treatment technology, simultaneously achieving oxidative wall breaking and alkaline hydrolysis of sludge. Through water splitting, CaO_2 dissolves and decomposes into H_2O_2,·OH, and O_2^- species, thereby facilitating the degradation of ECs in sludge (Equations 1–5).

$$CaO_2 + H_2O \rightarrow Ca(OH)_2 + H_2O_2 \quad (3.1)$$

$$2CaO_2 + 2H_2O \rightarrow 2Ca(OH)_2 + O_2 \quad (3.2)$$

$$H_2O_2 + e^- \rightarrow \cdot OH + OH^- \quad (3.3)$$

$$H_2O_2 + \cdot OH \rightarrow H_2O + OH_2 \quad (3.4)$$

$$HO_2 \cdot \rightarrow O_2^- + H^+ \quad (3.5)$$

3.4.2.4 Other treatments

With the continuous progress of technology, in addition to the common technologies mentioned above, hydrothermal carbonization and microwave technology can also remove ECs from sludge. Below is a brief introduction to these two methods. Hydrothermal carbonization refers to the process of hydrolysis, dehydration, decarboxylation, condensation, and aromatization of sludge into high-value-added products (hydrothermal carbon) and a small amount of gas under appropriate temperature, pressure, and pH. During this process, pathogenic bacteria and other microorganisms in sludge are killed and ECs are removed, achieving sludge reduction and harmless treatment. In WWPTs, sludge can better combine with ECs in wastewater through extracellular polymers. Therefore, adding microbial cells with their cellular structures to sludge can release high-molecular-weight polymers, which are combined to form complex extracellular polymers. Due to cellular polymers, extracellular polymers can directly come into contact with ECs, making it easier for sludge to become colloids for EC sedimentation treatment. Because there is a strong attraction between extracellular polymers and ECs, effectively removing ECs from sludge. For example, a study showed that bisphenol A can bind with extracellular polymers contained in sludge through hydrophobic interactions, thereby achieving a higher removal rate of bisphenol A (Yan *et al.*, 2019).

Microwave technology can evenly and quickly heat samples to decompose, regulate, and destroy pathogens. Combined with chemical oxidation, this technology can also be used as an independent pre-treatment process and potentially improve sludge decomposition. The chemical oxidants H_2O_2 and PS are widely used in sludge disposal, and they can also be combined with microwave technology to produce synergistic effects. Research has found that microwave/hydrogen peroxide and microwave/persulfate technologies have certain effects on sludge dissolution, degradation of ECs in sludge, and removal of pathogenic bacteria (Bilgin Oncu & Akmehmet Balcioglu, 2013).

3.5 CONCLUSION AND FUTURE PERSPECTIVES

ECs in municipal sewage/sludge have become an important ecological and public health problem. This chapter reviews and analyzes their sources, risk assessment, and treatment techniques. ECs in urban sewage/sludge originate from medical institutions, pharmaceutical industries, and PPCPs. ECs entering the sewage/sludge through the municipal sewage networks pose potential ecosystem risks and negatively impact human health. The risk assessment of ECs in sewage/sludge is based on the ecological risk assessment framework. The focus is to identify the risk sources of ecosystems and their components, quantitatively predict the probability of risk occurrence and its harmful effects, and take corresponding control measures. Considering the removal of ECs in sewage/sludge, common treatment technologies include adsorption, biological methods (aerobic composting or anaerobic digestion), and advanced oxidation processes. Although positive results have been obtained in the identification, risk assessment, and treatment technology research of ECs in sewage/sludge, there are also new challenges that need to be addressed simultaneously, including but not limited to the following:

(1) With the continuous improvement of water quality management requirements, the toxicity testing of ECs should cover the entire life cycle of the tested species while considering regional research. The selection of testing indicators should not only be a simple acute toxicity endpoint but also require the addition of more sensitive testing endpoints.
(2) Currently, research focuses on evaluating specific single ECs, while multiple ECs and their metabolites often coexist in actual environments. The mixture of ECs may have synergistic or antagonistic effects on aquatic organisms, and the toxic effects generated by the combined effects should also be a key aspect of risk assessment. Therefore, comprehensive evaluation methods for multiple ECs and their metabolites should be thoroughly studied.
(3) There are many uncertainties in each risk assessment stage, such as screening of risk sources, defining risk receptors, determining evaluation endpoints, and selecting risk assessment methods. Especially for risk assessment methods, there is significant uncertainty in the selection of evaluation factors, statistical methods or models, the setting of various elements in simulated ecosystems, the construction of risk assessment models, and the determination of parameters. Therefore, establishing uncertainty analysis methods and reducing risk assessment uncertainty are important research topics for future risk assessment.
(4) Research has shown that multiple methods are available for reducing and removing ECs from sewage/sludge. Biological treatment technologies, such as aerobic composting and bioreactors, have proven effective in degrading and transforming pollutants. Physical and chemical treatment technologies such as adsorption, oxidation, and membrane separation also show

potential. However, it should be noted that different ECs may require targeted treatment methods, and by-products or conversion products may be generated during the treatment process, requiring further research on their environmental behaviors and potential risks.

(5) Traditional treatment techniques have certain treatment effects on ECs, but considering the limitations of a single process and removal efficiency issues. Therefore, it is inevitable to develop integrated technologies for the efficient removal of ECs, such as coupling membrane bioreactors with advanced oxidation processes or coupling membrane bioreactors with forward/reverse osmosis.

REFERENCES

Alonso J. J. S., El Kori N., Melián-Martel N. and Del Río-Gamero B. (2018). Removal of ciprofloxacin from seawater by reverse osmosis. *Journal of Environmental Management*, **217**, 337–345, https://doi.org/10.1016/j.jenvman.2018.03.108

Barret M., Delgadillo-Mirquez L., Trably E., Delgenes N., Braun F., Cea-Barcia G., Steyer J. P. and Patureau D. (2012). Anaerobic removal of trace organic contaminants in sewage sludge: 15 years of experience. *Pedosphere*, **22**(4), 508–517, https://doi.org/10.1016/S1002-0160(12)60035-6

Bedia J., Belver C., Ponce S., Rodriguez J. and Rodriguez J. J. (2018). Adsorption of antipyrine by activated carbons from $FeCl_3$-activation of Tara gum. *Chemical Engineering Journal*, **333**, 58–65, https://doi.org/10.1016/j.cej.2017.09.161

Bilgin Oncu N. and Akmehmet Balcioglu I. (2013). Microwave-assisted chemical oxidation of biological waste sludge: simultaneous micropollutant degradation and sludge solubilization. *Bioresource Technology*, **146**, 126–134, https://doi.org/10.1016/j.biortech.2013.07.043

Bruce G. M., Pleus R. C. and Snyder S. A. (2010). Toxicological relevance of pharmaceuticals in drinking water. *Environmental Science & Technology*, **44**(14), 5619–5626, https://doi.org/10.1021/es1004895

Cai C., Duan X., Xie X., Kang S., Liao C., Dong J., Liu Y., Xiang S. and Dionysiou D. D. (2021). Efficient degradation of clofibric acid by heterogeneous catalytic ozonation using $CoFe_2O_4$ catalyst in water. *Journal of Hazardous Materials*, **410**, 124604, https://doi.org/10.1016/j.jhazmat.2020.124604

Cai Y., Mitrano D. M., Heuberger M., Hufenus R. and Nowack B. (2020). The origin of microplastic fiber in polyester textiles: the textile production process matters. *Journal of Cleaner Production*, **267**, 121970, https://doi.org/10.1016/j.jclepro.2020.121970

Camotti Bastos M., Soubrand M., Le Guet T., Le Floch É, Joussein E., Baudu M. and Casellas M. (2020). Occurrence, fate and environmental risk assessment of pharmaceutical compounds in soils amended with organic wastes. *Geoderma*, **375**, 114498, https://doi.org/10.1016/j.geoderma.2020.114498

Carneiro R. B., Gonzalez-Gil L., Londoño Y. A., Zaiat M., Carballa M. and Lema J. M. (2020). Acidogenesis is a key step in the anaerobic biotransformation of organic micropollutants. *Journal of Hazardous Materials*, **389**, 121888, https://doi.org/10.1016/j.jhazmat.2019.121888

Cetecioglu Z., Ince B., Gros M., Rodriguez-Mozaz S., BarcelμË D., Ince O. and Orhon D. (2015). Biodegradation and reversible inhibitory impact of sulfamethoxazole on the utilization of volatile fatty acids during anaerobic treatment of pharmaceutical industry wastewater. *Science of the Total Environment*, **536**, 667–674, https://doi.org/10.1016/j.scitotenv.2015.07.139

Chang M. (2015). Reducing microplastics from facial exfoliating cleansers in wastewater through treatment versus consumer product decisions. *Marine Pollution Bulletin*, **101**(1), 330–333, https://doi.org/10.1016/j.marpolbul.2015.10.074

Cheng N., Wang B., Wu P., Lee X., Xing Y., Chen M. and Gao B. (2021). Adsorption of emerging contaminants from water and wastewater by modified biochar: A review. *Environmental Pollution*, **273**, 116448, https://doi.org/10.1016/j.envpol.2021.116448

Cheung P. K., Hung P. L. and Fok L. (2019). River microplastic contamination and dynamics upon a rainfall event in Hong Kong, *China. Environmental Processes*, **6**(1), 253–264, https://doi.org/10.1007/s40710-018-0345-0

Chu X., Shan G., Chang C., Fu Y., Yue L. and Zhu L. (2016). Effective degradation of tetracycline by mesoporous Bi_2WO_6 under visible light irradiation. *Frontiers of Environmental Science & Engineering*, **10**(2), 211–218, https://doi.org/10.1007/s11783-014-0753-y

Dubey M., Mohapatra S., Tyagi V. K., Suthar S. and Kazmi A. A. (2021). Occurrence, fate, and persistence of emerging micropollutants in sewage sludge treatment. *Environmental Pollution*, **273**, 116515, https://doi.org/10.1016/j.envpol.2021.116515

Gao P., Yin Z., Feng L., Liu Y., Du Z., Duan Z. and Zhang L. (2020). Solvothermal synthesis of multiwall carbon nanotubes/BiOI photocatalysts for the efficient degradation of antipyrine under visible light. *Environmental Research*, **185**, 109468, https://doi.org/10.1016/j.envres.2020.109468

Garner E., Benitez R., von Wagoner E., Sawyer R., Schaberg E., Hession W. C., Krometis L.-A. H., Badgley B. D. and Pruden A. (2017). Stormwater loadings of antibiotic resistance genes in an urban stream. *Water Research*, **123**, 144–152, https://doi.org/10.1016/j.watres.2017.06.046

Golet E. M., Strehler A., Alder A. C. and Giger W. (2002). Determination of fluoroquinolone antibacterial agents in sewage sludge and sludge-treated soil using accelerated solvent extraction followed by solid-phase extraction. *Analytical Chemistry*, **74**(21), 5455–5462, https://doi.org/10.1021/ac025762m

González-González R. B., Sharma A., Parra-Saldívar R., Ramirez-Mendoza R. A., Bilal M. and Iqbal H. M. N. (2022). Decontamination of emerging pharmaceutical pollutants using carbon-dots as robust materials. *Journal of Hazardous Materials*, **423**, 127145, https://doi.org/10.1016/j.jhazmat.2021.127145

Guo J., Huang M., Gao P., Zhang Y., Chen H., Zheng S., Mu T. and Luo X. (2020). Simultaneous robust removal of tetracycline and tetracycline resistance genes by a novel UiO/TPU/PSF forward osmosis membrane. *Chemical Engineering Journal*, **398**, 125604, https://doi.org/10.1016/j.cej.2020.125604

He W., Kong X., Qin N., He Q., Liu W., Bai Z., Wang Y. and Xu F. (2019). Combining species sensitivity distribution (SSD) model and thermodynamic index (exergy) for system-level ecological risk assessment of contaminates in aquatic ecosystems. *Environment International*, **133**, 105275, https://doi.org/10.1016/j.envint.2019.105275

Hu H., Mao L., Fang S., Xie J., Zhao M. and Jin H. (2020). Occurrence of phthalic acid esters in marine organisms from Hangzhou Bay, China: implications for human exposure. *Science of the Total Environment*, **721**, 137605, https://doi.org/10.1016/j.scitotenv.2020.137605

Huang S. and Jaffé P. R. (2019). Defluorination of perfluorooctanoic acid (PFOA) and perfluorooctane sulfonate (PFOS) by *Acidimicrobium* sp. strain A6. *Environmental Science & Technology*, **53**(19), 11410–11419, https://doi.org/10.1021/acs.est.9b04047

Jiang T., Wang B., Gao B., Cheng N., Feng Q., Chen M. and Wang S. (2023). Degradation of organic pollutants from water by biochar-assisted advanced oxidation processes: mechanisms and applications. *Journal of Hazardous Materials*, **442**, 130075, https://doi.org/10.1016/j.jhazmat.2022.130075

Johnson A. C., Belfroid A. and Di Corcia A. (2000). Estimating steroid oestrogen inputs into activated sludge treatment works and observations on their removal from the effluent. *Science of the Total Environment*, **256**(2), 163–173, https://doi.org/10.1016/S0048-9697(00)00481-2

Kamal N., Sindhu R. and Chaturvedi Bhargava P. (2023). Biodegradation of emerging organic pollutant gemfibrozil: mechanism, kinetics and pathway modelling. *Bioresource Technology*, **374**, 128749, https://doi.org/10.1016/j.biortech.2023.128749

Khadra A., Ezzariai A., Merlina G., Capdeville M.-J., Budzinski H., Hamdi H., Pinelli E. and Hafidi M. (2018). Fate of antibiotics present in a primary sludge of WWTP during their co-composting with palm wastes. *Waste Management*, **84**, 13–19, https://doi.org/10.1016/j.wasman.2018.11.009

Kong X., Jiang J., Qiao B., Liu H., Cheng J. and Yuan Y. (2019). The biodegradation of cefuroxime, cefotaxime and cefpirome by the synthetic consortium with probiotic Bacillus clausii and investigation of their potential biodegradation pathways. *Science Total Environment*, **651**, 271–280, https://doi.org/10.1016/j.scitotenv.2018.09.187

Kooistra L., Huijbregts M. A. J., Ragas A. M. J., Wehrens R. and Leuven R. S. E. W. (2005). Spatial variability and uncertainty in ecological risk assessment: a case study on the potential risk of cadmium for the little owl in a dutch river flood plain. *Environmental Science Technology*, **39**, 2177–2187, https://doi.org/10.1021/es049814w

Li W., Shi Y., Gao L., Liu J. and Cai Y. (2013). Occurrence, distribution and potential affecting factors of antibiotics in sewage sludge of wastewater treatment plants in China. *Science of the Total Environment*, **445–446**, 306–313, https://doi.org/10.1016/j.scitotenv.2012.12.050

Lin H., Zhang J., Chen H., Wang J., Sun W., Zhang X., Yang Y., Wang Q. and Ma J. (2017). Effect of temperature on sulfonamide antibiotics degradation, and on antibiotic resistance determinants and hosts in animal manures. *Science of the Total Environment*, **607–608**, 725–732, https://doi.org/10.1016/j.scitotenv.2017.07.057

Liu H., Pu C., Yu X., Sun Y. and Chen J. (2018). Removal of tetracyclines, sulfonamides, and quinolones by industrial-scale composting and anaerobic digestion processes. *Environmental Science and Pollution Research*, **25**, 35835–35844, https://doi.org/10.1007/s11356-018-1487-3

Liu D., Yan Z., Liao W., Bai Y. and Feng C. (2020). The toxicity effects and mechanisms of tris(1,3-dichloro-2-propyl) phosphate (TDCPP) and its ecological risk assessment for the protection of freshwater organisms. *Environmental Pollution*, **264**, 114788, https://doi.org/10.1016/j.envpol.2020.114788

Liu F., Zhang S., Xu D., Sun F., Wang W., Li X., Yu W., Dong X., Liu G. and Yu H. (2022). Columnar cactus-like 2D/1D BiOBr/$BiVO_4$:Yb^{3+},Er3+ heterostructure photocatalyst with ultraviolet-visible-near infrared response for photocatalytic degrading dyes, endocrine disruptors and antibiotics. *Journal of Alloys and Compounds*, **929**, 167330, https://doi.org/10.1016/j.jallcom.2022.167330

Marce M., Palacios O., Bartolomé A., Caixach J., Baig S. and Esplugas S. (2017). Application of ozone on activated sludge: micropollutant removal and sludge quality. *Ozone: Science & Engineering*, **39**(5), 319–332, https://doi.org/10.1080/01919512.2017.1334535

Margenat A., You R., Cañameras N., Carazo N., Díez S., Bayona J. M. and Matamoros V. (2020). Occurrence and human health risk assessment of antibiotics and trace elements in Lactuca sativa amended with different organic fertilizers. Environmental Research, **190**, 109946, https://doi.org/10.1016/j.envres.2020.109946

Martín J., Santos J. L., Aparicio I. and Alonso E. (2015). Pharmaceutically active compounds in sludge stabilization treatments: anaerobic and aerobic digestion, wastewater stabilization ponds and composting. *Science of the Total Environment*, **503–504**, 97–104, https://doi.org/10.1016/j.scitotenv.2014.05.089

Mendes G., Faria M., Carvalho A., Gonçalves M. C. and de Pinho M. N. (2018). Structure of water in hybrid cellulose acetate-silica ultrafiltration membranes and permeation properties. *Carbohydrate Polymers*, **189**, 342–351, https://doi.org/10.1016/j.carbpol.2018.02.030

Mohammad A., Khan M. E., Cho M. H. and Yoon T. (2021). Adsorption promoted visible-light-induced photocatalytic degradation of antibiotic tetracycline by tin oxide/cerium oxide nanocomposite. *Applied Surface Science*, **565**, 150337, https://doi.org/10.1016/j.apsusc.2021.150337

Mpongwana N. and Rathilal S. (2022). Exploiting biofilm characteristics to enhance biological nutrient removal in wastewater treatment plants. *Applied Sciences*, **12**(15), 7561, https://doi.org/10.3390/app12157561

Murugalakshmi M., Mamba G. and Muthuraj V. (2020). A novel In_2S_3/Gd_2O_3 p-n type visible light-driven heterojunction photocatalyst for dual role of Cr(VI) reduction and oxytetracycline degradation. *Applied Surface Science*, **527**, 146890, https://doi.org/10.1016/j.apsusc.2020.146890

Pinlova B., Hufenus R. and Nowack B. (2022). Systematic study of the presence of microplastic fibers during polyester yarn production. *Journal of Cleaner Production*, **363**, 132247, https://doi.org/10.1016/j.jclepro.2022.132247

Pirsaheb M., Hossaini H. and Janjani H. (2020). Reclamation of hospital secondary treatment effluent by sulfate radicals based-advanced oxidation processes (SR-AOPs) for removal of antibiotics. *Microchemical Journal*, **153**, 104430, https://doi.org/10.1016/j.microc.2019.104430

Pokkiladathu H., Farissi S., Sakkarai A. and Muthuchamy M. (2022). Degradation of bisphenol A: a contaminant of emerging concern, using catalytic ozonation by activated carbon impregnated nanocomposite-bimetallic catalyst. *Environmental Science and Pollution Research*, **29**(48), 72417–72430, https://doi.org/10.1007/s11356-022-19513-3

Praveena S. M., Shaifuddin S. N. M. and Akizuki S. (2018). Exploration of microplastics from personal care and cosmetic products and its estimated emissions to marine environment: an evidence from Malaysia. *Marine Pollution Bulletin*, **136**, 135–140, https://doi.org/10.1016/j.marpolbul.2018.09.012

Qin K., Wei L., Li J., Lai B., Zhu F., Yu H., Zhao Q. and Wang K. (2020). A review of ARGs in WWTPs: sources, stressors and elimination. *Chinese Chemical Letters*, **31**(10), 2603–2613, https://doi.org/10.1016/j.cclet.2020.04.057

Ren B., Shi X., Jin X., Wang X. C. and Jin P. (2021). Comprehensive evaluation of pharmaceuticals and personal care products (PPCPs) in urban sewers: degradation, intermediate products and environmental risk. *Chemical Engineering Journal*, **404**, 127024, https://doi.org/10.1016/j.cej.2020.127024

Ren J., Shi H., Liu J., Zheng C., Lu G., Hao S., Jin Y. and He C. (2023). Occurrence, source apportionment and ecological risk assessment of thirty antibiotics in farmland system. *Journal of Environmental Management*, **335**, 117546, https://doi.org/10.1016/j.jenvman.2023.117546

Shakak M., Rezaee R., Maleki A., Jafari A., Safari M., Shahmoradi B., Daraei H. and Lee S.-M. (2020). Synthesis and characterization of nanocomposite ultrafiltration membrane (PSF/PVP/SiO_2) and performance evaluation for the removal of amoxicillin from aqueous solutions. *Environmental Technology & Innovation*, **17**, 100529, https://doi.org/10.1016/j.eti.2019.100529

Stevens-Garmon J., Drewes J. E., Khan S. J., McDonald J. A. and Dickenson E. R. V. (2011). Sorption of emerging trace organic compounds onto wastewater sludge solids. *Water Research*, **45**(11), 3417–3426, https://doi.org/10.1016/j.watres.2011.03.056

Taşkan B., Hanay Ö, Taşkan E., Erdem M. and Hasar H. (2016). Hydrogen-based membrane biofilm reactor for tetracycline removal: biodegradation, transformation products, and microbial community. *Environmental Science and Pollution Research*, **23**(21), 21703–21711, https://doi.org/10.1007/s11356-016-7370-1

Thomaidi V. S., Stasinakis A. S., Borova V. L. and Thomaidis N. S. (2016). Assessing the risk associated with the presence of emerging organic contaminants in sludge-amended soil: a country-level analysis. *Science of the Total Environment*, **548–549**, 280–288, https://doi.org/10.1016/j.scitotenv.2016.01.043

Van Thuan D., Nguyen T. B. H., Pham T. H., Kim J., Hien Chu T. T., Nguyen M. V., Nguyen K. D., Al-onazi W. A. and Elshikh M. S. (2022). Photodegradation of ciprofloxacin antibiotic in water by using ZnO-doped g-C_3N_4 photocatalyst. *Chemosphere*, **308**, 136408, https://doi.org/10.1016/j.chemosphere.2022.136408

Varsha M., Senthil Kumar P. and Senthil Rathi B. (2022). A review on recent trends in the removal of emerging contaminants from aquatic environment using low-cost adsorbents. *Chemosphere*, **287**, 132270, https://doi.org/10.1016/j.chemosphere.2021.132270

Vieira W. T., Bispo M. D., De Melo Farias S., De Almeida A.D. S. V., Da Silva T. L., Vieira M. G. A., Soletti J. I. and Balliano T. L. (2021). Activated carbon from macauba endocarp (*Acrocomia aculeate*) for removal of atrazine: experimental and theoretical investigation using descriptors based on DFT. *Journal of Environmental Chemical Engineering*, **9**(2), 105155, https://doi.org/10.1016/j.jece.2021.105155

Wang P., Yap P.-S. and Lim T.-T. (2011). C–N–S tridoped TiO_2 for photocatalytic degradation of tetracycline under visible-light irradiation. *Applied Catalysis A*, **399**(1), 252–261, https://doi.org/10.1016/j.apcata.2011.04.008

Wang X., Lu W., Zhao Z., Zhong H., Zhu Z. and Chen W. (2020). In situ stable growth of β-FeOOH on g-C_3N_4 for deep oxidation of emerging contaminants by photocatalytic activation of peroxymonosulfate under solar irradiation. *Chemical Engineering Journal*, **400**, 125872, https://doi.org/10.1016/j.cej.2020.125872

Wang N., Peng L., Gu Y., Liang C., Pott R. W. M. and Xu Y. (2023). Insights into biodegradation of antibiotics during the biofilm-based wastewater treatment processes. *Journal of Cleaner Production*, **393**, 136321, https://doi.org/10.1016/j.jclepro.2023.136321

Wu Y., Song S., Chen X., Shi Y., Cui H., Liu Y. and Yang S. (2023). Source-specific ecological risks and critical source identification of PPCPs in surface water: comparing urban and rural areas. *Science of the Total Environment*, **854**, 158792, https://doi.org/10.1016/j.scitotenv.2022.158792

Xu F.-L., Li Y.-L., Wang Y., He W., Kong X.-Z., Qin N., Liu W.-X., Wu W.-J. and Jorgensen S. E. (2015). Key issues for the development and application of the species sensitivity distribution (SSD) model for ecological risk assessment. *Ecological Indicators*, **54**, 227–237, https://doi.org/10.1016/j.ecolind.2015.02.001

Xu Q., Lai D., Xing Z., Liu X. and Wang Y. (2023). Strengthened removal of emerging contaminants over S/Fe codoped activated carbon fabricated by a mild one-step thermal transformation scheme. *Chemosphere*, **310**, 136897, https://doi.org/10.1016/j.chemosphere.2022.136897

Yan Q., Gao X., Chen Y.-P., Peng X.-Y., Zhang Y.-X., Gan X.-M., Zi C.-F. and Guo J.-S. (2014). Occurrence, fate and ecotoxicological assessment of pharmaceutically active compounds in wastewater and sludge from wastewater treatment plants in Chongqing, the Three Gorges Reservoir Area. *Science of the Total Environment*, **470–471**, 618–630, https://doi.org/10.1016/j.scitotenv.2013.09.032

Yan Z.-R., Zhu Y.-Y., Meng H.-S., Wang S.-Y., Gan L.-H., Li X.-Y., Xu J. and Zhang W. (2019). Insights into thermodynamic mechanisms driving bisphenol A (BPA) binding to extracellular polymeric substances (EPS) of activated sludge. *Science of the Total Environment*, **677**, 502–510, https://doi.org/10.1016/j.scitotenv.2019.04.413

Yang Y., Chen T., Liu X., Wang S., Wang K., Xiao R., Chen X. and Zhang T. (2022a). Ecological risk assessment and environment carrying capacity of soil pesticide residues in vegetable ecosystem in the Three Gorges Reservoir Area. *Journal of Hazardous Materials*, **435**, 128987, https://doi.org/10.1016/j.jhazmat.2022.128987

Yang Y., Khan H., Gao S., Khalil A. K., Ali N., Khan A., Show P. L., Bilal M. and Khan H. (2022b). Fabrication, characterization, and photocatalytic degradation potential of chitosan-conjugated manganese magnetic nano-biocomposite for emerging dye pollutants. *Chemosphere*, **306**, 135647, https://doi.org/10.1016/j.chemosphere.2022.135647

You R., Margenat A., Lanzas C. S., Cañameras N., Carazo N., Navarro-Martín L., Matamoros V., Bayona J. M. and Díez S. (2020). Dose effect of Zn and Cu in sludge-amended soils on vegetable uptake of trace elements, antibiotics, and antibiotic resistance genes: human health implications. *Environmental Research*, **191**, 109879, https://doi.org/10.1016/j.envres.2020.109879

Zhang X., Ma Y., Xi L., Zhu G., Li X., Shi D. and Fan J. (2019a). Highly efficient photocatalytic removal of multiple refractory organic pollutants by $BiVO_4/CH_3COO(BiO)$ heterostructured nanocomposite. *Science of the Total Environment*, **647**, 245–254, https://doi.org/10.1016/j.scitotenv.2018.07.450

Zhang Y., An Y., Liu C., Wang Y., Song Z., Li Y., Meng W., Qi F., Xu B., Croue J.-P., Yuan D. and Ikhlaq A. (2019b). Catalytic ozonation of emerging pollutant and reduction of toxic by-products in secondary effluent matrix and effluent organic matter reaction activity. *Water Research*, **166**, 115026, https://doi.org/10.1016/j.watres.2019.115026

Zhao W., Liu H., Liu Y., Jian M., Gao L., Wang H. and Zhang X. (2018). Thin-film nanocomposite forward-osmosis membranes on hydrophilic microfiltration support with an intermediate layer of graphene oxide and multiwall carbon nanotube. *ACS Applied Materials & Interfaces*, **10**(40), 34464–34474, https://doi.org/10.1021/acsami.8b10550

Zhi S., Shen S., Zhou J., Ding G. and Zhang K. (2020). Systematic analysis of occurrence, density and ecological risks of 45 veterinary antibiotics: focused on family livestock farms in Erhai Lake basin, Yunnan, China. *Environmental Pollution*, **267**, 115539, https://doi.org/10.1016/j.envpol.2020.115539

Zhou H., Cao Z., Zhang M., Ying Z. and Ma L. (2020). Zero-valent iron enhanced in-situ advanced anaerobic digestion for the removal of antibiotics and antibiotic resistance genes in sewage sludge. *Science of the Total Environment*, **754**, 142077, https://doi.org/10.1016/j.scitotenv.2020.142077

doi: 10.2166/9781789063752_0055

Chapter 4

Recent advances in treatment of microplastics in wastewater

*Surya Singh**

Division of Environmental Monitoring and Exposure Assessment (Water & Soil), ICMR – National Institute for Research in Environmental Health, Bhopal 462 030, India

**Corresponding author: suryasingh.nireh@icmr.gov.in*

ABSTRACT

Microplastics are newly emerged contaminants having a ubiquitous presence in almost every kind of environmental matrix. Water is one of the most important reservoirs of microplastics, which also serves as an efficient medium for the transfer of these particles to various abiotic matrices and biotic lives. As per recent research, microplastics in water are an emerging public health issue that needs attention, and hence, suitable treatment methodologies are needed to reduce the contamination potential of water/wastewater. Although conventional wastewater treatment methods are able to remove microplastics to some extent, complete removal is challenging to date. Therefore, new techniques are being explored, among which the use of bioinspired molecules, metal organic frameworks, and biological materials is important. Further, the mechanical removal of microplastics through engineered micromotors and chemical degradation through various techniques have also been investigated. While some of these techniques are attractive and provide suitable solutions, their wide-scale applicability and cost-effectiveness are issues. Moreover, the techniques that can be suitably incorporated into conventional wastewater treatment systems are more preferred. Considering all these issues, this chapter will discuss the recent technological advances in the removal of microplastics from wastewater.

Keywords: emerging contaminants, microplastics, removal techniques, treatment, wastewater

4.1 INTRODUCTION

Microplastics are one of the emerging contaminants that are found to be present in almost all kinds of environmental matrices, such as water, soil, air, snow/glaciers, etc. (Singh *et al.*, 2022). Apart from that, microplastics are also found in various food items, sewage/wastewater treatment plants, landfill leachates, and so on (Acarer, 2023; Conti *et al.*, 2020; Egea-Corbacho *et al.*, 2023; Makhdoumi *et al.*, 2021; Sekar & Sundaram, 2023). The global presence of these microplastics in a variety of matrices is primarily a result of inappropriate disposal and management of plastic waste. Accumulation of plastic waste in the environment for a long time results in the breakdown of larger particles into smaller pieces, resulting in microplastics (Liu *et al.*, 2023). In order to be defined as microplastics, the size of the plastic particles should lie in the range of 1 μm–5 mm (Frias & Nash, 2019). These small-size plastic particles are able to invade biotic species, including humans, through ingestion and inhalation, which poses a serious threat to the health of living beings. The presence of plastic particles has already been reported in various fish, animal, and plant species (Hariharan *et al.*, 2021, 2022; Veen *et al.*, 2022; Yu *et al.*, 2021); and also in human blood, lungs, placenta, semen, urine, and so on (Jenner *et al.*, 2022; Leslie *et al.*, 2022; Pironti *et al.*, 2023; Ragusa *et al.*, 2021; Zhao *et al.*, 2023). Consequently, genotoxic and cytotoxic impacts have also been reported (Cobanoglu *et al.*, 2021; Kaur *et al.*, 2022). Therefore, the effective management of microplastic particles is essential.

Microplastics primarily originate through two different means – the first mode is their intentional manufacture in the defined size range for various applications (known as primary microplastics), such as in cosmetics, pharmaceutical industries, and daily-use products (*viz.*, toothpaste, shaving creams, shampoo, moisturizers, scrubs, etc.) (Osman *et al.*, 2023); second mode of microplastics' origin (known as secondary microplastics) is the breakdown of large-size plastic material/plastic waste in the environment upon the action of various biotic and abiotic agencies (Osman *et al.*, 2023; Singh *et al.*, 2021a). Between these two, the second mode of microplastics' origin contributes more (69–81%) to the environment and therefore calls for the appropriate disposal of plastic waste (EU, 2018; Singh *et al.*, 2022). As far as the fate of microplastics is concerned, primary microplastics do find their way into our daily lives for various reasons and then become part of the sullage or domestic sewage, which is ultimately directed towards the sewage treatment plants (STPs) (Koyuncuoglu & Erden, 2021). Similarly, industrial effluent with varying amounts of plastic waste, depending on the type of industry, finds its way into effluent treatment plants (ETPs) (Umar *et al.*, 2023). Secondary microplastics, on the other hand, reach the water bodies through direct disposal of plastic waste in the aquatic bodies, illegitimate discharge of effluent from industries, leaching and percolation of leachate from municipal or hazardous dumpsites/landfills, and so on (Koyuncuoglu & Erden, 2021).

As a significant amount of microplastics reaches the STPs and ETPs, it is essential to have an effective removal system for this contaminant (Ridall *et al.*, 2023). Moreover, since these treatment plants provide an environment where microplastics can breakdown into smaller pieces at a faster rate owing to the presence of various chemicals and microbial moieties, the resulting microplastics become more fragmented and smaller. Thus, it makes microplastics more prone to adsorbing various contaminants, such as metals, chemicals, pigments, microbes, and so on, making these particles more pernicious (Gao *et al.*, 2023). Therefore, efficient removal of microplastics in various treatment plants is crucial. However, as of date, there is no mechanism in these units for the targeted removal of microplastics (Patil *et al.*, 2023). The conventional treatment units help to remove some amount of microplastic, but efficiency needs to be improved. For understanding the removal of microplastics, STPs and ETPs may be clubbed into one unit, namely, wastewater treatment plants (WWTPs). In this chapter, the details of various treatment processes for WWTPs will be outlined. The efficiency and limitations of these processes will also be discussed. Further, major emphasis would be laid on the advanced technologies in the field of microplastics' removal from wastewater.

4.2 CHALLENGES IN THE MICROPLASTICS REMOVAL

The removal of microplastics from wastewater becomes challenging owing to various factors. Among them, some of the factors are the different sources of origin of microplastics, the varied size range (1 μm–5 mm), and the different shapes such as fibers, foam, granules, fragments, pellets, and so on. The variable chemical composition of various microplastic particles also sometimes makes it challenging to adopt any single technology (Lu *et al.*, 2023). The chemical composition of microplastic particles may be of several types, but the majority of the particles are composed of polyethylene (PE), polypropylene (PP), polyethylene terephthalate (PET), polyvinyl chloride (PVC), polystyrene (PS), polyamide (PA), polyester, nylon, polyacrylates, and so on (Singh *et al.*, 2023). It is noteworthy to mention here that a specific industry may discharge any particular composition of microplastics; however, mixing effluents from a number of industries in the common effluent treatment plants results in the mingling of different types of microplastics. Further, microplastic particles are not wholesome plastic materials; rather they also contain a significant amount of chemicals in the form of additives and plasticizers, which are generally mixed during their manufacturing process (Hahladakis *et al.*, 2018). Moreover, these particles also serve as an efficient vector for various contaminants occurring in the environment, such as heavy metals, chemicals, microbes, and so on (Singh & Bhagwat, 2022; Upadhyay *et al.*, 2022). Thus, the removal of microplastics becomes crucial considering their inherent as well as associated hazardous chemicals.

4.3 OVERVIEW OF CONVENTIONAL TREATMENT TECHNIQUES AND SHORTCOMINGS

Any conventional wastewater treatment process involves mainly three-step treatment, namely, primary, secondary, and tertiary. Based on the technologies involved at each of these stages, the removal efficiency keeps improving from primary to tertiary treatment. The efficiency of different treatment steps evaluated by researchers is given in Table 4.1. Primary wastewater treatment generally involves physical mechanisms to remove dirt, suspended particles, and various other solid materials. At the initial level, different-size screens are applied, which grab the solid materials. This is basically an exclusion process based on size. Screening is then followed by sedimentation. The basic principle involved here is the suspension of wastewater in an idle position for a certain specified duration so that heavier particles may settle down owing to the gravitational force. The efficiency of the primary step specifically for microplastic removal ranges from 50 to 98% (Sun *et al.*, 2019). It is notable here to mention that the shape of the microplastics does affect the efficiency, as fiber-shaped particles are removed more compared to fragments (Magnusson & Noren, 2014; Ziajahromi *et al.*, 2017). The limitation of this treatment process is that it cannot remove plastic particles with micrometer and nanometer sizes.

The secondary step of the WWTPs generally involves biological treatment using attached growth systems (e.g., trickling filters and rotating biological contactors) or suspended growth systems (e.g., activated sludge processes, aerated lagoons, aerobic digestion, etc.). Further modifications in the attached growth systems involve the development of moving-bed biofilm reactors (MBBRs). Similarly, suspended growth systems involve the development of sequencing batch reactors (SBRs) and membrane bio-reactors (MBRs). It has been reported that approximately 88% of the microplastic removal may be achieved once the water undergoes a secondary treatment process (Sun *et al.*, 2019). Lares *et al.* studied the removal of microplastics in conventional activated sludge processes as well as in advanced MBR systems. Upon comparison of both systems, it was concluded that conventional systems resulted in approximately 1 microplastic particle per litre of effluent, while MBR technology reduced this amount to 0.4 microplastic particle per litre of effluent (Lares *et al.*, 2018). Though this removal efficiency is significant, the remaining amount of microplastics in the effluent is finally discharged into surface water bodies. Thus, WWTPs have become one of the most important sources of microplastic pollution in the aquatic environment.

The tertiary step provides another higher degree of treatment for the wastewater, generally involving filtration and disinfection. Talvitie *et al.* compared the microplastics' removal efficiency of different tertiary-level treatments of WWTPs, *viz.* disc filter, rapid sand filtration, and dissolved air flotation (Talvitie *et al.*, 2017). Along with these, the efficiency of MBR in treating primary effluent was also compared. It was found that 99.9% of the microplastics could be removed through MBR technology. Moreover, the removal efficiencies of disc filter, rapid sand filtration, and dissolved air flotation lied in the range of 40–98.5%, 97%, and 95%, respectively. The highest removal efficiency of MBR technology among all the studied methods was attributed to the smallest pore size of membranes (0.4 μm) used in MBR. However, there is scope for microplastics' presence in the effluent even after tertiary treatment. On average, tertiary treatment can enhance microplastic removal by up to 97% (Sun *et al.*, 2019). At first instance, it might appear that up to 97% treatment efficiency is good enough for the removal of microplastics. However, it is to be understood that even 1% of microplastics per liter of treated water discharged will result in millions of microplastic particles in the environment, considering the amount of wastewater treated in WWTPs. Therefore, treatment methods need to be much more efficient.

4.4 ADVANCED TECHNIQUES FOR REMOVAL OF MICROPLASTICS

Considering the amount and risks imposed by microplastics, researchers are attempting to develop advanced techniques that can ensure their complete removal from WWTPs (Table 4.1). These techniques may be clubbed into physical, chemical, biological, and miscellaneous techniques (Figure 4.1).

4.4.1 Physical techniques

4.4.1.1 Adsorption

The adsorption process is one of the most important physical means of removal reported for microplastics. Adsorption refers to the entanglement and diffusion of plastic particles onto a surface through various forces, such as hydrogen bond/electrostatic/π–π interactions, and so on. A sponge-type surface made up of graphene oxide and chitin has been synthesized and employed for microplastics' removal. By using this, an efficiency of as high as 90% could be achieved for virgin polystyrene (Sun *et al.*, 2020). Another low-cost adsorption material, *viz.* biochar, made up of bio-based materials, has also been synthesized. Basically, biochar is a type of charcoal that can be synthesized by pyrolysis of a variety of feed materials, such as biowaste, cattle litter, crop residues, and so on. (Abuwatfa *et al.*, 2021). Therefore, it is characterized by a porous structure rich in carbon content. Removal of microplastics has been studied using biochar synthesized from corn straw and hardwood. This biochar was integrated with the filtration system of the WWTP to assess the change in removal efficiency. It was found that efficiency increased from 60 to 80% to more than 95% through this biochar integration (Wang *et al.*, 2020a). Siipola *et al.* synthesized steam-activated porous biochar from the bark of coniferous trees, *viz.* pine and spruce. This biochar was used for the removal of polyethylene microplastics of different shapes. The shapes of the particles included spheres, beads, cylindrical fiber, and fleece fiber. It was found that the removal efficiency of biochar varied for different shapes of particles. While 100% removal was achieved for cylindrical and fleece fibers, beads and spheres were not removed completely. This difference was attributed to the entanglement of different shapes of microplastics with biochar (Siipola *et al.*, 2020). Another study carried out modifications into the biochar surface properties through the addition of iron nanoparticles. A comparison was made between the unmodified biochar and iron-modified biochar, and it was deduced that microplastics' removal efficiency increased from 75% to 100% upon integration of iron (Singh *et al.*, 2021b). Moreover, the pH of the solution was found to have no effect on the iron-modified biochar adsorption properties. Similarly, magnesium/zinc-modified biochar was also used for the adsorptive removal of polystyrene microplastics, and efficiency up to 99.5% could be achieved (Wang *et al.*, 2021).

4.4.1.2 Filtration

Filtration is the process of physical removal of microplastics from wastewater by creating a barrier using a membrane. Ultrafiltration membranes (pore size 1–100 nm) have been used for removing microplastics (polyethylene) by Ma *et al.*, and it was found that removal efficiency increased from 13% to 91% (Ma *et al.*, 2019). Generally, filtration processes are applied in combination with other water treatment processes, such as sedimentation and coagulation. Reverse osmosis (RO) is another technique through which particles are captured through the application of semi-permeable membranes. Ziajahromi *et al.* reported the performance of RO employed in a WWTP. It was reported that even after the RO process, few microplastic fibers remained in the effluent (Ziajahromi *et al.*, 2017). It was deduced that it might be due to some defects, either in the membrane or in the supply pipe system. The performance of polycarbonate, cellulose acetate, and polytetrafluoroethylene membranes

Table 4.1 Removal methods of microplastics.

Process Description	Major Mechanism	Lowest Size of Microplastic Particle Removed/Finest Mesh	Efficiency (%)	Advantages	Challenges	References
Wastewater treatment plant processes	Skimming, settling of the entrapped microplastics	300 μm	99.9	Conventional process, no additional cost	Not possible to remove MPs of size <300 μm	Magnusson and Noren (2014)
Wastewater treatment plant processes	Primary, secondary, and tertiary	100 μm	99.9	Conventional process, no additional cost	Not possible to remove MPs of size <100 μm	Carr *et al.* (2016)
Wastewater treatment plant processes	Secondary treatment	20 μm	95.6	MBR process exhibited greater overall efficiency	Complete retention is not possible	Michielssen *et al.* (2016)
	Tertiary treatment		97.2			
	MBR		99.4			
Wastewater treatment plant processes	MBR	250 μm	99.3	MBR process helped to retain more microplastics compared to conventional activated sludge process	Not possible to remove MPs of size <250 μm	Lares *et al.* (2018)
Al and Fe salt	Coagulation	<0.5 mm	45.34 ± 3.93	Simple process, does not require additional set-up	Low efficiency	Ma *et al.* (2019)
Adsorption using biochar	Morphologically controlled mechanism (stuck, trapped, and entangled)	10 μm	>95%	Low cost and efficient	Process is slow and results in obstruction of the pores with time, costly, regeneration is tough	Wang *et al.* (2020a)
Adsorption using steam-activated porous biochar made up of the bark of coniferous trees	Retention in the surface pores	>10 μm ≤10 μm	~100% Could not retain	Low cost and efficient for particles >10 μm	Not possible to remove MPs of size ≤10 μm	Siipola *et al.* (2020)

(*Continued*)

Table 4.1 Removal methods of microplastics (*Continued*).

Process Description	Major Mechanism	Lowest Size of Microplastic Particle Removed/ Finest Mesh	Efficiency (%)	Advantages	Challenges	References
Adsorption and thermal degradation using:	Electrostatic interaction and chemical bonding interaction	1 μm		Low cost, eco-friendly, and efficient Stable performance, keeping the efficiency intact even after 5 cycles of adsorption-pyrolysis	Efficiency decreases in alkalescent condition, in water having high chemical oxygen demand, and in presence of anions such as $H_2PO_4^-$	Wang *et al.* (2021)
Magnetic biochar			94.81%			
Mg modified magnetic biochar			98.75%			
Zn modified magnetic biochar			99.46%			
Filtration with GAC (combined with coagulation and sedimentation)	Physical properties (size and shape)	1–5 μm	56.8–60.9%	Efficient to remove plastic particles of nano-size range	Process is slow and results in obstruction of the pores with time, costly, regeneration is tough	Wang *et al.* (2020b)
Pulse clarification with filtration	Entrapment in sludge blanket formed due to coagulation floats	<100 μm	85%	Removal efficiency is comparable to the other treatment plants having advanced processes	Complete retention is not possible	Sarkar *et al.* (2021)
Bioinspired molecules	Mechanical capture mechanism driven by the hydrophobic and van der Waals interactions	–	–	Flexible, possibility to remove different types of plastic particles in wastewater stream	Method yet to be established for practical purposes	Herbort and Schuhen (2017)
Photocatalytic micromotors	Phoretic interaction and shovelling/pushing interactions	–	–	Self-propelled devices, works efficiently independent of the fuel	Selectivity of micromotors for microplastics is crucial	Wang *et al.* (2019)
Zr MOF-based foams	Entrapment	–	95.5 ± 1.2%	High performance, excellent durability	Flexibility and robustness of MOF-based foams, removal efficiency affected by the particle size and zeta potential	Chen *et al.* (2020)

(*Continued*)

MOF (MIL-100 Fe) nanoparticles incorporated into polysulfone matrix	Membrane separation	40 μm	–	Removal of MPs (PE and PVC) from textile wastewater	Tested for only textile wastewater	Gnanasekaran *et al.* (2021)
Photocatalytic degradation	Photocatalysis using *n*-TiO_2 semiconductor	MPs extracted from commercial exfoliating scrub	Max. mass loss of 6.4%	Environment-friendly technology	High cost, requirement of specific set-up	Ariza-Tarazona *et al.* (2019)
Algal degradation	Electrostatic charge on the microplastic particles and algal surfaces	20 μm	94.5%	No chemical, electrical, and mechanical operations	Efficiency will vary owing to physiological and topographical differences on the seaweed surface	Sundbæk *et al.* (2018)
Fungal degradation	Biodegradation	–	–	Biological process, no toxic side-products	Study investigated only low-density polyethylene. Method yet to be established for other polymers	Kunlere *et al.* (2019)
Bacterial degradation	Biodegradation	–	50–61%	Biological process, no toxic side-products	Method yet to be established for wide variety of polymers	Rajandas *et al.* (2012)
Electrocoagulation	Charge neutralization, flocculation	–	90 (pH 3–10) 99.24 (pH 7.5)	Does not rely on chemicals or microorganisms, energy efficient	Operation time needs to lowered down	Perren *et al.* (2018)
Combinatorial method	Electrocoagulation, electroflotation, and membrane filtration	MPs collected from WWTPs filtered through mesh dia 26 μm	100%	Short retention time, high efficiency	Application of pressure creates additional cost, Membrane rupture results in high replacement frequency	Akarsu *et al.* (2021)

Adapted from Singh *et al.* (2021a).

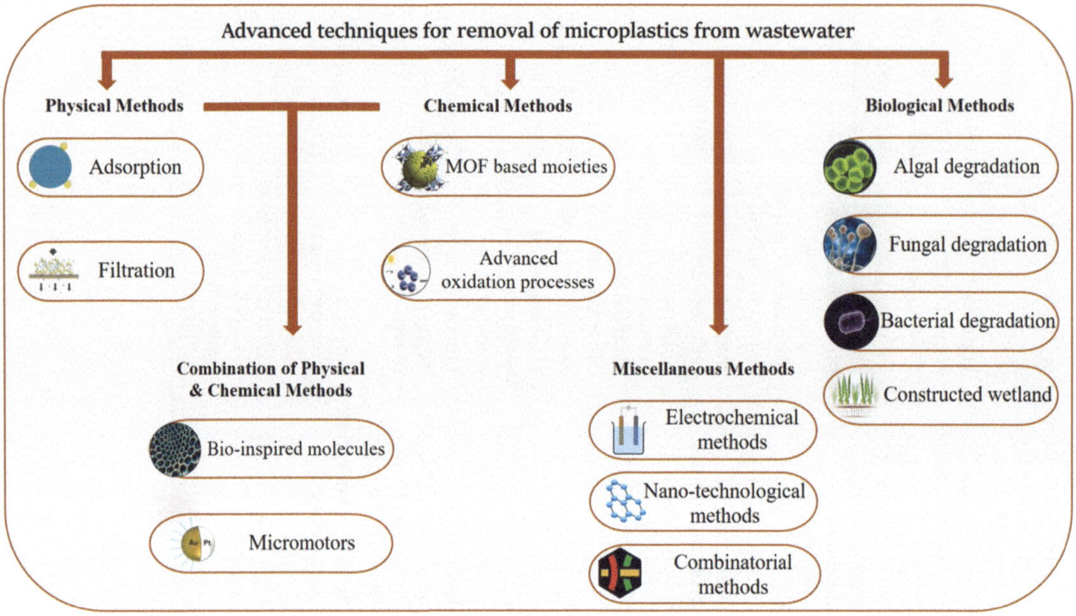

Figure 4.1 Illustration of advanced techniques for removal of microplastics from wastewater.

was also evaluated for the removal of polystyrene and polyamide microplastic particles (size range of 20–300 μm). More than 94% of mass removal efficiency could be achieved from all three membranes (Pizzichetti *et al.*, 2021). However, major disadvantages of membrane technology are fouling and abrasion with time.

4.4.1.3 Agglomeration and sol–gel process using bioinspired molecules

A new concept of bioinspired molecules has been proposed for the possible removal of microplastics (Herbort & Schuhen, 2017). This concept involves the agglomeration of microplastics facilitated through silicone-based chemicals in a water-induced sol–gel process (Herbort *et al.*, 2018), thus combining the attributes of both physical and chemical techniques. Since microplastics are small particles that remain scattered in the aquatic system, agglomeration thereof helps in size enhancement, making the trapping of these particles easier (Sturm *et al.*, 2023). Sturm *et al.* studied the removal of microplastics in a WWTP and compared the efficiency of three mechanisms, *viz.* advanced oxidation process (AOP), granular activated carbon (GAC), and GAC combined with organosilanes. It was found that individual mechanisms of AOP and GAC could not result in significant removal of microplastics; however, reduction reached up to 61% upon connecting the organosilane molecules in series with the GAC (Sturm *et al.*, 2023). The principle involved here was the chemical binding of organosilanes with microplastics through the sol–gel process, resulting in three-dimensional agglomerates. These agglomerates can then be easily skimmed from the water. The polarity of the microplastic polymers plays an important role here, as non-polar polymers (e.g., polyethylene and polypropylene) are easier to remove compared to polar polymers (e.g., polyvinyl chloride and polyamide). Further, the effects of temperature and water composition were also studied. It was found that variation in temperature and water composition does not affect the removal of microplastics through the agglomeration-fixation reactions of organosilanes (Sturm *et al.*, 2021).

4.4.1.4 Micromotors

Micromotors are one of the recent developments in the field of environmental remedial applications, which are basically a combination of physical and chemical processes. These micromotors are energy-driven devices in which their motion direction can also be controlled (Zhang *et al.*, 2017). Micromotors have been found to be useful for the removal of a variety of pollutants, such as dyes (Zhang *et al.*, 2017), oil (Mou *et al.*, 2015), heavy metals (Villa *et al.*, 2018), suspended matter (Wang *et al.*, 2019), and so on. Considering their efficacy, researchers have applied micromotors for microplastics removal as well (Wang *et al.*, 2019). In this case, photocatalytic micromotors were designed to make them self-propelled under ultraviolet (UV) light illumination. The propulsion was made possible due to the photocatalytic reactions taking place in the particles, *viz.* TiO_2. These photocatalytic reactions involve the creation of conduction band electrons and valence band holes, which ultimately lead to fluid flow in the direction of propulsion. Wang *et al.* showed that micromotors help in the removal of microplastics through two mechanisms, *viz.* phoretic interactions and shoveling. Phoretic interactions result in the collection and simultaneous travel of microplastic particles along with the micromotors, thus resulting in their removal. On the other hand, shoveling involves the removal of a large number of microplastic particles through a moving assemblage of magnetic catalytic particles. This magnetic assemblage was constituted by the interaction between the magnetic nickel layers inside the core-shell structure of $Au@Ni@TiO_2$ micromotors (Wang *et al.*, 2019).

4.4.2 Chemical techniques

4.4.2.1 Metal organic framework (MOF)-based moieties

MOF-based moieties have been in use for various environmental applications, such as air pollution control (Zhang *et al.*, 2016), water contaminants removal (Kobielska *et al.*, 2018; Mon *et al.*, 2018), volatile organic carbon removal (Vikrant *et al.*, 2020), and so on. Considering their wide-scale utility, researchers have employed MOFs for microplastics' removal as well (Chen *et al.*, 2020). Gnanasekaran *et al.* synthesized hydrophilic MOF, *viz.* MIL-100 (Fe) nanoparticles, which were incorporated into a polysulfone matrix to design a composite membrane. This MOF-incorporated membrane was successfully utilized for the removal of microplastics in textile wastewater (Gnanasekaran *et al.*, 2021). This membrane also performed well for a range of microplastics' concentrations without losing its permeability. Another study reported the synthesis of zinc-based MOF for the removal of microplastics (Dongyu *et al.*, 2022). It was developed by cultivating the zeolitic imidazolate frameworks (ZIF), *viz.* ZIF-8, on the wood aerogel. The efficiency for removing the polystyrene particles of size 90–140 nm was reportedly 86% using ZIF-8@Aerogel. The removal efficiency of this MOF was attributed to the strong electrostatic interaction between negatively charged microplastics and positively charged MOF (Dongyu *et al.*, 2022). Similarly, the removal of polystyrene microplastics was also reported using the MOF material ZIF-67 (Wan *et al.*, 2022). In this study, the adsorption ratio of polystyrene microplastics to ZIF-67 reached approximately 92% with an equilibrium time of 20 min. Moreover, apart from electrostatic interactions, hydrogen bond interactions and π–π interactions were also reported to be significant in the microplastics' removal mechanism (Wan *et al.*, 2022). In a recent study, another MOF structure was reported to remove microplastics along with other dissolved contaminants from water (Haris *et al.*, 2023). This nano-pillared structure was synthesized by growing the two-dimensional MOF on the core-shell-structured carbon and FeO (C@FeO) nano-pillars. The advantage of adding FeO to the structure was obtained by developing magnetic properties. Moreover, this structure had a high surface area and ample active sites for the adsorption process. Approximately 100% removal efficiency was achieved through this MOF structure for microplastics' removal (Haris *et al.*, 2023).

4.4.2.2 Advanced oxidation processes

Advanced oxidation processes (AOPs), such as the Fenton reaction, photocatalysis, persulfate oxidation, photolysis, and so on are some of the most suitable methods for the mineralization of various recalcitrant organic contaminants (Santos *et al.*, 2023). Reactive oxygen species (ROS), such

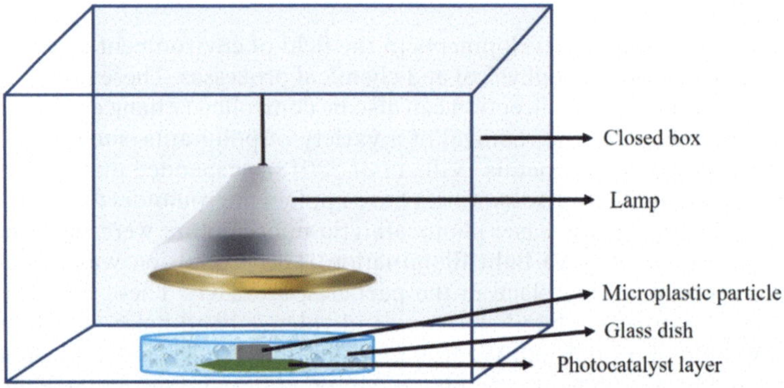

Figure 4.2 Schematic representation of photocatalytic degradation of microplastic particles.

as hydroxyl radical (•OH), sulfate radical ($SO_4^{\bullet-}$), superoxide ($O_2^{\bullet-}$), and so on are usually employed to mineralize a variety of organic moieties. Therefore, microplastic degradation was also studied using single- as well as hybrid-AOPs (Kim *et al.*, 2022). Ariza-Tarazona *et al.* demonstrated the photocatalytic degradation of polyethylene microplastics in an aqueous and solid matrix using the *n*-TiO_2 semiconductor (Figure 4.2). The degradation was estimated by the weight loss (Ariza-Tarazona *et al.*, 2019). Similarly, Tofa *et al.* used zinc oxide nanorods for the photocatalytic degradation of low-density polyethylene (LDPE) residues (Tofa *et al.*, 2019). The degradation was confirmed by the increased brittleness and the presence of cavities, wrinkles, and cracks on the surface of LDPE after the photocatalysis. The degree of degradation was found to be directly proportional to the catalyst surface area. Some investigators also used ozone (O_3) and O_3/H_2O_2 and reported the degradation and removal of microplastics up to 99.2% (Hidayaturrahman & Lee, 2019). Easton *et al.* reported polyester microfiber degradation using UV/H_2O_2 with mass loss efficiency of 52.7% after 48 hours of reaction (Easton *et al.*, 2023).

4.4.3 Biological techniques

4.4.3.1 Algal degradation

Researchers have shown that algal masses affect the nature of microplastics in the environment and thus may help in their removal. Algae either degrade the polymer matrix biologically or alter the density of the polymer particle, resulting in a change in its floatation behavior (Priya *et al.*, 2022). Sundbæk *et al.* reported that polystyrene microplastics get sorbed on marine microalgae, namely, *Fucus vesiculosus*, with an efficiency as high as 94.5% (Sundbæk *et al.*, 2018). The alginate compounds released from the cell walls of the microalgae further assist in the sorption of microplastics. Adsorption of polystyrene microplastics has also been studied on the green alga – *Chlorella*, *Scenedesmus*, and *Pseudokirchneriella subcapitata* (Bhattacharya *et al.*, 2010; Nolte *et al.*, 2017). The mechanism of this interaction between the microplastic particles and algal cells was attributed to the electrostatic charges on the surface. It was deduced that positive and/or neutral charges on the surface of microplastic particles show greater affinity towards the algal cells (Bhattacharya *et al.*, 2010; Nolte *et al.*, 2017). Further, blue-green algal species such as *Oscillatoria subbrevis* and *Phormidium lucidum* have also been found to degrade polyethylene plastics (Sarmah & Rout, 2018). The fundamental requisite for the degradation of plastics through the algal community has been identified as the formation of biofilms over the surface of polymers. With the advancement in the field of biotechnology, some genetically modified algae have also been produced, which help in degrading microplastics. For example, *Chlamydomonas reinhardtii* was modified to produce 'polyethylene terephthalate hydrolase' in order to degrade polyethylene terephthalate films and terephthalic acid (Kim *et al.*, 2020).

4.4.3.2 Fungal degradation

Fungi are known to be tolerant to a variety of chemicals, metalloids, metals, and so on owing to their ability to produce various extracellular enzymes and/or bio-surfactants that can efficiently breakdown polymers into monomeric units (Straub *et al.*, 2017). Fungi also help in catalyzing the oxidation of aromatic/non-aromatic substances, owing to the production of lignin-degrading enzyme 'laccase' (Straub *et al.*, 2017). This ability of fungi may be utilized to degrade microplastic polymers as well. Genera that can perform the degradation of common polymers (such as polyethylene, polypropylene, and polyethylene terephthalate) include *Aspergillus*, *Penicillium*, *Cladosporium*, and so on since these utilize microplastic particles as the only source of energy (Oliveira *et al.*, 2020). Kunlere *et al* recently reported that *Mucor circinelloides* and *Aspergillus flavus* can efficiently breakdown low-density polyethylene (Kunlere *et al.*, 2019).

4.4.3.3 Bacterial degradation

Bacterial degradation of plastic polymers was reported long ago in 1993, when a consortium of bacteria was used to degrade polypropylene (Cacciari *et al.*, 1993). Since then, a number of bacterial species have been isolated from a variety of habitats that possess the potential for plastic degradation (Awasthi *et al.*, 2020). It has been seen that bacteria generally develop an efficient enzymatic mechanism to degrade microplastics for survival in polluted habitats. These bacterial enzymes usually increase the hydrophilicity of microplastic particles, thus converting them into alcoholic or carbonyl residues. Such bacteria can be used to degrade microplastics in other habitats as well. Degradation of low-density polyethylene within two months of incubation has been reported in the presence of *Microbacterium paraoxydans* and *Pseudomonas aeruginosa*, with efficiency as high as 61% and 50.5%, respectively (Rajandas *et al.*, 2012). Similarly, polypropylene-degrading bacterial strains include *Chelatococcus*, *Bacillus*, *Pseudomonas*, and so on, which were obtained from areas contaminated with plastics (Anand *et al.*, 2023). Bacterial enzymes that were found to be substantial for microplastics degradation include lipases, laccases, esterases, lignin peroxides, manganese peroxides, and so on. Recently, bacterial cellulose hydrogels have been used as potential bioflocculants for trapping microplastics (Mendonca *et al.*, 2023). Microplastics were found to be adsorbed as well as embedded in the fibrillar and porous gel-like hydrogel, which resulted in flocculation. A total flocculation rate of 88.59% could be achieved at the optimized conditions of temperature, immersion time, and hydrogel:microplastics ratio (Mendonca *et al.*, 2023).

Bacterial degradation generally involves physico-chemical alterations in the polymers, such as polymer chain length reduction, polymeric functional group alterations, and so on which helps to degrade polymers through their enzymatic action. However, the major drawback of this method is its extremely slow rate of degradation. Further, in order to minimize the formation of any noxious end-product after the bacterial degradation process, it is considerable to involve a consortium of bacteria (Singh & Wahid, 2015).

4.4.3.4 Constructed wetlands

Constructed wetlands are one of the most cost-effective and efficient wastewater treatment techniques, which involve naturally occurring geochemical and biological processes to treat the wastewater (Parashar *et al.*, 2022). Microplastics removal through constructed wetlands has recently been identified (Long *et al.*, 2023; Rozman *et al.*, 2023). In one of the studies, fibers and bead-shaped microplastic particles were introduced into the wetland system with *Iris* vegetation. A hydraulic retention time of four days was provided for microplastics. Upon testing the effluent, it was found that only 0.296% and 0.003% of the total microbeads and fibers, respectively, were present in the effluent. Moreover, longer fibers were comparatively retained more in the wetland system compared to smaller fibers (Rozman *et al.*, 2023). Multi-combination and multi-stage constructed wetlands were also utilized for estimating the microplastics' removal efficiency. It was found that a combination of systems can efficiently trap a considerable amount of microplastics. Vertical flow, horizontal subsurface flow, and surface flow types of constructed wetlands resulted in 25.71%, 32%, and 23.53% removal of microplastics, respectively (Long *et al.*, 2023).

4.4.4 Miscellaneous techniques

4.4.4.1 Electrochemical methods

Electrochemical methods are used for the removal of a variety of pollutants from water and wastewater. Electrocoagulation is a technique in which direct current is supplied to the anode and cathode. As electrodes make contact with water, metal gets oxidized and releases ions into water, resulting in the formation of agglomerates along with the impurities settling at the bottom of the tank, which can be removed later. In the electroflotation process, these agglomerates float at the top of the tank. Electrodecantation is another process that involves the phenomena of gravity as well as electrodialysis. These processes are efficient as well as environment friendly since they do not involve the addition of chemicals. Commonly used metal ions for the electrocoagulation process are Fe^{2+} and Al^{3+}. In a study conducted by Perren *et al.*, Al^{3+} ions were utilized in the electrocoagulation process for the removal of sphere-shaped polyethylene particles (Perren *et al.*, 2018). The obtained removal efficiency was more than 90%. Moreover, the highest removal was found at a pH of 7.5.

4.4.4.2 Nanotechnological methods

Kang *et al.* employed the combination of oxidation and hydrolysis of microplastics over carbon nanotubes (CNTs). In this study, manganese carbide nanoparticles were encapsulated in nitrogen-doped CNTs using the pyrolytic method. This synthesized material was then used for peroxymonosulfate activation (for producing ROS) and microplastics' mineralization under the hydrothermal condition. The outcome demonstrated that up to 50% removal of microplastics could be realized using this method (Kang *et al.*, 2019).

4.4.4.3 Combinatorial methods

Akarsu *et al.* used a combination of electrocoagulation, electroflotation, and membrane filtration techniques for the removal of polyethylene and polyvinyl chloride microplastics from wastewater (Akarsu *et al.*, 2021). The electrode combination of Al–Fe was found to be optimal at a pH of 7 and a current density of 20 A/m^2. The membrane used was a polyvinylidene fluoride microfiltration membrane with a pore size of 0.22 μm. With this combination, 100% removal efficiency for microplastics could be achieved.

Wang *et al.* employed the chemical coagulation process with sedimentation and filtration using GAC for microplastics' removal in water treatment plants (Wang *et al.*, 2020b). When coagulation was combined with only sedimentation, only 40–54% removal efficiency could be achieved. Nevertheless, larger-sized microplastics could be removed more efficiently using coagulation–sedimentation compared to smaller-sized microplastics. However, upon integrating GAC filtration in the treatment chain, 57–61% of microplastics' removal could be achieved. Further, GAC filtration enhanced the removal of smaller-sized microplastics (1–5 μm) up to 74–98%. Similar findings were reported by Zhang *et al.* after exploring the removal efficiency of the combined processes of coagulation/flocculation, sedimentation, and granular filtration. Without the granular filtration process, the efficiency was not enough to remove micro- as well as nanoplastics. However, upon adding filtration after the sedimentation, efficiency increased from 87% to almost 100%, especially for the particle size range of >100 μm (Zhang *et al.*, 2020).

4.5 FUTURE PERSPECTIVES

Literature shows the possibility of removing microplastics through various techniques; however, there is still a long way to go. Since WWTPs receive microplastics of mixed size, shape, and composition, and the removal techniques employed in various WWTPs are also different, performance comparison among different techniques becomes challenging. Moreover, from sample collection to the final analysis and identification of microplastics, uniformity is also required in procedures. Therefore, one of the most important aspects of future research in the area of microplastics' removal is to develop a consensus over universally accepted definitions and uniform analytical procedures. Further, it has been seen across the studies that PE, PP, PVC, PET, and PA are the most common types of microplastics in the WWTPs. Therefore, removal mechanisms for these polymers need to be focused on. Another area that requires attention is the monitoring of microplastics of size less than 1 μm (*viz*., nanoplastics). Owing to the limitations of instrumentation techniques, smaller-sized microplastics are often left out of analytical identification. Apart from this, it is also important to target the microplastics present in sludge. Usually, microplastics trapped in the WWTPs result in their accumulation in the sludge, which poses a risk for the soil and groundwater contamination in the long run. Therefore, future studies should also take into account the removal of microplastics from the sludge produced from WWTPs.

4.6 CONCLUSION

Efficient removal of microplastics from wastewater is critical in order to ascertain the quality of natural water reservoirs (surface/ground), as effluent is ultimately discharged into freshwater resources after treatment. As of date, conventional wastewater treatment procedures do not involve any specific mechanism to remove microplastics; rather, this contaminant gets removed simultaneously with other organic and inorganic contaminants in wastewater. However, considering the increasing concern over the microplastics' presence in the treated effluent and their repercussions on human health, advanced treatment methods need to evolve.

In this chapter, various advanced wastewater treatment techniques developed so far have been discussed. These techniques offer significant potential to reduce the microplastics' concentration in the treated effluent. Nevertheless, operational cost, practical applicability, efficacy for most of the types of microplastic particles, and energy efficiency need to be explored and improved further in the advanced techniques.

REFERENCES

Abuwatfa W. H., Al-Muqbel D., Al-Othman A., Halalsheh N. and Tawalbeh M. (2021). Insights into the removal of microplastics from water using biochar in the era of COVID-19: a mini review. *Case Studies in Chemical and Environmental Engineering*, **4**, 100151, https://doi.org/10.1016/j.cscee.2021.100151

Acarer S. (2023). Microplastics in wastewater treatment plants: sources, properties, removal efficiency, removal mechanisms, and interactions with pollutants. *Water Science & Technology*, **87**, 685, https://doi.org/10.2166/wst.2023.022

Akarsu C., Kumbur H. and Kideys A. E. (2021). Removal of microplastics from wastewater through electrocoagulation – electroflotation and membrane filtration processes. *Water Science & Technology*, **84**, 1648–1662, https://doi.org/10.2166/wst.2021.356

Anand U., Dey S., Bontempi E., Ducoli S., Vethaak A. D., Dey A. and Federici S. (2023). Biotechnological methods to remove microplastics: a review. *Environmental Chemistry Letters*, **21**, 1787–1810, https://doi.org/10.1007/s10311-022-01552-4

Ariza-Tarazona M. C., Villarreal-Chiu J. F., Barbieri V., Siligardi C. and Cedillo-Gonzalez E. I. (2019). New strategy for microplastic degradation: green photocatalysis using a protein-based porous N-TiO_2 semiconductor. *Ceramic International*, **45**, 9618–9624, https://doi.org/10.1016/j.ceramint.2018.10.208

Awasthi A. K., Tan Q. and Li J. (2020). Biotechnological potential for microplastic waste. *Trends in Biotechnology*, **38**, 1196–1199, https://doi.org/10.1016/j.tibtech.2020.03.002

Bhattacharya P., Lin S., Turner J. P. and Ke P. C. (2010). Physical adsorption of charged plastic nanoparticles affects algal photosynthesis. *Journal of Physical Chemistry C*, **114**, 16556–16561, https://doi.org/10.1021/jp1054759

Cacciari I., Quatrini P., Zirletta G., Mincione E., Vinciguerra V., Lupattelli P. and Sermanni G. G. (1993). Isotactic polypropylene biodegradation by a microbial community: physicochemical characterization of metabolites produced. *Applied and Environmental Microbiology*, **59**, 3695–3700, https://doi.org/10.1128/aem.59.11.3695-3700.1993

Carr S. A., Liu J. and Tesoro A. G. (2016). Transport and fate of microplastic particles in wastewater treatment plants. *Water Research*, **91**, 174–182, https://doi.org/10.1016/j.watres.2016.01.002

Chen Y. J., Chen Y., Miao C., Wang Y. R., Gao G. K., Yang R. X., Zhu H. J., Wang J. H., Li S. L. and Lan Y. Q. (2020). Metal-organic framework-based foams for efficient microplastic removal. *Journal of Materials Chemistry A*, **8**, 14644–14652, https://doi.org/10.1039/D0TA04891G

Cobanoglu H., Belivermis M., Sikdokur E., Kilic O. and Cayir A. (2021). Genotoxic and cytotoxic effects of polyethylene microplastics on human peripheral blood lymphocytes. *Chemosphere*, **272**, 129805, https://doi.org/10.1016/j.chemosphere.2021.129805

Conti G. O., Ferrante M., Banni M., Favara C., Nicolosi I., Cristaldi A., Fiore M. and Zuccarello P. (2020). Micro- and nano-plastics in edible fruit and vegetables. The first diet risks assessment for the general population. *Environmental Research*, **187**, 109677, https://doi.org/10.1016/j.envres.2020.109677

Dongyu Y., Yujuan Z., Weiting Y., Qinhe P. and Jiyang L. (2022). Metal-organic framework-based wood aerogel for effective removal of micro/nano plastics. *Chemical Research in Chinese Universities*, **38**, 186–191, https://doi.org/10.1007/s40242-021-1317-x

Easton T., Koutsos V. and Chatzisymeon E. (2023). Removal of polyester fibre microplastics from wastewater using a UV/H_2O_2 oxidation process. *Journal of Environmental Chemical Engineering*, **11**, 109057, https://doi.org/10.1016/j.jece.2022.109057

Egea-Corbacho A., Martin-Garcia A. P., Franco A. A., Quiroga J. M., Andreasen R. R., Jorgensen M. K. and Christensen M. L. (2023). Occurrence, identification, and removal of microplastics in a wastewater treatment plant compared to an advanced MBR technology: full-scale pilot plant. *Journal of Environmental Chemical Engineering*, **11**, 109644, https://doi.org/10.1016/j.jece.2023.109644

EU. (2018). Microplastics: sources, effects and solutions. Available online at https://www.europarl.europa.eu/news/en/headlines/society/20181116STO19217/microplastics-sources-effects-and-solutions (accessed on June 27th, 2023)

Frias J. P. G. L. and Nash R. (2019). Microplastics: finding a consensus on the definition. *Marine Pollution Bulletin*, **138**, 145–147, https://doi.org/10.1016/j.marpolbul.2018.11.022

Gao Z., Chen L., Cizdziel J. and Huang Y. (2023). Research progress on microplastics in wastewater treatment plants: a holistic review. *Journal of Environmental Management*, **325**, 116411, https://doi.org/10.1016/j.jenvman.2022.116411

Gnanasekaran G., Arthanareeswaran G. and Mok Y. S. (2021). A high-flux metal-organic framework membrane (PSF/MIL-100(Fe)) for the removal of microplastics adsorbing dye contaminants from textile wastewater. *Separation and Purification Technology*, **277**, 119655, https://doi.org/10.1016/j.seppur.2021.119655

Hahladakis J. N., Velis C. A., Weber R., Iacovidou E. and Purnell P. (2018). An overview of chemical additives present in plastics: migration, release, fate, and environmental impact during their use, disposal, and recycling. *Journal of Hazardous Materials*, **344**, 179–199, https://doi.org/10.1016/j.jhazmat.2017.10.014

Hariharan G., Purvaja R., Anandavelu I., Robin R. S. and Ramesh R. (2021). Accumulation and ecotoxicological risk of weathered polyethylene (wPE) microplastics on green mussel (Perna viridis). *Ecotoxicology and Environmental Safety*, **208**, 111765, https://doi.org/10.1016/j.ecoenv.2020.111765

Hariharan G., Purvaja R., Anandavelu I., Robin R. S. and Ramesh R. (2022). Ingestion and toxic impacts of weathered polyethylene (wPE) microplastics and stress defensive responses in whiteleg shrimp (Penaeus vannamei). *Chemosphere*, **300**, 134487, https://doi.org/10.1016/j.chemosphere.2022.134487

Haris M., Khan M. W., Zavabeti A., Mahmood N. and Eshtiaghi N. (2023). Self-assembly of C@FeO nanopillars on 2D-MOF for simultaneous removal of microplastic and dissolved contaminants from water. *Chemical Engineering Journal*, **455**, 140390, https://doi.org/10.1016/j.cej.2022.140390

Herbort A. F. and Schuhen K. (2017). A concept for the removal of microplastics from the marine environment with innovative host-guest relationships. *Environmental Science and Pollution Research*, **24**, 11061–11065, https://doi.org/10.1007/s11356-016-7216-x

Herbort A. F., Sturm M. T., Fiedler S., Abkai G. and Schuhen K. (2018). Alkoxy-silyl induced agglomeration: a new approach for the sustainable removal of microplastic from aquatic systems. *Journal of Polymers and the Environment*, **26**, 4258–4270, https://doi.org/10.1007/s10924-018-1287-3

Hidayaturrahman H. and Lee T. G. (2019). A study on characteristics of microplastic in wastewater of South Korea: identification, quantification, and fate of microplastics during treatment process. *Marine Pollution Bulletin*, **146**, 696–702, https://doi.org/10.1016/j.marpolbul.2019.06.071

Jenner L. C., Rotchell J. M., Bennett R. T., Cowen M., Tentzeris V. and Sadofsky L. R. (2022). Detection of microplastics in human lung tissue using μFTIR spectroscopy. *Science of the Total Environment*, **831**, 154907, https://doi.org/10.1016/j.scitotenv.2022.154907

Kang J., Zhou L., Duan X., Sun H., Ao Z. and Wang S. (2019). Degradation of cosmetic microplastics via functionalized carbon nanosprings. *Matter*, **1**, 745–758, https://doi.org/10.1016/j.matt.2019.06.004

Kaur M., Xu M. and Wang L. (2022). Cyto-genotoxic effect causing potential of polystyrene microplastics in terrestrial plants. *Nanomaterials*, **12**, 2024, https://doi.org/10.3390/nano12122024

Kim J. W., Park S. B., Tran Q. G., Cho D. H., Choi D. Y., Lee Y. J. and Kim H. S. (2020). Functional expression of polyethylene terephthalate degrading enzyme (PETase) in green microalgae. *Microbial Cell Factories*, **19**, 97, https://doi.org/10.1186/s12934-020-01355-8

Kim S., Sin A., Nam H., Park Y., Lee H. and Han C. (2022). Advanced oxidation processes for microplastics degradation: a recent trend. *Chemical Engineering Journal Advances*, **9**, 100213, https://doi.org/10.1016/j.ceja.2021.100213

Kobielska P. A., Howarth A. J., Farha O. K. and Nayak S. (2018). Metal-organic frameworks for heavy metal removal from water. *Coordination Chemistry Reviews*, **358**, 92–107, https://doi.org/10.1016/j.ccr.2017.12.010

Koyuncuoglu P. and Erden G. (2021). Sampling, pre-treatment, and identification methods of microplastics in sewage sludge and their effects in agricultural soils: a review. *Environmental Monitoring and Assessment*, **193**, 175, https://doi.org/10.1007/s10661-021-08943-0

Kunlere I. O., Fagade O. E. and Nwadike B. I. (2019). Biodegradation of low density polyethylene (LDPE) by certain indigenous bacteria and fungi. *International Journal of Environmental Studies*, **76**, 428–440, https://doi.org/10.1080/00207233.2019.1579586

Lares M., Ncibi M. C., Sillanpaa M. and Sillanpaa M. (2018). Occurrence, identification, and removal of microplastic particles and fibers in conventional activated sludge process and advanced MBR technology. *Water Research*, **133**, 236–246, https://doi.org/10.1016/j.watres.2018.01.049

Leslie H. A., van Velzen M. J. M., Brandsma S. H., Vethaak A. D., Garcia-Vallejo J. J. and Lamoree M. H. (2022). Discovery and quantification of plastic particle pollution in human blood. *Environment International*, **163**, 107199, https://doi.org/10.1016/j.envint.2022.107199

Liu F., Zhang C., Li H., Offiong N. A. O., Bi Y., Zhou R. and Ren H. (2023). A systematic review of electrocoagulation technology applied for microplastics removal in aquatic environment. *Chemical Engineering Journal*, **456**, 141078, https://doi.org/10.1016/j.cej.2022.141078

Long Y., Zhou Z., Wen X., Wang J., Xiao R., Wang W., Li X., Lai X., Zhang Y., Deng C., Cao J. and Yin L. (2023). Microplastics removal and characteristics of a typical multi-combination and multi-stage constructed wetlands wastewater treatment plant in Changsha, China. *Chemosphere*, **312**, 137199, https://doi.org/10.1016/j.chemosphere.2022.137199

Lu Y., Li M. C., Lee J., Liu C. and Mei C. (2023). Microplastic remediation technologies in water and wastewater treatment processes: current status and future perspectives. *Science of the Total Environment*, **868**, 161618, https://doi.org/10.1016/j.scitotenv.2023.161618

Ma B., Xue W., Hu C., Liu H., Qu J. and Li L. (2019). Characteristics of microplastic removal *via* coagulation and ultrafiltration during drinking water treatment. *Chemical Engineering Journal*, **359**, 159–167, https://doi.org/10.1016/j.cej.2018.11.155

Magnusson K. and Noren F. (2014). Screening of Microplastic Particles in and Downstream A Wastewater Treatment Plant. IVL Swedish Environmental Research Institute, Stockholm, Sweden, *C*55, 1–22.

Makhdoumi P., Naghshbandi M., Ghaderzadeh K., Mirzabeigi M., Yazdanbakhsh A. and Hossini H. (2021). Micro-plastic occurrence in bottled vinegar: qualification, quantification, and human risk exposure. *Process Safety and Environmental Protection*, **152**, 404–413, https://doi.org/10.1016/j.psep.2021.06.022

Mendonca I., Sousa J., Cunha C., Faria M., Ferreira A. and Cordeiro N. (2023). Solving urban water microplastics with bacterial cellulose hydrogels: leveraging predictive computational models. *Chemosphere*, **314**, 137719, https://doi.org/10.1016/j.chemosphere.2022.137719

Michielssen M. R., Michielssen E. R., Ni J. and Duhaime M. B. (2016). Fate of microplastics and other small anthropogenic litter (SAL) in wastewater treatment plants depends on unit processes employed. *Environmental Science: Water Research & Technology*, **2**, 1064–1073, https://doi.org/10.1039/C6EW00207B

Mon M., Bruno R., Ferrando-Soria J., Armentano D. and Pardo E. (2018). Metal-organic framework technologies for water remediation: towards a sustainable ecosystem. *Journal of Materials Chemistry A*, **6**, 4912–4947, https://doi.org/10.1039/C8TA00264A

Mou F., Pan D., Chen C., Gao Y., Xu L. and Guan J. (2015). Magnetically modulated pot-like $MnFe_2O_4$ micromotors: nanoparticles assembly fabrication and their capability for direct oil removal. *Advanced Functional Materials*, **25**, 6173–6181, https://doi.org/10.1002/adfm.201502835

Nolte T. M., Hartmann N. B., Kleijn J. M., Garnæs J., van de Meent D., Hendriks A. J. and Baun A. (2017). The toxicity of plastic nanoparticles to green algae as influenced by surface modification, medium hardness, and cellular adsorption. *Aquatic Toxicology*, **183**, 11–20, https://doi.org/10.1016/j.aquatox.2016.12.005

Oliveira J., Belchior A., da Silva V. D., Rotter A., Petrovski Z., Almeida P. L., Lourenco N. D. and Gaudencio S. P. (2020). Marine environmental plastic pollution: mitigation by microorganism degradation and recycling valorization. *Frontiers in Marine Science*, **7**, 567126, https://doi.org/10.3389/fmars.2020.567126

Osman A. I., Hosny M., Eltaweil A. S., Omar S., Elgarahy A. M., Farghali M., Yap P. S., Wu Y. S., Nagandran S., Batumalaie K., Gopinath S. C. B., John O. D., Sekar M., Saikia T., Karunanithi P., Hatta M. H. M. and Akinyede K. A. (2023). Microplastic sources, formation, toxicity, and remediation: a review. *Environmental Chemistry Letters*, **21**, 2129–2169, https://doi.org/10.1007/s10311-023-01593-3

Parashar V., Singh S., Purohit M. R., Tamhankar A. J., Singh D., Kalyanasundaram M., Lundborg C. S. and Diwan V. (2022). Utility of constructed wetlands for treatment of hospital effluent and antibiotic resistant bacteria in resource limited settings: a case study in Ujjian, India. *Water Environment Research*, **94**, e10783, https://doi.org/10.1002/wer.10783

Patil S., Kamdi P., Chakraborty S., Das S., Bafana A., Krishnamurthi K. and Sivanesan S. (2023). Characterization and removal of microplastics in a sewage treatment plant from urban Nagpur, India. *Environmental Monitoring and Assessment*, **195**, 47, https://doi.org/10.1007/s10661-022-10680-x

Perren W., Wojtasik A. and Cai Q. (2018). Removal of microbeads from wastewater using electrocoagulation. *ACS Omega*, **3**, 3357–3364, https://doi.org/10.1021/acsomega.7b02037

Pironti C., Notarstefano V., Ricciardi M., Motta O., Giorgini E. and Montano L. (2023). First evidence of microplastics in human urine, a preliminary study of intake in the human body. *Toxics*, **11**, 40, https://doi.org/10.3390/toxics11010040

Pizzichetti A. R. P., Pablos C., Alvarez-Fernandez C., Reynolds K., Stanley S. and Marugan J. (2021). Evaluation of membranes performance for microplastic removal in a simple and low-cost filtration system. *Case Studies in Chemical and Environmental Engineering*, **3**, 100075, https://doi.org/10.1016/j.cscee.2020.100075

Priya A. K., Jalil A. A., Dutta K., Rajendran S., Vasseghian Y., Karimi-Maleh H. and Soto-Moscoso M. (2022). Algal degradation of microplastic from the environment: mechanism, challenges, and future prospects. *Algal Research*, **67**, 102848, https://doi.org/10.1016/j.algal.2022.102848

Ragusa A., Svelato A., Santacroce C., Catalano P., Notarstefano V., Carnevali O., Papa F., Rongioletti M. C. A., Baiocco F., Draghi S., D'Amore E., Rinalso D., Matta M. and Giorgini E. (2021). Plasticenta: first evidence of microplastics in human placenta. *Environment International*, **146**, 106274, https://doi.org/10.1016/j.envint.2020.106274

Rajandas H., Parimannan S., Sathasivam K., Ravichandran M. and Yin L. S. (2012). A novel FTIR-ATR spectroscopy based technique for the estimation of low-density polyethylene biodegradation. *Polymer Testing*, **31**, 1094–1099, https://doi.org/10.1016/j.polymertesting.2012.07.015

Ridall A., Farrar E., Dansby M. and Ingels J. (2023). Influence of wastewater treatment plants and water input sources on size, shape, and polymer distributions of microplastics in St. Andrew Bay, Florida, USA. *Marine Pollution Bulletin*, **187**, 114552, https://doi.org/10.1016/j.marpolbul.2022.114552

Rozman U., Klun B. and Kalcikova G. (2023). Distribution and removal of microplastics in a horizontal sub-surface flow laboratory constructed wetland and their effects on the treatment efficiency. *Chemical Engineering Journal*, **461**, 142076, https://doi.org/10.1016/j.cej.2023.142076

Santos N. O. D., Busquets R. and Campos L. C. (2023). Insights into the removal of microplastics and microfibers by advanced oxidation processes. *Science of the Total Environment*, **861**, 160665, https://doi.org/10.1016/j.scitotenv.2022.160665

Sarkar D. J., Sarkar S. D., Das B. K., Praharaj J. K., Mahajan D. K., Purokait B., Mohanty T. R., Mohanty D., Gogoi P., Kumar V. S., Behera B. K., Manna R. K. and Samanta S. (2021). Microplastics removal efficiency of drinking water treatment plant with pulse clarifier. *Journal of Hazardous Materials*, **413**, 125347, https://doi.org/10.1016/j.jhazmat.2021.125347

Sarmah P. and Rout J. (2018). Efficient biodegradation of low-density polyethylene by cyanobacteria isolated from submerged polyethylene surface in domestic sewage water. *Environmental Science and Pollution Research*, **25**, 33508–33520, https://doi.org/10.1007/s11356-018-3079-7

Sekar V. and Sundaram B. (2023). Preliminary evidence of microplastics in landfill leachate, Hyderabad, India. *Process Safety and Environmental Protection*, **175**, 369–376, https://doi.org/10.1016/j.psep.2023.05.070

Siipola V., Pflugmacher S., Romar H., Wendling L. and Koukkari P. (2020). Low-cost biochar adsorbents for water purification including microplastics removal. *Applied Science*, **10**, 788, https://doi.org/10.3390/app10030788

Singh S. and Bhagwat A. (2022). Microplastics: a potential threat to groundwater resources, *Groundwater for Sustainable Development*, **19**, 100852, https://doi.org/10.1016/j.gsd.2022.100852

Singh L. and Wahid Z. A. (2015). Methods for enhancing bio-hydrogen production from biological process: a review. *Journal of Industrial and Engineering Chemistry*, **21**, 70–80, https://doi.org/10.1016/j.jiec.2014.05.035

Singh S., Kalyanasundaram M. and Diwan V. (2021a). Removal of microplastics from wastewater: current practices and way forward, *Water Science & Technology*, **84**, 3689–3704, https://doi.org/10.2166/wst.2021.472

Singh N., Khandelwal N., Ganie Z. A., Tiwari E. and Darbha G. K. (2021b). Eco-friendly magnetic biochar: an effective trap for nanoparticles of varying surface functionality and size in the aqueous environment. *Chemical Engineering Journal*, **418**, 129405, https://doi.org/10.1016/j.cej.2021.129405

Singh S., Trushna T., Kalyanasundaram M., Tamhankar A. J. and Diwan V. (2022). Microplastics in drinking water: a macro issue. *Water Supply*, **22**, 5650–5674, https://doi.org/10.2166/ws.2022.189

Singh S., Chakma S., Alawa B., Kalyanasundaram M. and Diwan V. (2023). Identification, characterization, and implications of microplastics in soil: a case study of Bhopal, *Central India. Journal of Hazardous Materials Advances*, **9**, 100225, https://doi.org/10.1016/j.hazadv.2022.100225

Straub S., Hirsch P. E. and Burkhardt-Holm P. (2017). Biodegradable and petroleum-based microplastics do not differ in their ingestion and excretion but in their biological effects in a freshwater invertebrate *Gammarus fossarum. International Journal of Environmental Research and Public Health*, **14**, 774, https://doi.org/10.3390/ijerph14070774

Sturm M. T., Horn H. and Schuhen K. (2021). Removal of microplastics from waters through agglomeration-fixation using organosilanes – effects of polymer types, water composition, and temperature. *Water*, **13**, 675, https://doi.org/10.3390/w13050675

Sturm M. T., Myres E., Schober D., Korzin A., Thege C. and Schuhen K. (2023). Comparison of AOP, GAC, and novel organosilane-based process for the removal of microplastics at a municipal wastewater treatment plant. *Water*, **15**, 1164, https://doi.org/10.3390/w15061164

Sun J., Dai X., Wang Q., Loosdrecht M. C. M. V. and Ni B. J. (2019). Microplastics in wastewater treatment plants: detection, occurrence, and removal. *Water Research*, **152**, 21–37, https://doi.org/10.1016/j.watres.2018.12.050

Sun C., Wang Z., Chen L. and Li F. (2020). Fabrication of robust and compressive chitin and graphene oxide sponges for removal of microplastics with different functional groups. *Chemical Engineering Journal*, **393**, 124796, https://doi.org/10.1016/j.cej.2020.124796

Sundbæk K. B., Koch I. D. W., Villaro C. G., Rasmussen N. S., Holdt S. L. and Hartmann N. B. (2018). Sorption of fluorescent polystyrene microplastic particles to edible seaweed *Fucus vesiculosus. Journal of Applied Phycology*, **30**, 2923–2927, https://doi.org/10.1007/s10811-018-1472-8

Talvitie J., Mikola A., Koistinen A. and Setala O. (2017). Solutions to microplastic pollution – removal of microplastics from wastewater effluent with advanced wastewater treatment technologies. *Water Research*, **123**, 401–407, https://doi.org/10.1016/j.watres.2017.07.005

Tofa T. S., Kunjali K. L., Paul S. and Dutta J. (2019). Visible light photocatalytic degradation of microplastic residues with zinc oxide nanorods. *Environmental Chemistry Letters*, **17**, 1341–1346, https://doi.org/10.1007/s10311-019-00859-z

Umar M., Singdahl-Larsen C. and Ranneklev S. B. (2023). Microplastics removal from a plastic recycling industrial wastewater using sand filtration. *Water*, **15**, 896, https://doi.org/10.3390/w15050896

Upadhyay R., Singh S. and Kaur G. (2022). Sorption of pharmaceuticals over microplastics' surfaces: interaction mechanisms and governing factors. *Environmental Monitoring and Assessment*, **194**, 803, https://doi.org/10.1007/s10661-022-10475-0

Veen I. V., Mourik L. M., Velzen M. J. M., Groenewoud Q. R. and Leslie H. A. (2022). Plastic Particles in Livestock Feed, Milk, Meat, and Blood. Vrije Universiteit Amsterdam, Netherlands. Available online at https://www.plasticsoupfoundation.org/wp-content/uploads/2022/07/Final-Report-pilot-study-plastic-particles-in-livestock-feed-milk-meat-and-blood-SIGNED.pdf (accessed on June 28th, 2023)

Vikrant K., Kim K. H., Kumar V., Giannakoudakis D. A. and Boukhvalov D. W. (2020). Adsorptive removal of an eight-component volatile organic compound mixture by Cu-, Co-, and Zr-metal-organic frameworks: experimental and theoretical studies. *Chemical Engineering Journal*, **397**, 125391, https://doi.org/10.1016/j.cej.2020.125391

Villa K., Palenzuela L. M., Sofer Z., Matejkova S. and Pumera M. (2018). Metal-free visible-light photoactivated C_3N_4 bubble-propelled tubular micromotors with inherent fluorescence and on/off capabilities. *ACS Nano*, **12**, 12482–12491, https://doi.org/10.1021/acsnano.8b06914

Wan H. Wang J., Sheng X., Yan J., Zhang W. and Xu Y. (2022). Removal of polystyrene microplastics from aqueous solution using the metal-organic framework material of ZIF-67. *Toxics*, **10**, 70, https://doi.org/10.3390/toxics10020070

Wang L., Kaeppler A., Fischer D. and Simmchen J. (2019). Photocatalytic TiO_2 micromotors for removal of microplastics and suspended matter. *ACS Applied Materials & Interfaces*, **11**, 32937–32944, https://doi.org/10.1021/acsami.9b06128

Wang Z. Sedighi M. and Lea-Langton A. (2020a). Filtration of microplastic spheres by biochar: removal efficiency and immobilisation mechanisms. *Water Research*, **184**, 116165, https://doi.org/10.1016/j.watres.2020.116165

Wang Z., Lin T. and Chen W. (2020b). Occurrence and removal of microplastics in an advanced drinking water treatment plant (ADWTP). *Science of the Total Environment*, **700**, 134520, https://doi.org/10.1016/j.scitotenv.2019.134520

Wang J., Sun C., Huang Q. X., Chi Y. and Yan J. H. (2021). Adsorption and thermal degradation of microplastics from aqueous solutions by Mg/Zn modified magnetic biochars. *Journal of Hazardous Materials*, **419**, 126486, https://doi.org/10.1016/j.jhazmat.2021.126486

Yu Z., Song S., Xu X., Ma Q. and Lu Y. (2021). Sources, migration, accumulation, and influence of microplastics in terrestrial plant communities. *Environmental and Experimental Botany*, **192**, 104635, https://doi.org/10.1016/j.envexpbot.2021.104635

Zhang Y., Yuan S., Feng X., Li H., Zhou J. and Wang B. (2016). Preparation of nanofibrous metal-organic framework filters for efficient air pollution control. *Journal of American Chemical Society*, **138**, 5785–5788, https://doi.org/10.1021/jacs.6b02553

Zhang Q., Dong R., Wu Y., Gao W., He Z. and Ren B. (2017). Light-driven Au-WO_3@C Janus micromotors for rapid photodegradation of dye pollutants. *ACS Applied Materials & Interfaces*, **9**, 4674–4683, https://doi.org/10.1021/acsami.6b12081

Zhang Y., Diehl A., Lewandowski A., Gopalakrishnan K. and Baker T. (2020). Removal efficiency of micro- and nanoplastics (180 nm–125 μm) during drinking water treatment. *Science of the Total Environment*, **720**, 137383, https://doi.org/10.1016/j.scitotenv.2020.137383

Zhao Q., Zhu L., Weng J., Jin Z., Cao Y., Jiang H. and Zhang Z. (2023). Detection and characterization of microplastics in the human testis and semen. *Science of the Total Environment*, **877**, 162713, https://doi.org/10.1016/j.scitotenv.2023.162713

Ziajahromi S., Neale P. A., Rintoul L. and Leusch D. L. (2017). Wastewater treatment plants as a pathway for microplastics: development of a new approach to sample wastewater-based microplastics. *Water Research*, **112**, 93–99, https://doi.org/10.1016/j.watres.2017.01.042

doi: 10.2166/9781789063752_0073

Chapter 5

A brief account of the antibiotics and antibiotic resistance genes in an aquatic environment

Nikita Yadav[1], Ashootosh Mandpe[2] and Sudeep Shukla[3]*

[1]Amity School of Earth and Environmental Sciences, Amity University Haryana, Gurugram 122 413, India
[2]Department of Civil Engineering, Indian Institute of Technology Indore, Indore 453 552, Madhya Pradesh, India
[3]Environment Pollution Analysis Lab, Bhiwadi, Alwar 301 019, Rajasthan, India
**Corresponding Author: Dr. Ashootosh Mandpe, Assistant Professor, Department of Civil Engineering, Indian Institute of Technology Indore, Indore 453 552, India*
**Corresponding author: as_mandpe@iiti.ac.in*

ABSTRACT

Inappropriate disposal of pharmaceutical waste containing antibiotics into water bodies developed the catastrophe of antibiotic resistance. The occurrence and, thereafter, bioaccumulation of antibiotics in the aquatic ecosystem have raised concern about their associated toxicity toward aquatic organisms. Their presence in aquatic bodies has been widely discussed in developed nations. However, in many developing countries like India, minimal research has been reported on the status of antibiotics in the aquatic environment. This chapter summarizes the global distribution of antibiotics in the aquatic environment, their effects on the microbial community, the evaluation and assessment of antibiotic risks, and the source tracking of these emerging pollutants. Another key objective of this chapter is to investigate the trends in consumption, distribution, and fate of major antibiotic classes in India and their dynamic relation to the biotic world. The eco-toxicity of these emerging contaminants towards different flora and fauna communities at different trophic levels with targeted assessment and remediation technologies, including percent removal efficiencies, has been discussed. Legislative and regulatory measures carried out for the effective management of this growing epidemic by various governments have also been discussed.

Keywords: antibiotics, antibiotic resistance genes, bioaccumulation, emerging contaminants, remediation

5.1 INTRODUCTION

Mankind has witnessed a paradigm shift in the pathogenic/microbiological world after the sudden discovery of penicillin by Alexander Fleming in the 1920s. Among the most significant achievements of the 20th century was the discovery of antibiotics, which are produced by microbes. Several microorganisms produce these metabolites, which can kill or delay the onset of further microbial growth and development. This property of antibiosis provided them with a significant tool to conquer the epidemic world by reducing mortality and morbidity for many significant diseases

(Nkoh *et al.*, 2023). Considerable growth has been reported in pharmaceutical industries (Aitken, 2020), which also gave rise to a dramatic increase in the utilization of antibiotics, not only for humans but for animal husbandry practices too. The dumping of antibiotics in the aquatic waterways due to improper handling and disposal from pharmaceutical industries, municipality runoff, hospitals and dispensaries runoff, and effluent from animal husbandry practices led to the growing traces of these emerging contaminants (ECs) into the environmental matrices. Both lentic and lotic water bodies face the dilemma of antibiotic resistance due to the pseudo-persistent nature of these ECs (Wang *et al.*, 2023). This puts the ecosystem and healthcare services in a compromising position. Different aquatic bodies with different levels of accumulation have other effects on the existing micro-communities of flora and fauna, that is, they have adverse effects on the micro-biomes of humans as well as physiological aberrations in the plant world. The lakes and ponds (i.e., lentic) system offers an essential setting for studying the fate of antibiotics and the haulage of antibiotic resistance genes (ARGs) with a diverse variety of niche areas, such as bacterial populations and associated floral and faunal communities (Dong *et al.*, 2022). Due to the low hydraulic coefficient, the lentic system is likely to succumb to the accumulation of these new-age contaminants (Zhang *et al.*, 2023). Hydraulic properties led to the effective dispersal of pollutants among the water bodies. Since lentic system has a quantitatively higher freshwater portion than lotic system, previous ones have a higher bioaccumulation factor and serve as a standing reservoir for acquiring new resistances among the microbes of human interest.

Antibiotics are chemically complex substances with several functional moieties in their molecular structure that belong to different specific classifications (Table 5.1) based on their non-selective mode of action and their wide range of potential. Their varied and unique chemical structure provides them with lipophilicity (Gevao *et al.*, 2022). The presence of specific functional groups gives them a peculiar feature for targeting specific bacterial genomes. Since bacteria have labile and less stable genetic materials, it aids them in overcoming the hindrances posed by antibiotics. Antibiotics can be resisted by microbes through a wide variety of their associated metabolic strategies. Spontaneous mutations in prokaryotic genomes lead to antibiotic resistance epidemic with horizontal gene transfer (HGT) (Deng *et al.*, 2020). The evolution of AR processes accumulated by adaptive and malignant microbes involves Darwinian factors, that is, alterations occurring in antecedent functional genes of the prokaryotic genome, chosen by selective environmental stressors. Due to selection processes, the defensive mechanism has been reported in transportation systems such as efflux pumps, mutations in intracellular proteins like porins that affect bacterial cell porosity, and barriers to these drugs' entry. The lateral transmission of genes for antibiotic resistance from source organisms is accelerated by adjustment to the evolutionary pressure of antibiotics (Grenni *et al.*, 2018). When confronted with the antibiotics generated, bacteria within the same ecosystem may modify their innate mechanisms – for example, by over-expressing efflux systems, they introduce new pathways through HGT, that is, the transduction of heterologous genes, the resistance mechanism from the producers (Lupo *et al.*, 2012) (Figure 5.1).

Other varied genetic materials and mobile genetic elements are ubiquitous in aquatic bodies, such as integrons, phages, transposons, and plasmids, which help acquire new resistances by serving as platforms for gene aggregation and transmitting it to other microbes of human interest by subsequent mixing. Henceforth, aquatic life was impacted, from the micro to macro level, as these chemicals bioaccumulated at trophic levels. Bioaccumulated concentrations ultimately pass onto human consumption via the supplies of veggies and fruits irrigated by such contaminated recycled water, dairy products, and water drinking supplies, resulting in Multi Drug Resistance (MDR), as shown in Figure 5.2.

Based on pre-finding studies, the possible sources or occurrences, their fate, exposure to drinking water, possible assessment of human health risk, removal by available treatment methodologies, and regulatory preventive measures are taken into account in this study.

5.1.1 Antibiotics as emerging pollutants

The inception of rapid industrialization and altered patterns in consumer goods led to the emergence of a new class of pollutants in the different environmental matrices. New-age contaminants are

Table 5.1 Characteristics of different antibiotic classes.

S. No.	Type of Abs (Interest, 2020; Map *et al.*, 2020; Moore, 2016)	Category	Discovery (Interest, 2020)	Common Examples	Mode of Action (Map *et al.*, 2020; Moore, 2016)	Toxicity (Moore, 2016)
01.	β-lactams	Bactericidal and broad spectrum	1928	Penicillin, amoxicillin, cephalosporin	Interfere with the synthesis of peptidoglycan layer of cell wall of bacteria	Hypersensitivity, hemolytic anemia
02.	Sulfonamides	Broad spectrum and bacteriostatic	1932	Prontosil, sulfadiazine, sulfisoxazole	Inhibiting the bacterial synthesis instead of killing them	Thrombocytopenia
03.	Aminoglycosides	Bacteriostatic	Early 1940s	Streptomycin, kanamycin, neomycin	Inhibit the synthesis of protein in bacteria	Nephrotoxicity, ototoxicity
04.	Tetracyclines	Broad spectrum and bacteriostatic	Late 1940s	Doxycycline, tetracycline, oxytetracycline	Inhibit the protein synthesis, so as the growth and reproduction	Hepatotoxicity, tooth discoloration, impaired growth
05.	Chloramphenicol	Broad spectrum and bacteriostatic	Late 1940s	Levomycetin, Chlornitromycin, chloromycetin	Inhibit the protein synthesis, so as the growth and reproduction	Aplastic anemia, gray baby syndrome
06.	Macrolides	Bacteriostatic	1950	Erythromycin	Same as lactams but target more number of species than lactams	Coumadin interaction
07.	Glycopeptides	Bactericidal	Late 1950s	Vancomycin (drug of last resort), teicoplanin	Inhibit bacterial cell wall synthesis	Red man syndrome, nephrotoxicity, ototoxicity
08.	Oxazolidinones	Broad spectrum and bactericidal	Late 1970s	Linezolid, cycloserine, posizolid	Inhibiting the protein synthesis at P-site of ribosomal subunit 50S	Pharmokinetic profile with low toxicity
09.	Ansamycins	Bactericidal and narrow spectrum	Late 1950s	Rifamycin, isoniazid	Inhibit the production of RNA, anti-viral activity reported	Metabolic acidosis, hematological disorders
10.	Quinolones	Bactericidal and broad spectrum	Early 1960s	Ciproflaxacin, levoflaxacin, travoflaxcin	Inhibits replication and transcription of DNA	Phototoxicity, impaired fracture healing, Achillies tendon rupture
11.	Streptogramins	Narrow spectrum with bacteriostatic properties	Early 1960s	Quinupristin dalfopristin	Streptogramins A & B synergistically inhibit cell growth	Gastrointestinal disturbance
12.	Lipopeptides	Bactericidal	1987	Daptomycin, surfactin	Multiple cell- membrane functions in bacteria	Low toxic and high biodegradability reported

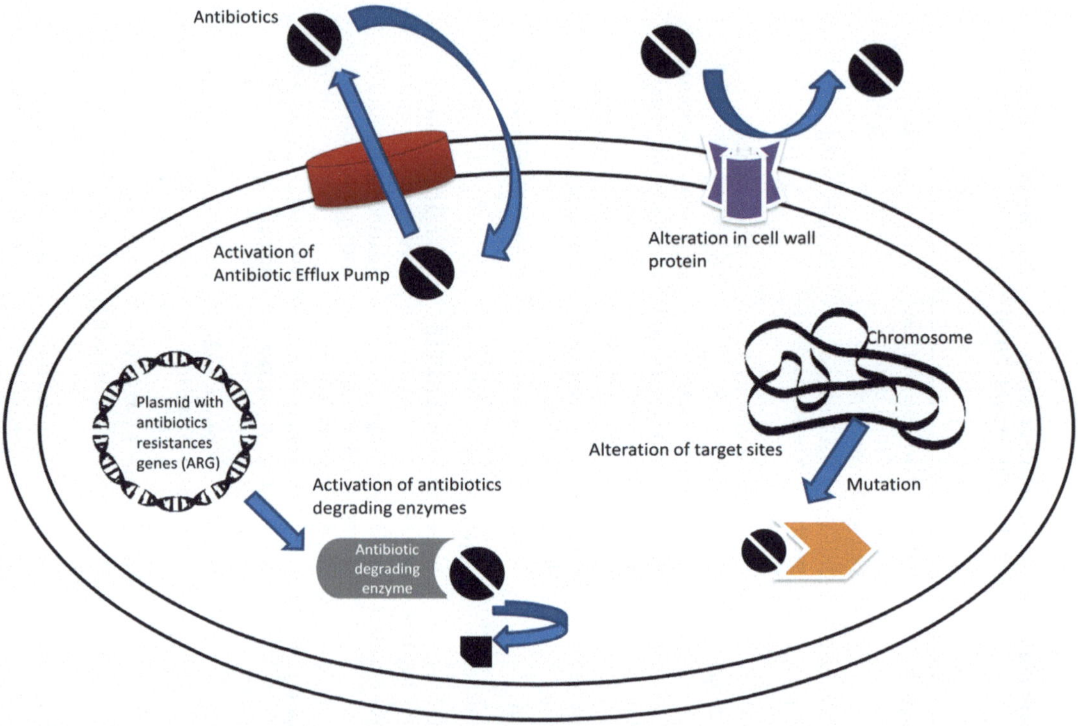

Figure 5.1 Antibiotic resistance mechanism through different pathways: efflux pumps, alteration of cell wall proteins, plasmid transfer, mutational events in the bacterial genome, and activation of antibiotics-degrading enzymes.

artificial or naturally present compounds that are not regularly detected in the ecosystem but possess the propensity to infiltrate and have documented or anticipated harmful environmental consequences and health impacts. Examples include compounds like pharmaceuticals, personnel care products, corticosteroids, pheromones, flame retardants, insecticides, hormone-disrupting substances, cleansers, lingering organic pollutants (the dirty dozen), and so on (Stefanakis & Becker, 2015). Pharmaceutical compounds, especially antibiotics, have been identified as emerging contaminants across the globe. Based on people's consumption, these chemicals tend to be present in water bodies as conventional treatment facilities are unable to remove them effectively. However, their detection in the environmental matrices can be attributed to recent advancements in sophisticated analytical tools and techniques that can quantify them at trace levels. Further sections evaluated the consumption trends in antibiotic usage worldwide, particularly in India.

5.1.2 Occurrence of antibiotics and ARGs in water bodies

According to research that is presently accessible, the levels of antibiotics present in rivers and streams, aquifers, and improperly disposed wastewater are generally within the acceptable range of 0.05–0.1 μg/L, as presented in Table 5.2. However, these concentrations can be identified with the course of evolution in analytical techniques such as Liquid Chromatography Mass Spectrometry (LC-MS), High Performance Liquid Chromatography (HPLC), and so on (Danner *et al.*, 2019).

The possible significant pathways and hotspots for the contamination of aquatic bodies reported in various studies with high anthropogenic activities are: effluents from wastewater treatment plants

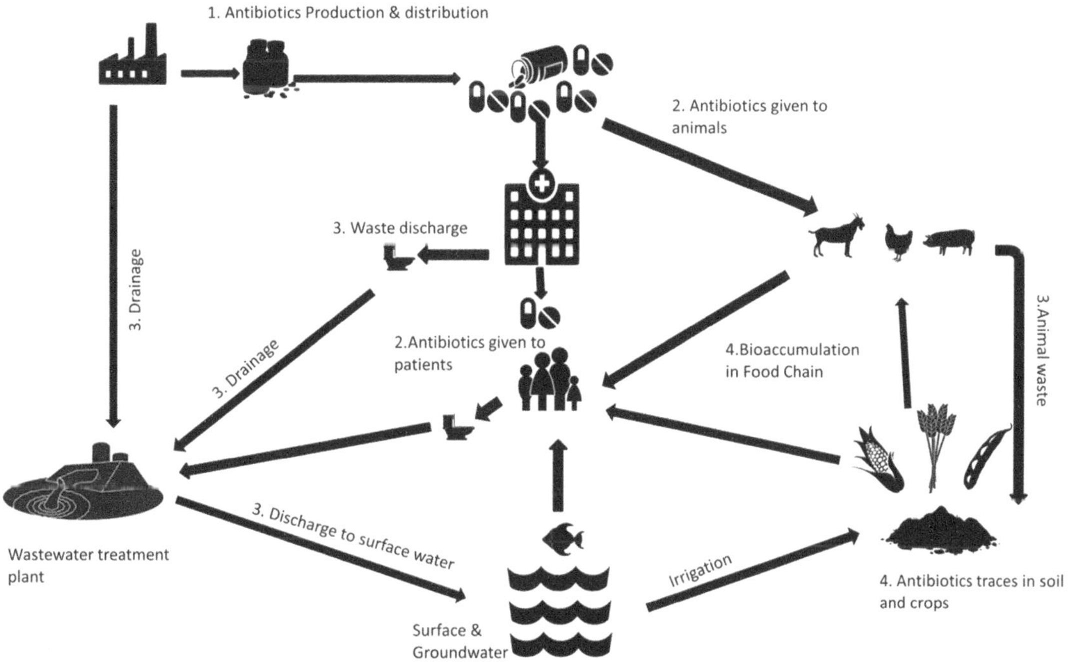

Figure 5.2 Lifecycle analysis and processes associated with the fate of antibiotics.

(WWTPs), which include household usage and municipal runs; pharmaceutical manufacturing plants (Kumar *et al.*, 2020); and animal husbandry practices, which include poultry and shrimp farms; and aquaculture, as shown in Table 5.2. The utilization of antibiotics in livestock farming creates a conduit for these substances to enter adjacent aquifers and waterways, as these are persistent and enter the environment either by direct effluent discharge or from the excretory pathway, as shown in Figure 5.2 (Gray *et al.*, 2019). Unused antibiotics, which are discarded in garbage, end up in landfills and finally in the soil (Marathe *et al.*, 2016).

It has become widely accepted that the broad expansion of ARGs imposed by the improper application of antibiotics poses a severe threat to the well-being of people and the environment. Wastewater treatment facilities (WWTFs) encounter sewage comprising not merely antibiotics but also ARGs, which can develop into an epicenter for the dissemination of ARGs and associated bacteria since they provide favorable circumstances for bacterial development and have a higher selective constraint for the dissemination of ARGs among various bacterial genera. Numerous ARGs have been identified to date in samples connected to WWTFs, including wastewater influent prior to remediation and effluent despite sterilization. WWTFs serve as important repositories for the numerous pervasive and widely distributed ARGs found in ecological soil, water, and other ecosystems.

ARGs have been detected in WWTPs around the world. WWTPs are an important point source of antibiotics and ARGs, as they receive and treat wastewater from households, hospitals, and industries. Studies have shown that the concentration and diversity of ARGs in WWTPs can vary depending on factors such as the type of treatment technology used, the size of the plant, and the source of the wastewater. For example, the presence of sub-inhibitory concentrations of antibiotics in WWTPs can select for antibiotic-resistant bacteria. Some studies have found that WWTPs that receive wastewater from hospitals or animal farms have higher concentrations of ARGs compared to those that treat mainly domestic wastewater.

Table 5.2 Trend of antibiotics consumption in the global context.

S. No.	Continents	Country-wise Location	Antibiotics	Type River/Lake/TTP	Concentration (μg/L)	References
01.	AFRICA	Ghana	(i) Ampicillin (ii) Ciprofloxacin (iii) Erythromycin (iv) Sulfamethoxazole	Hospital waste water effluent and river water	0.027 11.35–15.73 7.94–10.61 7.194	Azanu *et al.* (2018)
		Kenya	(i) Sulfamethoxazole (ii) Ciprofloxacin (iii) Trimethoprim	Nairobi River	13.8 0.509 2.650	Ngumba *et al.* (2016)
		South Africa	(i) Ampicillin (ii) Ciprofloxacin (iii) Nalidixic acid	WWTP influent and effluent	6.57–8.92 27.1 25.2–29.9	Agunbiade and Moodley (2016)
02.	AMERICA	USA	(i) Ampicillin (ii) Cloxacilin (iii) Erythromycin (iv) Oxytetracycline (v) Sulfadimethoxine (vi) Sulfathiazole	Poudre River, WWTP (influent and effluent), ground water	1.969 1.993 0.18 0.07–1.34 0.24–15 0.08	Cha *et al.* (2006); Lindsey *et al.* (2001); Kolpin *et al.* (2002)
		Argentina	(i) Monensin (ii) Lascolid (iii) Salinomycin	Del Plata hydrological basin	0.288–4.67 – 1.150	Alonso *et al.* (2019)
03.	ASIA	India	(i) Trimethoprim (ii) Ciprofloxacin (iii) Enrofloxacin (iv) Norfloxacin (v) Ofloxacin (vi) Lomefloxacin	Patancheru Enviro Tech Ltd. Treatment plant undiluted effluent, Karnataka	4.4 14000 210 25 55 8.8	Fick *et al.* (2009)
		Bangladesh	(i) Trimethoprim (ii) Sulfadiazine (iii) Sulfamethiazine (iv) Penicillin	Finfish aquaculture	0.041 0.017 0.011 0.007	Hossain *et al.* (2017)
		Iraq	(i) Levofloxacin (ii) Amoxicillin (iii) Ciprofloxacin	WWTPs (raw water)	0.177–0.414 1.50 1.270–1.344	Mahmood *et al.* (2019)
		China	(i) Chlortetracycline (ii) Doxycline (iii) Enroflaxacin (iv) Erythromycin (v) Sulfadiazine (vi) Tetracycline	Filtered tap water	0.017 0.047 0.136 2.91 0.726 0.114	Ben *et al.* (2020)
		Iran	(i) Azithromycin (ii) Cefalexin (iii) Ciprofloxacin	WWTP	0.563 0.184 0.657	Mirzaei *et al.* (2019)
04.	EUROPE	Spain	(i) Chlortetracycline (ii) Clarithromycin (iii) Ciprofloxacin (iv) Enrofloxacin (v) Lincomycin (vi) Norfloxacin (vii) Trimethoprim (viii) Ofloxacin	Surface, waste waters and hospital effluent	0.059 0.01 0.74, 2.292,13.78 0.07,0.22 0.047,0.142 0.054,0.310 0.151,0.232 14.38	Díaz-Cruz *et al.* (2008); Rodriguez-Mozaz *et al.* (2015)
		North Portugual	(i) Ampicillin (ii) Azythromycin (iii) Clarithromycin	Urban wastewater	0.552 0.184–0.358 0.433–0.474	Iakovides *et al.* (2019)

ARGs can persist in WWTPs even after treatment, as some of the ARG-carrying bacteria may be resistant to the disinfectants and antibiotics used in the treatment process. In addition, some ARGs can be transferred between bacteria through mechanisms such as horizontal gene transfer, which can lead to the development and spread of antibiotic-resistant bacteria.

Researchers are exploring various strategies to reduce the occurrence of ARGs in WWTPs. For example, advanced treatment technologies such as membrane filtration and ozonation have been shown to be effective in removing ARGs from wastewater. Additionally, optimizing the treatment process to promote the growth of bacteria that are not resistant to antibiotics may also help to reduce the occurrence of ARGs in WWTPs. Overall, the detection of ARGs in WWTPs highlights the need for improved wastewater management practices to minimize the environmental impact of antibiotics and antibiotic-resistant bacteria. Continued monitoring and research will be necessary to better understand the occurrence and fate of ARGs in WWTPs and to develop effective strategies for their removal.

5.1.3 Global distribution of antibiotics as emerging pollutants

This study aims at the available literature for antibiotic utilization and their concentrations in various aquatic bodies reported worldwide and presented in Table 5.2. The study formulated 16 studies among all the continents of our globe, in which we reported some trends in the utilization of antibiotics in the different corners of this small global world. Ciprofloxacin and fluoroquinolones had been reported as the most persistent antibiotics in the studies.

Developing nations have a burden of microbial-induced mortality and, thereby, the utilization of antibiotics. Agunbiade and Moodley (2016) reported a study in which nalidixic acid and ciprofloxacin levels were detected in South African riverine waterways in the range of 25.2–29.5 μg/L and 27.1 μg/L, respectively. These high concentrations are reported due to rivers receiving high contamination from agricultural activities and lacking poor management of effluents. Another study from Ghana, reported by Azanu *et al.* (2018), showed the highest concentrations in the order of ciprofloxacin > erythromycin > sulfmethoxazole > ampicillin in the hospital wastewater effluents. The highest concentration of sulfamethoxazole, with a concentration of 7.194 μg/L, had been reported in WWTP effluent. Ampicillin was found to be the least in the concentration of both hospital and WWTP facility effluent. In Kenya, high concentrations of sulfamethoxazole (13.8 μg/L) and trimethoprim (2.650 μg/L) have been reported due to the high disease prevalence, especially HIV/AIDS (Ngumba *et al.*, 2016). Similar trends were observed in Asian nations. As reported in Iraq, considerably high levels are reported in the effluent from a drinking water treatment facility (Mahmood *et al.*, 2019). High concentrations were reported in China and India's aquatic bodies, as reported in Tables 5.2 and 5.3, respectively.

The aquatic antibiotic levels in American countries ranged from 2 g/L or less, except for sulfadimethoxine, which was detected at 15 g/L in Kansas, USA (Cha *et al.*, 2006; Lindsey *et al.*, 2001). Ampicillin and oxacillin levels from WWTP wastewaters to the Cache la Poudre River in northern Colorado have been documented at 86 and 95 g/L, respectively (Cha *et al.*, 2006). Monensin turned out to be the most often found molecule and was also discovered in larger quantities than salinomycin and lascolid, in line with the various levels and types of livestock farming in the Del Plata basin, Argentina (Alonso *et al.*, 2019).

Based on the data available, European nations have reported a considerable amount of antibiotics in various studies, whereas Spain reported the highest antibiotic consumption (Díaz-Cruz *et al.*, 2008). Healthcare facility effluents along the Ter River in Spain have been discovered to contain substantial amounts of microchemical pollutants, including ofloxacin and ciprofloxacin, at quantities above 13 g/L (Rodriguez-Mozaz *et al.*, 2015), which may corroborate the high consumption in Spain concerning other European nations. In France, clarithromycin had been reported as the maximum concentration with a 2.33 μg/L value (Feitosa-Felizzola & Chiron, 2009). The authors reported this due to the highly persistent nature of this chemical in the water bodies. Similarly,

Table 5.3 Studies reported in different sectors with varied concentration of antibiotics from several states in India.

S.No.	Type River/Lake/TTP	States	Antibiotics	Concentration (µg/L)	References
1.	Kshipra River	Madhya Pradesh	(i) Norfloxacin (ii) Sulfamethoxazole	0.66 1.59	Diwan *et al.* (2018)
2.	STP in South India	Tamil Nadu and Kerala	(i) Chloramphenicol (ii) Trimethoprim (iii) Sulfamethoxazole (iv) Ofloxacin	<0.01 0.043–0.285 0.040–0.637 0.00–0.2469	Akiba *et al.* (2015)
3.	Tamil Nadu Shrimp farms		(i) Chloramphenicol (ii) Sulfonamides (iii) Erythromycin	0.085–0.123 44.12–72.65 1.32–2.45	Swapna *et al.* (2012)
4.	Shrimp farm	Karnataka	(i) Chloramphenicol (ii) Sulfonamides (iii) Erythromycin	0–0.0135 58.36–69.65 53.69–56.98	Swapna *et al.* (2012)
5.	Patancheru Enviro Tech Ltd. Treatment plant undiluted effluent		(i) Ciprofloxacin (ii) Enrofloxacin (iii) Norfloxacin (iv) Ofloxacin (v) Lomefloxacin	4.4 14000 210 25 55 8.8	Fick *et al.* (2009)
6.	Musi River		(i) Ciprofloxacin (ii) Oxofloxacin (iii) Norfloxacin	5015 542.4 251	Gothwal and Shashidhar (2017)
7.	Yamuna River	Delhi	(i) Ampicillin (ii) Ciprofloxacin (iii) Gatifloxacin (iv) Sparfloxacin (v) Cefuroxin	0.2–13.75 ND–1.44 ND–0.48 ND–2.09 ND–1.7	Monitoring *et al.* (2014)

ciprofloxacin reported 1.674 µg/L in influents and 0.626 µg/L in effluents at Varese Olona STP, Italy, representing efficient reduction (Castiglioni *et al.*, 2008). North Portugal had reported a trend of ampicillin $>$ clarithromycin $>$ azithromycin $>$ trimethoprim antibiotics in urban wastewater (Iakovides *et al.*, 2019).

Almost all of the concentration of the antibiotic has been reported in our farthest destination, that is, the ice caps of Antarctica, where ciprofloxacin (in the range 0.37–1.86 µg/L) and norfloxacin (0.58–1.23 µg/L) have been reported in the highest amount, followed by metronidazole $>$ clandinomycin $>$ erythromycin $>$ others. These traces can be assigned to significant transport processes of the hydrological cycle, tourism, and international research stations in the area (Hernández *et al.*, 2019).

5.1.4 Studies for antibiotic distribution in Indian aquatic bodies

A brief literature review was conducted on the Indian subcontinent for the occurrence of these persistent chemicals. Studies conducted by Philip *et al.* (2017) stated that 39% of antibiotic contamination was found in the northern rivers Ganga and Yamuna. Both rivers are the lifeline of Northern India, contaminated with pollutants from sources like effluents from industries, households, and other agricultural activities carried out in the upper and lower stretches of the river. It is reported that the sewage treatment plant (STP) effluent enters the river with reportedly high concentrations of fluoroquinolones ranging from Not in Detection Limit (NDL) to 2.09 μg/L (Monitoring *et al.*, 2014). Although values fluctuated due to temporal variations reported least in monsoonal climate followed by summers and maximum in winters. This can be attributed to the hydraulic gradient of the riverine system and also to more diseases in the winter seasons.

Arsson (2009) reported an unusually high amount of persistent antibiotics, particularly trimethoprim (4.4 μg/L), ciprofloxacin (14,000 μg/L), enrofloxacin (210 μg/L), norfloxacin (25 μg/L), ofloxacin (55 μg/L), and lomefloxacin (8.8 μg/L), found in the undiluted effluents from the Patancheru Enviro Tech Ltd. (PETL) wastewater treatment plant in Hyderabad, which is the pharmaceutical capital of India. Recent studies also reported high contamination in the Musi River, Hyderabad, as PETL outlets drained it from 2009 onwards (Gothwal & Shashidhar, 2017). The situation is worst reported in soil sediments, groundwater (well studies), effluent, and influent scenarios. Another study from shrimp farms located in Hyderabad by Swapna *et al.* (2012) reported high levels of sulfonamide in the range of 58.36–69.65 μg/L, erythromycin within 53.69–56.98 μg/L, and a mild concentration of chloramphenicol ranged from 0 to 0.0135 μg/L. High concentrations can be attributed to animal husbandry practices, poor implementation of water regulation policies, and negligence by government authorities. Antibiotic medicines and their derivatives were discovered in wastewater and sludge from three residential STPs in India's southern states, conducted by Akiba *et al.* (2015). Comparatively temporally varied concentrations were reported in the Kshipra River. Norfloxacin was at 0.66 μg/L, and ofloxacin was at 0.99 μg/L. According to reports, sulfamethoxazole was found at higher concentrations in the fall and winter seasons (2.75 and 2.18 g/L, respectively) than in the summer (1.39 and 0.04 g/L, respectively) (Diwan *et al.*, 2018).

5.2 TRENDS IN CONSUMPTION OF ANTIBIOTIC POLLUTANTS

5.2.1 Antibiotic consumption trend at the global level

Universalizing the economies and the developed medical sector with better survival reports can be accredited to the availability of these useful drugs to humanity. Estimates from 76 countries reveal that the predicted global antibiotic use grew by 39% to 42.3 billion designated regular doses up to 15 years, that is, from 2000 and 2015 (Klein *et al.*, 2018). Defined daily dose (DDD) is the estimated daily sustaining dosage for a medication used in humans for its primary indication. By 2020, global pharmaceutical spending would have increased by 29–32% from 2015 versus 35% in the previous five years (Aitken, 2020) .

The predicted calculated global antibiotic consumption levels (such as in nations not listed in the IQVIA inventory (IMS Health Quintiles)) significantly decreased between 2000 and 2015 in High-Income Countries (HICs), ranging from 27.0 to 25.7 DDDs per 1000 residents per day, but increased by approximately 77% in Low-Middle Income Countries (LMICs), from 8.6 to 13.9 DDDs per 1000 inhabitants per day (Klein *et al.*, 2018). High-income nations used antibiotics more frequently per person than low- and middle-income nations, including India, China, and Brazil, which had a massive growth in antibiotic consumption. However, some fluctuating and considerate trends were reported in the consumption of antibiotics. While the usage rates of the subsequent three categories of antibiotics – cephalosporins (20% of all DDDs), quinolones (12% of all DDDs), and macrolides (12% of all DDDs) – all increased globally between 2000 and 2015, the overall usage of antibiotics in HICs reduced dramatically during that time. Usage of antibiotics surged 399%, 125%, and 119% in LMICs, respectively, for cephalosporins, quinolones, and macrolides. These three medications' antimicrobial use rates in HICs declined by 18%, 1%, and 25%, respectively. Five countries with major emerging economies, that is, Brazil, Russia, India, China, and South Africa (a group known as BRICS) (State & The, 2015), recorded the highest increase in consumption of antibiotics in the first decade of the 21st century (2000–2010). Between 2000 and 2010, these nations' antibiotic usage increased by 68%, 19%, 66%, 37%, and 219%, respectively. Although these BRICS countries accounted for nearly 75% of world consumption growth, per capita spending in these nations remained lower than in the developed USA

(CDDEP, 2015; State & The, 2015). Despite this, as their populations grow, it is anticipated that in BRICS nations, usage will double by 2030, presuming no legislative changes.

5.2.2 Antibiotic consumption trend in India

With nearly 1.3 billion inhabitants, India is the fastest-growing second-most populated nation on the planet and has half a billion animals, accounting for 20% of the world's total livestock population. Being an agrarian and developing nation, the use of antibiotics has been on a tremendous rise to meet the requirement of eatables (dairy products) and healthcare facilities for such a huge population expansion, which is estimated to rise by two-thirds between 2010 and 2030 (Laxminarayan & Chaudhury, 2016) . India also has the world's largest cattle and poultry population with a sizable aquaculture industry, which consumes vast quantities of antibiotics to maximize yield. The use of prophylactic antibiotics in animal husbandry and aquaculture is increasing antibiotic-resistant pathogens, which remains a grave concern (Ghosh & Mandal, 2010) . Technological innovation, formal investigation, trained employees, relatively low cost, and a paucity of patent protection make the Indian market an appealing alternative for global multinational businesses looking to outsource their medication manufacturing.

In terms of pharmaceutical output and consumption, India is ranked third and thirteenth in the world, respectively. Rapid urbanization has been reported in India during the last three decades, which synergistically degrades and upgrades the quality of life. Bacterial infections, enteric fever, diarrhea, cholera, and acute respiratory infections are the undemanded repercussions of the teeming ambient of urban land. One of the main contributing factors to the inappropriate and unregulated use of antibiotics in India is the purchase of antibiotics without a prescription from pharmacies.

Antibiotic usage in India increased by 103% (3.2–6.5 billion specified daily doses) in the previous 15 years (2000–2015) (Klein *et al.*, 2018). In LMICs, India saw the highest rise in antibiotic use (65% between 2000 and 2015) and continued economic development (a 10% annual increase in Gross Domestic Product [GDP] during the 2000s. India's per capita usage of antibiotics (10.7 units per capita) was less than that of numerous other countries during this moment (e.g., 22 units per capita in the USA) ('Scoping Report on Antimicrobial Resistance in India', 2017). In 2010, the utilization of antimicrobial compounds in food animals was projected to be ~63,000 (1560) units; India holds 3% of the global demand and is the fourth largest exporter, after China (23%), the USA (13%), and Brazil (9%). Antimicrobial usage in India's food animal industry is predicted to quadruple by 2030.

However, in India, there were some reported trends in this particular class of antibiotics from 2000 to 2015. After quinolones (34%), cephalosporins (32%), macrolides (14%), and tetracyclines (6%), penicillins were the third most regularly given antibiotics in 28% of cases (Kotwani and Holloway, 2011). In 2012, India exceeded expectations as the world's largest user of oxazolidinone antibiotics.

5.3 ECOLOGICAL RISK POSED BY ANTIBIOTICS

Previous sections described the occurrence and possible sources of antibiotic channeling into our ecosystem components, mainly water and soil, as shown in Figure 5.2. Industrial activities, household waste effluent, and agricultural activities are the sources of the most intensive utilization of antibiotics. These sources contaminate the surface water bodies through improper disposal and runoff, physio-chemically trapped traces into soil sediments, and en-route into the groundwater table via leaching. Subsequently, these contaminated sources are used for irrigation purposes and bioaccumulate in the edible crop plants, ultimately reaching our plates at a much higher level. The vicious cycle of biotransformation, bioaccumulation, and biogeochemical cycling processes tends to aggravate the issue of increased risk. Antibiotics released without sufficient treatment have led to substantial levels of antibiotic remnants in the aquatic system, as mentioned in earlier sections. Many cases of antibiotic resistance have been reported in the last two decades, which has given rise to serious health concerns for livestock and the human population. The widespread antibiotic tetracycline, which is routinely used to cure both humans and animals, has been discovered to cause horizontal translocation of resistance genes in *Escherichia coli* at low doses in the microgram per liter range (Grenni *et al.*, 2018). Low concentrations mean non-lethal or sub-inhibitory ones tend to act in different ways, such as in selecting resistances, by generating genetic and phenotypic alterations to promote adaptive evolution, and as signaling molecules to produce other physiological functions (Müller *et al.*, 2002).

Although there was less research on antibiotic risk evaluation in India's aquatic systems, numerous studies have found a prevalence of antibiotic resistance in areas where antibiotic pollution is a problem. The persistent nature of antibiotics tends to be antibiotic resistance, ARGs, and antimicrobial resistance (AMR), which developed due to rampant and unregulated usage of antibiotics. Bacterial resistance is a significant adjustment or systemic response of microbes to substances that attempt to stop them from growing, and it is developed through many processes, as previously stated. According to a recent analysis, drug-resistant diseases kill 700,000 people every year, with 10 million people dying annually by 2050 (Proia *et al.*, 2018). Antibiotic resistance is likely to impede the battle against HIV and malaria, with 490,000 people developing multi-drug-resistant TB worldwide in 2016. *E. coli* has a high resilience rate to fluoroquinolone therapies (one of the most regularly used drugs for treating urinary incontinence). In at least 10 nations, treatment failure with the final resort of gonorrhea medication (third-generation cephalosporin antibiotics) has been proven. As of July 2016, five countries in the Greater Mekong subregion had demonstrated resistance to the first-line therapy for *Plasmodium falciparum* malaria (artemisinin-based combo regimens).

Being the antibiotic manufacturer capital of the world, India suffers from this catastrophe of drug resistance. India has one of the highest proportions of bacterial illnesses in the world. In India, approximately 410,000 children aged five and under perish from pneumonia each year, accounting for about a quarter of all child mortality. The crude death rate from contagious diseases in India is now 417 per 100,000 people ('Scoping Report on Antimicrobial Resistance in India', 2017).

Antibiotic residues can alter ecological integrity in general. Antibiotics, which are substances that may damage or limit the development of microorganisms, are likely to upset the environment's microbiome equilibrium. The microbiological community of aquatic bodies plays a significant role in channeling their ecological health. Due to the influence of persistent antibiotics, agricultural fields watered with wastewater and polluted with antibiotic residues may deplete the key microbial group for nitrification. On the other hand, antibiotic residues in wastewater can disturb the activity of microorganisms engaged in the disposal of both residential (septic tanks) and industrial effluents (WWTPs).

Antibiotic cocktails synergistically adversely and significantly affect the food web of the aquatic body, rather than the presence of singlets (Danner *et al.*, 2019). Chen *et al.* (2018) reported the synergistic actions of the combination of tetracycline and enrofloxacin. The authors reported the increased risk of biochemical transformation of reaction products after the interaction of two or more antibiotics instead of their existence, which may or may not be harmful to humans (Kumari & Kumar, 2020).

5.4 ASSESSMENT AND REMEDIATION METHODOLOGIES

Antibiotics are non-biodegradable, and several of them have been found to linger in the soil. Various studies have been reported by Grenni *et al.* (2018) on the newly introduced concept of the biodegradability of antibiotics by the available microbial community. The other noted effects of antibiotics can be seen in an altered pattern of ecological functions like methanogenesis, sulfate reduction, nutrient cycling, nitrogen transformation, and the rate of humus formation (Grenni *et al.*, 2018). Apart from that, the basal metabolic rate of an aquatic body also fluctuates due to changes in physio-chemical and biological parameters.

Although the persistence of antibiotics in the effluents of different residential, commercial, and other institutional sectors creates a nuisance for their proper disposal and treatment for sustainable medical management, fewer available technologies are there to solve this challenge. Although conventional treatment technologies are not designed for emerging contaminants, conventional treatment processes such as coagulation, flocculation, and sedimentation can be partially used for the remediation of antibiotics in the aquatic system. Various studies support the fact that the combination of different advanced treatment processes efficiently removes them to a satisfactory level. Hence, it emphasizes the necessity of a synergistic and integrative approach to treating wastewater containing antibiotics.

Each methodology has its own set of dimensions, both positive (effective removal, low cost) and negative (cost factor, high maintenance). There is a practical need for an integrative approach to applying the treatment methodologies for better management and application processes. A conventional (activated sludge) and cutting-edge (membrane filtration/osmosis) effluent treatment facility in Brisbane, Australia, which excludes antibiotics, showed that both treatment processes substantially lowered antibiotic quantities, with an average removal efficiency of 92% from the liquid stage (Watkinson *et al.*, 2007). Antibiotics had been discovered in both effluents in the low to mid-nanogram per liter range. But the concentration findings in the advanced methodology's treated effluent were lower than the conventional ones.

5.4.1 Conventional treatment processes

The goal of conventional water treatment is to guarantee that water is safe to drink by removing physiological, tangible, and microbiological pollutants such as heavy metals, suspended solids, and spoilage microorganisms. These water-treatment methods are usually not designed to extract trace contaminants. Although some researchers have confirmed that traditional WWTPs may remove Abs, they must use processes including coagulation, precipitation, sand filtering, and clarifying.

5.4.1.1 Activated sludge process (ASP)

In municipal WWTPs, activated sludge is mingled with wastewater and microbes to remove micronutrients that oxidize carbon-containing biological matter and other compounds throughout the treatment process. High removal efficiency varied from 30% to 70% reported, with a very unusually long retention time reported for the removal of sulfonamides, macrolides, and trimethoprim by the activated sludge process (Göbel *et al.*, 2005).

5.4.1.2 Membrane biological reactor (MBR)

MBR involves a series of actions required to achieve higher efficacy. Ab elimination through wastewater treatment mainly incorporates microbial degradation, affinity to sludge, photocatalytic degradation, and evaporation. When combined with ASP, this approach produces less sediment, lower suspended particles, and higher germ elimination, making it ideal for domestic wastewater. Longer retention times allow denitrification, increased bacterial nitrification activity, and improved micropollutant elimination.

5.4.2 Advanced emerging treatment techniques

Antibiotic pollution in water is a significant environmental and public health concern. The emergence of antibiotic-resistant bacteria due to the presence of antibiotics in water has led to a growing need for effective treatment methods. Emerging techniques such as advanced oxidation processes (AOPs), adsorption, and membrane filtration have shown promise in removing antibiotics from water. AOPs involve the generation of highly reactive oxidizing species that can degrade antibiotics. Adsorption involves the use of materials with a high surface area and adsorption capacity to remove antibiotics from water. Membrane filtration uses membranes with small pores to physically remove antibiotics from water. The following section discusses these emerging techniques, which have the potential to improve the quality of water and reduce the risk of antibiotic-resistant bacteria, but further research is needed to optimize their effectiveness and feasibility for large-scale implementation.

The use of a composite of modern methods of treatment such as hydroxyl radicals, catalyst supports, photovoltaic, or ultraviolet (UV) light to eliminate antibiotics from wastewater has lately received interest. AOPs generate the hydroxyl radical (•OH), which is a potent antioxidant or oxidative radical that reacts with molecules (Anjali & Shanthakumar, 2019). The desired substances can then be processed and oxidized into CO_2, H_2O, and inorganic materials.

Ultraviolet Photolysis of Hydrogen Peroxide (UV/H_2O_2), Photo-Fenton Process ($UV/H_2O_2/Fe^{2+}$), and ozonation methods have a high clearance rate (Bianculllo *et al.*, 2019). Except for metronidazole (92%) and ciprofloxacin (100%), antibiotics are destroyed up to 100% under various experimental settings using UV/H_2O_2 and ozonation approaches (93%). In antibiotics, $UV/H_2O_2/Fe^{2+}$ reveals a disintegration rate of up to 100% (Anjali & Shanthakumar, 2019).

5.4.2.1 Ozonation

By inducing oxidative stress, ozonation is used as a first-line solution to improve the processability of contaminants (Blaney, 2014). Ozonation can utilize minimal energy and recover up to 99% of the water without producing trash. Antibiotics are often removed from water and effluent via ozonation and certain other oxidizing techniques. Iakovides *et al.* (2019) had undertaken research using ozonation to eliminate antibiotics from unpasteurized milk for the first time (amoxicillin, doxycycline, ciprofloxacin, and sulfadiazine).

5.4.2.2 UV irradiation

UV irradiation is used for decontamination and is commonly employed in WWTPs as part of the tertiary treatment. UV therapy breaks down chemical bonds in contaminants using UV light, a process known as 'photocatalytic degradation.' On contrary, intense UV photocatalysis does not work efficiently. To improve the efficacy of the treatment, a variety of UV lamps, moderate or reduced mercury vapor lamps generating Ultraviolet C (UV-C) light, and oxidants or catalysts (e.g., H_2O_2, $Fe^{2+/3+}$, TiO_2) were added. UV irradiation coupled with advanced peroxidation (including UV/chlorine, UV/H2O2,UV/O3, and H2O2/Fe2+/UV (photo-Fenton)) may accelerate Antibiotic Resistant Bacteria (ARB) and Antibiotic Resistance Gene (ARG) elimination in potable water. UV irradiation is used for decontamination, and UV light-emitting diodes have recently attracted a lot of interest in their development, particularly for industrial wastewater treatment, due to their eco-friendliness (by substituting mercury) and sustained period (Bianculllo *et al.*, 2019).

5.4.2.3 Adsorption-based removal

The aggregation of substances from a gaseous stage to the sorbent surface, which might be physical or chemical, is referred to as adsorption (Mahmoud *et al.*, 2020). Four stages are involved in the removal of impurities: (i) mass solute mobility, (ii) adsorbate layer propagation, (iii) adsorbate diffusion through porosity, and (iv) adsorption – interplay among adsorbate and permeable morphology. The primary mechanisms identified for carbon-based nanomaterials include hydrogen bonding, separation into uncarbonized components, cavity loading, electrostatic effect (ionic attractions), hydrodynamic impact (water-insoluble action), and a few more processes like surface deposition and so on. The parameters of the adsorbent, such as effective surface area, permeability (macro or microporosity), channel width, and functional groups, are all intrinsically linked to adsorption effectiveness.

The effectiveness of various adsorbents in eliminating antibiotics from wastewater was assessed by looking at the adsorption coefficient values. Different sorbents for sulfamethoxazole followed a similar trend: charcoal Biochar (BC) > Multiwalled Carbon NanoTubes (MWCNT) > graphite = silicate minerals. The adsorptive materials for tetracycline continued the pattern: activated carbon (AC) = humic material = clay particles = Single Walled Carbon Nanotubes (SWCNT) > graphite > MWCNT (Ahmed *et al.*, 2015).

5.4.2.3.1 Activated carbon (AC)

It is one of the most commonly cited adsorbents for removing antibiotics. Recent research on using air conditioners (ACs) for antibiotic elimination has found that ACs or enhanced ACs may efficiently separate a wide range of antibiotics from wastewater, ranging from 74% to 100%. After 24 hours of adsorption, the most significant chemical oxygen demand (COD) removal of more than 90% was achieved at a pH of 5, the operating temperature of 20°C, 0.15 g of activated carbon, and a 100 ppm baseline tetracycline dosage (Profile, 2019).

5.4.2.3.2 Carbon nanotubes (CNTs)

CNTs are made up of tubular, stacked graphite sheets with a huge surface area and a significant van der Waals index. Because sp^2-hybridized carbon atoms are present, the benzenoid chains in graphene-layered sheets have a strong dipole moment. The interaction between CNTs and aromatic contaminants typically involves adsorption, where the hydrophobic nature of CNTs allows them to adsorb or capture the hydrophobic aromatic compounds. This process often relies on van der Waals forces or π-π stacking interactions between the carbon nanotubes and the aromatic rings of the contaminants. Lincomycin and sulfamethoxazole (sulfonamides) were reported to be removed from aqueous solutions by SWCNT and MWCNT in current antibiotic remediation experiments.

5.4.2.3.3 Ionic exchange resins

Ion exchange is a technique for transferring excited ions on a solid sorbent for metallic ions or negative ions in a liquid solution while retaining electro-neutrality throughout the process. Overall, it has been demonstrated that ion exchange resin hybrids are up to 90% efficient in eliminating antibiotics from wastewater and drinking water. Accordingly, tetracyclines and sulfonamides had adsorption reduction effectiveness of 80% and 90% on an ion exchanger.

5.5 REGULATIONS BY GLOBAL AUTHORITIES FOR ANTIBIOTICS UTILIZATION

With the introduction of new sophisticated and analytical tools and techniques, the detection of antibiotics in trace amounts, their eco-toxicological studies, and hence the formulation of regulatory measures, became possible recently. Governments across the globe are revising the standards for water quality parameters, and it has been shown by the steps procured by the Indian government by amending the Environment Protection Act (EPA), 1986, in January 2020. The Central Government amended the EPA of 1986 to formulate the standard for traces of antibiotic presence in STP/ETP effluents from bulk drug and manufacturer industries. The values suggested and disposed of nowadays are nowhere compared because of the unregulated and inappropriate washing of antibiotics in water bodies. This act will put a hold on and check measures on regulating the quality of effluents entering Indian rivers and other aquatic bodies (Authority, 2013).

The legal framework governing fish farming procedures differs from that of the poultry and dairy industries. Antibiotics and numerous pharmacologically active compounds have been outlawed in fisheries by the Food Safety and Standards Authority of India (FSSAI). On the other hand, there are no regulations in the chicken business, and many commercially available blended meals include antibiotics. Since March 2014, the Drugs and Cosmetic Rules, 1945 has included a distinct Schedule H-1 to control the sale of antimicrobials nationwide. The schedule covers about 24 antimicrobials, including third and fourth-generation cephalosporins and carbapenems. Antimicrobials cannot be dispensed without a legitimate medical recommendation, and the medicine container must have the following phrase with a red border: 'Schedule H1 Drug–Warning: Taking the drug without professional counsel is harmful; not to be distributed by retail without a licensed medical practitioner's authorization.' The pharmacist must keep a separate register with information on the physician, the patient, and the medicine sold.

As part of the study, 30 possible policy options disseminated throughout the pollutant life cycle's 10 primary probable action regions have been identified to inform the development of the European

Union's (EU) strategic approach. The objectives of these options are to (i) obtain a better knowledge of the issue and (ii) establish circular economy-compliant, more efficient, and sustainable strategies for manufacturing, utilization, and dumping.

The Food & Agricultural Organization (FAO) Action Plan on AMR contains five strategic priorities that are in accordance with the international action plan having common goals: (i) enhance consciousness of drug resistance; (ii) develop stronger expertise through vigilance and investigations; (iii) reduce the incidence of infection; (iv) improve the use of antimicrobial agents in the wellness, animal, and food sectors; and (v) establish the monetary justification for sustainable investing that caters to the needs of all nations, and boost investment in new medicines, diagnostics, and vaccines. The actions have been started by the various administrative units across the globe, but a comprehensive, integrative, and realistic strategy to combat the problem of antibiotic resistance is the need of the hour. Strict punitive actions and regulations must be directed at pharmaceutical companies to avoid more severe and lethal repercussions that are on their way.

5.6 CURRENT ADVANCES AND FUTURE OUTLOOK

Antibiotic resistance is a major public health concern worldwide. It is well known that the overuse and misuse of antibiotics in human and animal health care can lead to the emergence and spread of antibiotic-resistant bacteria. However, research has shown that the presence of antibiotics and ARGs in the aquatic environment can also contribute to the development and spread of antibiotic-resistant bacteria. Recent research has focused on understanding the fate and behavior of antibiotics and ARGs in the aquatic environment. Studies have shown that antibiotics can persist in the environment for long periods of time, even after they have been discontinued. Antibiotics can enter aquatic systems through a variety of sources, including wastewater discharge from hospitals, farms, and households. Current research in this area has focused on several key areas:

(a) Identifying the sources and pathways of antibiotics and ARGs in aquatic environments: Studies have shown that the release of antibiotics and ARGs into aquatic environments can occur via a variety of pathways, including wastewater discharges, agricultural runoff, and the use of antibiotics in aquaculture.
(b) Developing methods for detecting and quantifying antibiotics and ARGs in aquatic environments: There is ongoing research into the development of more sensitive and accurate methods for detecting antibiotics and ARGs in water and sediment samples, such as Polymeric Chain Reaction (PCR) -based assays and metagenomics.
(c) Understanding the fate and behavior of antibiotics and ARGs in aquatic environments: Researchers are investigating the mechanisms by which antibiotics and ARGs are transported, transformed, and degraded in aquatic environments and how their persistence can be influenced by factors such as water chemistry, sediment characteristics, and microbial activity.
(d) Assessing the ecological and human health impacts of antibiotics and ARGs in aquatic environments: Studies have shown that exposure to antibiotics and ARGs can have negative effects on aquatic organisms, such as reducing their survival, growth, and reproduction. There are also concerns about the potential for antibiotic resistance to spread to human pathogens through the aquatic environment.

Researchers are exploring various strategies to mitigate the impact of antibiotics and ARGs in the aquatic environment. For example, WWTPs can use advanced treatment technologies to remove antibiotics and ARGs from wastewater before it is discharged into waterways. Additionally, some researchers are exploring the use of phages, which are viruses that can infect and kill bacteria, as a potential treatment for antibiotic-resistant bacteria in the aquatic environment. Looking to the future, it is likely that the issue of antibiotics and ARGs in the aquatic environment will continue to be an important area of research. Continued monitoring of water quality and the development of new treatment technologies will be necessary to reduce the spread of antibiotic-resistant bacteria and protect public health. Additionally, improved regulations and guidelines for the use and disposal of antibiotics can help minimize the environmental impact of these drugs. Future research in this area is likely to focus on several key areas:

(a) Developing strategies for reducing the release of antibiotics and ARGs into aquatic environments: This could involve improved wastewater treatment technologies, more sustainable agricultural practices, and a reduced use of antibiotics in aquaculture.
(b) Developing new approaches for removing antibiotics and ARGs from aquatic environments: Researchers are investigating the potential of various technologies, such as bioremediation, nanotechnology, and adsorption, for removing antibiotics and ARGs from water and sediment.
(c) Investigating the link between antibiotic use in humans and animals and the presence of antibiotics and ARGs in aquatic environments: Studies are needed to better understand the relationship between human and animal antibiotic use and the occurrence of antibiotics and ARGs in aquatic environments, as well as the potential for exposure and transmission of antibiotic resistance through the aquatic environment.
(d) Assessing the long-term effects of antibiotics and ARGs on aquatic ecosystems: There is a need for more long-term studies to understand the potential ecological impacts of exposure to antibiotics and ARGs over time and to identify the most vulnerable aquatic ecosystems and organisms.

5.7 CONCLUSION

In conclusion, antibiotics have transformed modern medicine and enhanced the quality of life for millions of people. However, the presence of antibiotics and ARGs in aquatic environments is a mounting concern that entails urgent consideration. The aquatic environment, including both freshwater and marine ecosystems, has been documented as a reservoir of ARGs and antibiotic-resistant bacteria. The discharge of antibiotics and their residues into the aquatic environment from various sources, including agricultural and medical activities, is believed to contribute to the dissemination and persistence of ARGs. These compounds can have harmful effects on aquatic organisms and contribute to the development of antibiotic-resistant bacteria, which pose a significant threat to human health. Thus, understanding the dynamics of antibiotic resistance in the aquatic environment is crucial to mitigating the spread of ARGs and developing effective strategies to combat antibiotic resistance. This can be achieved through the development and implementation of regulations, the promotion of responsible use of antibiotics, and the adoption of appropriate treatment technologies for wastewater. Further research is also needed to understand the extent and distribution of antibiotics and ARGs in different aquatic environments, as well as their impacts on ecosystem health and human well-being. By taking concerted actions, we can protect our precious aquatic resources and prevent the emergence and spread of antibiotic-resistant bacteria.

REFERENCES

Agunbiade F. O. and Moodley B. (2016). Occurrence and distribution pattern of acidic pharmaceuticals in surface water, wastewater, and sediment of the Msunduzi River, Kwazulu-Natal, South Africa. *Environmental Toxicology and Chemistry*, **35**(1), 36–46. Available at: https://doi.org/10.1002/etc.3144

Ahmed M. B., Zhou J. L., Ngo H. H. and Guo W. (2015). Science of the total environment adsorptive removal of antibiotics from water and wastewater: progress and challenges. *Science of the Total Environment*, **532**, 112–126. Available at: https://doi.org/10.1016/j.scitotenv.2015.05.130

Aitken M. (2020) 'Global Medicines Use in 2020 Outlook and Implications', (November 2015).

Akiba M., Senba H., Otagiri H., Prabhasankar V. P., Taniyasu S., Yamashita N., Lee K., Yamamoto T., Tsutsui T., Ian Joshua D., Balakrishna K., Bairy I., Iwata T., Kusumoto M., Kannan K. and Guruge K. S. (2015). Impact of wastewater from different sources on the prevalence of antimicrobial-resistant *Escherichia coli* in sewage treatment plants in South India. *Ecotoxicology and Environmental Safety*, **115**, 203–208. Available at: https://doi.org/10.1016/j.ecoenv.2015.02.018

Alonso L. L., Demetrio P. M., Capparelli A. L., and Marino, D. J. G. (2019). Behavior of ionophore antibiotics in aquatic environments in Argentina: the distribution on different scales in water courses and the role of wetlands in depuration. *Environment International*, **133**, 105144. Available at: https://doi.org/10.1016/j.envint.2019.105144

Anjali R. and Shanthakumar S. (2019). Insights on the current status of occurrence and removal of antibiotics in wastewater by advanced oxidation processes. *Journal of Environmental Management*, **246**, 51–62, https://doi.org/10.1016/j.jenvman.2019.05.090

Antimicrobial resistance. (2020). *February 2018*, 1–7.

Arsson D. G. J. O. L. (2009). 'Pharmaceuticals and Personal Care Products in the Environment CONTAMINATION OF SURFACE, GROUND, AND DRINKING WATER FROM', **28**(12), 2522–2527.

Authority P. B. Y. (2013). 'The Gazette of; LITEdia', **1935**(D).

Azanu D., Styrishave B., Darko G., Weisser J. J. and Abaidoo R. C. (2018). Occurrence and risk assessment of antibiotics in water and lettuce in Ghana. *Science of the Total Environment*, **622–623**, 293–305. Available at: https://doi.org/10.1016/j.scitotenv.2017.11.287

Ben Y., Hu M., Zhang X., Wu S., Wong M. H., Wang M., Andrews C. B. and Zheng C. (2020). Efficient detection and assessment of human exposure to trace antibiotic residues in drinking water. *Water Research*, **175**, 115699. Available at: https://doi.org/10.1016/j.watres.2020.115699

Biancullo F., Moreira N. F. F., Ribeiro A. R., Manaia C. M., Faria J. L., Nunes O. C., Castro-silva S. M. and Silva A. M. T. (2019). Heterogeneous photocatalysis using UVA-LEDs for the removal of antibiotics and antibiotic resistant bacteria from urban wastewater treatment plant effluents. *Chemical Engineering Journal*, **367**(November 2018), 304–313. Available at: https://doi.org/10.1016/j.cej.2019.02.012

Blaney L. (2014). *Ozone Treatment of Antibiotics in Water*. https://doi.org/10.1016/B978-0-12-411645-0.00012-2

Castiglioni S., Pomati F., Miller K., Burns B. P., Zuccato E., Calamari D. and Neilan B. A. (2008). Novel homologs of the multiple resistance regulator marA in antibiotic-contaminated environments. *Water Research*, **42**(16), 4271–4280. Available at: https://doi.org/10.1016/j.watres.2008.07.004

Cha J. M., Yang S. and Carlson K. H. (2006). Trace determination of β-lactam antibiotics in surface water and urban wastewater using liquid chromatography combined with electrospray tandem mass spectrometry. *Journal of Chromatography A*, **111 5**(1–2), 46–57. Available at: https://doi.org/10.1016/j.chroma.2006.02.086

Chen H., Jing L., Teng Y. and Wang, J. (2018). Science of the total environment characterization of antibiotics in a large-scale river system of China: occurrence pattern, spatiotemporal distribution and environmental risks. *Science of the Total Environment*, **618**, 409–418. Available at: https://doi.org/10.1016/j.scitotenv.2017.11.054

Danner M. C., Robertson A., Behrends V. and Reiss J. (2019). Antibiotic pollution in surface fresh waters: occurrence and effects. *Science of the Total Environment*, **664**, 793–804. Available at: https://doi.org/10.1016/j.scitotenv.2019.01.406

Deng C., Liu X., Li L., Shi J., Guo W. and Xue, J. (2020). Temporal dynamics of antibiotic resistant genes and their association with the bacterial community in a water-sediment mesocosm under selection by 14 antibiotics. *Environment International*, **137**, 105554. Available at: https://doi.org/10.1016/j.envint.2020.105554

Díaz-Cruz M. S., García-Galán, M. J. and Barceló, D. (2008). Highly sensitive simultaneous determination of sulfonamide antibiotics and one metabolite in environmental waters by liquid chromatography-quadrupole linear ion trap-mass spectrometry. *Journal of Chromatography A*, **1193**(1–2), 50–59, https://doi.org/10.1016/j.chroma.2008.03.029

Diwan V., Hanna N., Purohit M., Chandran S., Riggi E., Parashar V., Tamhankar A. J. and Lundborg C. S. (2018). Seasonal Variations in Water-Quality, Antibiotic Residues, Resistant Bacteria and Antibiotic Resistance Genes of Escherichia coli Isolates from Water and Sediments of the Kshipra River in Central India, pp. 1–16. Available at: https://doi.org/10.3390/ijerph15061281

Dong J., Yan D., Mo K., Chen Q., Zhang J., Chen Y. and Wang, Z. (2022). Antibiotics along an alpine river and in the receiving lake with a catchment dominated by grazing husbandry. *Journal of Environmental Sciences*, **115**, 374–382. Available at: https://doi.org/10.1016/J.JES.2021.08.007

Feitosa-Felizzola J. and Chiron S. (2009). Occurrence and distribution of selected antibiotics in a small Mediterranean stream (Arc River, Southern France). *Journal of Hydrology*, **364**(1–2), 50–57. Available at: https://doi.org/10.1016/j.jhydrol.2008.10.006

Fick J., Söderström H., Lindberg R. H., Phan C., Tysklind M. and Larsson D. G. J. (2009). Contamination of surface, ground, and drinking water from pharmaceutical production. *Environmental Toxicology and Chemistry*, **28**(12), 2522–2527, https://doi.org/10.1897/09-073.1

Gevao B., Uddin S., Krishnan D., Rajagopalan S. and Habibi N. (2022). Antibiotics in wastewater: baseline of the influent and effluent streams in Kuwait. *Toxics*, **10**(4), 1–14. Available at: https://doi.org/10.3390/toxics10040174

Ghosh K. and Mandal S. (2010). Antibiotic resistant bacteria in consumable fishes from Digha coast, West Bengal, India. *Proceedings of the Zoological Society*, **63**(1), 13–20, https://doi.org/10.1007/s12595-010-0002-8

Gothwal R. and Shashidhar (2017). Occurrence of high levels of fluoroquinolones in aquatic environment due to effluent discharges from bulk drug manufacturers. *Journal of Hazardous, Toxic, and Radioactive Waste*, **21**(3). Available at: https://doi.org/10.1061/(ASCE)HZ.2153-5515.0000346

Gray A. D., Todd D. and Hershey A. E. (2019). The seasonal distribution and concentration of antibiotics in rural streams and drinking wells in the piedmont of North Carolina. *Science of the Total Environment*, **136286**. Available at: https://doi.org/10.1016/j.scitotenv.2019.136286

Grenni P., Ancona V. and Barra Caracciolo A. (2018). Ecological effects of antibiotics on natural ecosystems: a review. *Microchemical Journal*, **136**, 25–39. Available at: https://doi.org/10.1016/j.microc.2017.02.006

Hernández F., Calısto-ulloa N., Gómez-fuentes C., Gómez M. and Ferrer, J. (2019). Occurrence of antibiotics and bacterial resistance in wastewater and sea water from the Antarctic. *Journal of Hazardous Materials*, **363**(July 2018), 447–456. Available at: https://doi.org/10.1016/j.jhazmat.2018.07.027

Hossain A., Nakamichi S., Habibullah-Al-Mamun M., Tani K., Masunaga S. and Matsuda H. (2017). Occurrence, distribution, ecological and resistance risks of antibiotics in surface water of finfish and shellfish aquaculture in Bangladesh. *Chemosphere*, **188**, 329–336. Available at: https://doi.org/10.1016/j.chemosphere.2017.08.152

Iakovides I. C., Michael-Kordatou I., Moreira N. F. F., Ribeiro A. R., Fernandes T., Pereira M.F.R., Nunes O.C., Manaia C.M., Silva A.M.T. and Fattta-Kassinos D. (2019). Continuous ozonation of urban wastewater: removal of antibiotics, antibiotic-resistant Escherichia coli and antibiotic resistance genes and phytotoxicity. *Water Research*, **159**, 333–347. Available at: https://doi.org/10.1016/j.watres.2019.05.025

Klein E. Y., Boeckel T. P. Van, Martinez E. M., Pant S., Gandra S. and Levin S. A. (2018). Global increase and geographic convergence in antibiotic consumption between 2000 and 2015, pp. 1–8. Available at: https://doi.org/10.1073/pnas.1717295115

Kolpin D. W., Furlong E. T., Meyer M. T., Thurman E. M., Zaugg S. D., Barber L. B. and Buxton H. T. (2002). Pharmaceuticals, hormones, and other organic wastewater contaminants in U.S. streams, 1999-2000: A national reconnaissance. *Environmental Science and Technology*, **36**(6), 1202–1211. Available at: https://doi.org/10.1021/es011055j

Kotwani A. and Holloway K. (2011). Trends in antibiotic use among outpatients in New Delhi, India. *BMC Infectious Diseases*, **11**, https://doi.org/10.1186/1471-2334-11-99

Kumar N. M., Sudha M. C., Damodharam T. and Varjani S. (2020). Chapter 3 – Micro-pollutants in surface water: Impacts on the aquatic environment and treatment technologies. In: Current Developments in Biotechnology and Bioengineering. Elsevier B.V. Available at: https://doi.org/https://doi.org/10.1016/B978-0-12-819594-9.00003-6

Kumari M. and Kumar A. (2020). Human health risk assessment of antibiotics in binary mixtures for finished drinking water. *Chemosphere*, **240**, 124864. Available at: https://doi.org/10.1016/j.chemosphere.2019.124864

Laxminarayan R. and Chaudhury R. R. (2016). Antibiotic resistance in India: drivers and opportunities for action. *PLoS Medicine*, **13**(3), 1–7, https://doi.org/10.1371/journal.pmed.1001974

Lindsey M. E., Meyer M. and Thurman E. M. (2001). Analysis of trace levels of sulfonamide and tetracycline antimicrobials in groundwater and surface water using solid-phase extraction and liquid chromatography/mass spectrometry. *Analytical Chemistry*, **73**(19), 4640–4646. Available at: https://doi.org/10.1021/ac010514w

Lupo A., Coyne S. and Berendonk T. U. (2012). Origin and evolution of antibiotic resistance: the common mechanisms of emergence and spread in water bodies. *Frontiers in Microbiology*, **3**(JAN), 1–13. Available at: https://doi.org/10.3389/fmicb.2012.00018

Mahmood A. R., Al-haideri H. H. and Hassan F. M. (2019). 'Detection of Antibiotics in Drinking Water Treatment Plants in Baghdad City, Iraq', 2019.

Mahmoud M. E., Al-Haideri H. H. and Hassan F. M. (2020). Enhanced adsorption of levofloxacin and ceftriaxone antibiotics from water by assembled composite of nanotitanium oxide/chitosan/nano-bentonite. *Materials Science and Engineering C*, **108**, 110199. Available at: https://doi.org/10.1016/j.msec.2019.110199

Marathe N. P., Shetty S. A., Shouche Y. S. and Larsson D. G. J. (2016). Limited bacterial diversity within a treatment plant receiving antibiotic containing waste from bulk drug production. *PLoS ONE*, **11**(11), 1–12. Available at: https://doi.org/10.1371/journal.pone.0165914

Mirzaei R., Mesdaghinia A., Hoseini S. S. and Yunesian M. (2019). Antibiotics in urban wastewater and rivers of Tehran, Iran: Consumption, mass load, occurrence, and ecological risk. *Chemosphere*, **221**, 55–66. Available at: https://doi.org/10.1016/j.chemosphere.2018.12.187

Monitoring E., Goi G. R. and Mittal A. K. (2014). 'Occurrences and fate of selected human antibiotics in influents and effluents of sewage treatment plant and effluent-receiving river Yamuna in Delhi Occurrences and fate of selected human antibiotics in influents and effluents of sewage treatment plant an', (January). Available at: https://doi.org/10.1007/s10661-013-3398-6

Müller A. K., Westergaard K., Christensen S. and Sørensen S. J. (2002). The diversity and function of soil microbial communities exposed to different disturbances. *Microbial Ecology*, **44**(1), 49–58. Available at: https://doi.org/10.1007/s00248-001-0042-8

Ngumba E., Gachanja A. and Tuhkanen T. (2016). Occurrence of selected antibiotics and antiretroviral drugs in Nairobi River Basin, Kenya. *Science of the Total Environment*, **539**, 206–213. Available at: https://doi.org/10.1016/j.scitotenv.2015.08.139

Nkoh J. N., Oderinde O., Oshogwue N., Kifle G. A., Okeke E. S., Ejeromedoghene O. and Mgbechidinma C. L. (2023). Recent perspective of antibiotics remediation: a review of the principles, mechanisms, and chemistry controlling remediation from aqueous media. *Science of the Total Environment*, **881**, 163469. Available at: https://doi.org/10.1016/J.SCITOTENV.2023.163469

Philip J. M., Aravind U. K. and Aravindakumar C. T. (2017). Emerging contaminants in Indian environmental matrices a review. *Chemosphere [Preprint]*. Available at: https://doi.org/10.1016/j.chemosphere.2017.09.120

Profile S. E. E. (2019). 'Removal of antibiotic wastes by adsorption method', (December 2018), pp. 2–6.

Proia L., Adriana A., Jessica S., Carles B., Marinella F., Marta L., Luis B. J. and Servais, P. (2018). Antibiotic resistance in urban and hospital wastewaters and their impact on a receiving freshwater ecosystem. *Chemosphere*, **206**, 70–82, https://doi.org/10.1016/j.chemosphere.2018.04.163

Rodriguez-Mozaz S., Chamorro S., Marti E., Huerta B., Gros M., Sànchez-Melsió A., Borrego C. M., Barceló D. and Balcázar J. L. (2015). Occurrence of antibiotics and antibiotic resistance genes in hospital and urban wastewaters and their impact on the receiving river. *Water Research*, **69**, 234–242. Available at: https://doi.org/10.1016/j.watres.2014.11.021

'Scoping Report on Antimicrobial Resistance in India'. (2017). (November).

State T. H. E. and The O. F. (2015). 'THE STATE OF THE WORLD ' S ANTIBIOTICS'.

Stefanakis A. I. and Becker J. A. (2015). A review of emerging contaminants in water: classification, sources, and potential risks. *Impact of Water Pollution on Human Health and Environmental Sustainability*, 55–80. Available at: https://doi.org/10.4018/978-1-4666-9559-7.ch003

Swapna K. M., Rajesh R. and Lakshmanan P. T. (2012). Incidence of antibiotic residues in farmed shrimps from the southern states of India, **41**(August), 344–347.

Wang Y., Dong X., Zang J., Zhao X., Jiang F., Jiang L., Xiong C., Wang N. and Fu C. (2023). Antibiotic residues of drinking-water and its human exposure risk assessment in rural eastern China. *Water Research*, **236**, 119940. Available at: https://doi.org/10.1016/J.WATRES.2023.119940

Watkinson A. J., Murby E. J. and Costanzo S. D. (2007). Removal of antibiotics in conventional and advanced wastewater treatment: implications for environmental discharge and wastewater recycling, **41**, 4164–4176. Available at: https://doi.org/10.1016/j.watres.2007.04.005

Zhang L., Bai J., Zhang K., Wang Y., Xiao R., Campos M., Acuña J. and Jorquera M. A. (2023). Occurrence, bioaccumulation and ecological risks of antibiotics in the water-plant-sediment systems in different functional areas of the largest shallow lake in North China: impacts of river input and historical agricultural activities. *Science of The Total Environment* **857**, 159260. Available at: https://doi.org/10.1016/J.SCITOTENV.2022.159260

doi: 10.2166/9781789063752_0093

Chapter 6

Function of nanomaterials in the treatment of emerging pollutants in wastewater

Paramjeet Dhull[1], Neha Saini[1], Mohd Aamir[2], Shama Parveen[3] and Samina Husain[4]*

[1]Department of Environmental Science & Engineering, Guru Jambheshwar University of Science & Technology, Hisar 125001, Haryana, India
[2]Division of Plant Pathology, ICAR-Indian Council of Agricultural Research, Pusa, New Delhi 110012, India
[3]Department of Physics, DPG Degree College, Gurugram, India
[4]Centre for Nanoscience and Nanotechnology, Jamia Millia Islamia, New Delhi 110025, India
**Corresponding author: shusain3@jmi.ac.in*

ABSTRACT

Freshwater accessibility has grown to be a serious global challenge. The naturally occurring freshwater reserves are contaminated by the increased demographic, industrialization, and climatic changes. The health, environment, economy, and daily life are all extremely harmed by water pollution. Emergent pollutants including microplastics, antibiotics, hormones, unregulated medicines, nano-based materials, endocrine disruptors, pesticides, and so on are detrimental to human health and the environment. The development of wastewater treatment methods that are quick, feasible, low-cost, efficient, and sustainable is a problem posed by the emergence of new pollutants in the water. The shortcomings of current traditional treatment methods can be reduced with nanotechnology's intervention as it can remediate the contaminants commonly found in traces within complex organic mineral compounds. Based on the types of pollutants and required level of treatment efficiency, several nanomaterials such as carbon nanotubes, nanocomposites, nano-sorbents, graphene, nanomembranes, nanofibers, and nano-catalysts and so on are employed for wastewater treatment. Nanomaterials have unique physcio-chemical properties like shape, size and structure, surface morphology, crystallinity, and so on. These special qualities make them ideal substitutes for wastewater cleanup, purification, and contamination detection using pollutant-specific nanosensors and detectors. This chapter covers the type of nanomaterials and nanotechnologies useful in wastewater treatment to remediate emerging pollutants of concern. It also discusses the toxicity associated with nanotechnology and its environmental concern. Further the recent trends of large-scale clean-up of wastewater using nanotechnology and the challenges and future perspective associated with it are discussed.

Keywords: emerging contaminants (ECs), nanomaterials, wastewater treatment, nanoparticles, green synthesis, eco-toxicity

6.1 INTRODUCTION

As we all know, the availability of water in a pure state is becoming very crucial for the survival of human beings and other life forms (Kurniawan *et al.*, 2023). It is regarded as a universal solvent because of its special properties such as solubility. Clean and unpolluted water access to everyone has become a major challenge for the whole world (Sufiani *et al.*, 2023). Further, the ever-increasing world population and industrialization demand more and more clean water for various activities like drinking, agriculture, industrial activities, transportation, and many more. As reported, the world's water requirement has doubled from a few decades ago and it has become a global issue and a major challenge for the life in 21st century (Hodges *et al.*, 2018). Due to reasons like the dumping of industrial wastes, sewage treatment, marine dumping issues, dumping of radioactive waste materials, agricultural run-offs directly into water bodies, and so on lead to the introduction of emerging contaminants (ECs) into the water bodies. These excessive amounts of pollutants and ECs render them unsuitable for usage (Yaqoob *et al.*, 2020). The organic pollutants, also known as ECs, have drawn the attention of the public and raised concerns since they not only worsen the quality of water but also present a serious risk to already-installed water treatment systems and have recently been identified as a major hazardous pollutant (Rathi *et al.*, 2021). These are natural or artificial compounds/substances that are mainly unregulated. Indeed, these are not necessarily of new use but are newly identified and the knowledge about their presence, sources, fate, and effects are not completely known. The long list includes pharmaceutical compounds like analgesics, antidepressants, hormones, antibiotics, lipid regulators, anti-inflammatory drugs, personal care products like fats, fragrances, sunscreens, detergents, oils, disinfectants, and insect repellants (Morin-Crini *et al.*, 2022); pesticides; fertilizers; sweeteners; polycyclic aromatic hydrocarbons (PAHs); dioxins; surfactants, and other chemicals, often not effectively removed by conventional wastewater treatment processes (Ahmed *et al.*, 2021; Rout *et al.*, 2021). These are often released into water bodies from point sources of industries, domestic effluents, agriculture, hospitals, aquaculture, and so on. They are introduced into groundwater through infiltration, leakage, leaching from landfills, leakage from septic tanks, or failure with sewage systems (Parida *et al.*, 2021). Due to the benefits provided by these products in the daily lifestyle, these are utilized and released continuously into the environment, even at very low concentrations (ng/L to μg/L). As a result, they can accumulate in water bodies, leading to potential ecological and human health risks like carcinogenicity and tissue degradation which are extremely hazardous, the development of antibiotic-resistant bacteria, and so on (Karpińska & Kotowska, 2021). Table 6.1 discusses the present emerging contaminant (EC) types, their sources, and the consequences which they pose to human and environmental health.

Currently, EC treatment technologies applied in wastewater treatment plants (WWTPs) are of two kinds that is, conventional treatment and advanced treatment techniques including physical, mechanical, chemical, and biological methods (Wen *et al.*, 2021). These treatment technologies are based on the fate of the water to be used for industry, agriculture, drinking, and domestic purpose. Most commonly used methods include filtration, ozonation, biochemical processes, chemical disinfection, dilution, photolysis, sorption, biodegradation, decontamination treatment, sedimentation, flocculation, volatilization, and so on (Rout *et al.*, 2021). Further, advanced processes like adsorption-oriented processes and advanced oxidation processes including biosorbents, biological-based technologies, activated carbon, and biochar have also occupied the focus of WWTPs research for the exclusion of ECs (Cheng *et al.*, 2021; Morin-Crini *et al.*, 2022). Regardless, existing WWTPs treatment technologies are not able to completely eliminate emerging contaminants in the wastewater because of their complex and non-biodegradability structure, polarity, and high-water solubility (Alvarino *et al.*, 2018). These treatment technologies pose several limitations like high operating and maintenance costs, lower efficiency, selective decontamination, large-scale applications, high energy requirement, production of harmful by-products, and so on (Mirzaei *et al.*, 2017). These drawbacks led researchers to explore improved and innovative technologies which are sustainable and eco-friendly.

Table 6.1 Types of emerging contaminants (ECs) and their consequences on human health and environment.

EC Types	Representative ECs	Major Sources	Adverse Effects	References
Pharmaceutically active compounds (PhACs)	Diazepam, Testosterone, Diclofenac, Clorfibric acid, Ciprofloxacin, Metoprolol, Carbamazepine,	Hospital effluent, pharmaceutical industry effluent, domestic wastewater, livestock, and aquaculture farms effluent	Induces antibiotic resistance in microbes, affects the structure of the microbial community, and reduces the population of nematodes, bacteria, algae, and so on.	Ahmed et al. (2021), Khan et al. (2021), Rout et al. (2021), Morin-Crini et al. (2022)
Personal care products (PCPs)	N′-diethyltoluamide (DEET), galaxolide, 4-benzophenone, N, musk xylene, musk ketone	Surface water, WWTP effluent, and landfill leachate	Enhances toxicity in the aquatic environment, causes oxidation stress to goldfish, is carcinogenic to rodents, and potentially causes damage to the human nervous system	
Perfluorinated alkylated substances (PFASs)	Perfluorooctanoic acid (PFOA) and perfluorooctanesulfonate (PFOS)	Surface water, wastewater, groundwater, and sediments	Kidney cancer, liver damage, thyroid disease, induces resistance to vaccines, and developmental effects on the unborn child.	
Endocrine-disrupting chemicals (EDCs)	Bisphenol A (BPA), phthalates, dioctyl phthalate (DOP), xenoestrogen, and bisphenol,	Secondary sludge, drinking water, soil, surface water, and sediments	Interfere with the endocrine system, estrogenic effects in rats, feminizing side effects in men, birth defects, and developmental delays.	
Regulated compounds (RCs)	Phenanthrene, anthracene, pyrene and chlorpyrifos	Surface water, soils, agricultural runoffs, sediment, effluent from sewage treatment plants,	Cardiovascular diseases, carcinogenic effects, poor fetal development	
Industrial chemicals	Tris (1-chloro–2-propyl) phosphate (TCPP), polybrominated diphenyl ethers (PBDEs) and dimethyl adipate (DMAD)	Industrial effluent and domestic wastewater	Affect hormonal activity, interfere with brain and nervous system, reproduction, and fertility	
Pesticides	Metaldehyde, butachlor, and epoxiconazole	Aquaculture effluent, agricultural runoff, and surface water	Carcinogenic effect, cardiovascular diseases, toxic to aquatic organisms	

In this scenario, it is possible to view nanomaterials and nanotechnologies as a feasible and successful way to get beyond the drawbacks of conventional and innovative wastewater treatment approaches (Karthigadevi *et al.*, 2021). These have emerged as a low-expense, more efficient option for the removal of ECs from wastewater. Because of their properties such as organophilic nature, sieve diameter, strong solution mobility, aspect ratio, sensitivity, selectivity, catalytic potential, tunable hydrophobicity, large surface area, absorptivity, antimicrobial properties, porosity, strong mechanical properties, recyclability, chemical stability, mobility, and many more makes them the better option to remove/treat emerging contaminants from wastewater (Wu *et al.*, 2019; Yaqoob & Ibrahim, 2019). Recently, there have been improvements in the research and development of nanomaterials such as nanomembranes, nanomotors, nanofiltration membranes, nanosorbents, engineered nanomaterials like nanocomposites, and so on which have shown effective removal of ECs from wastewater and have opened new opportunities to adapt towards both selectivity and capacity of particular ECs (Ollier *et al.*, 2020). Fang *et al.* (2017) found effective removal for Pb^{2+}, Cd^{2+}, and Cu^{2+} with an adsorption capacity of 20.23 mg Pb/g, 17.01 mg Cd/g, and 10.42 mg Cu/g, respectively, through PDA nanoparticle adhering on the walls of finger-like pores (PES/PDA-R). Adsorption capabilities were found to be 1.69, 2.25, and 1.91 times more, respectively, than in conventional PDA-decorated membranes (PES/PDA-F membranes). Other studies have also reported the efficiency of nanofiltration membrane-based nanomaterials for the elimination of ECs emerged from personal care goods, pesticides, and pharmaceuticals (Kollarahithlu & Balakrishnan, 2021).

This chapter provides a thorough explanation of how nanomaterials work for controlling emerging contaminants (ECs) in wastewater. It will discuss several kinds of nanomaterials, such as carbon-based nanomaterials, metal, and metal oxide nanoparticles, and hybrid nanocomposites, as well as their applications in the elimination and degradation of emerging contaminants. It will cover the latest advancements in the field, including the synthesis, characterization, mechanisms, and factors influencing their performance and application for the degradation and removal of emerging pollutants. It will provide insights into the latest advances and new developments in the application of nanomaterials to the pressing problem of emerging contaminants in wastewater and offer potential solutions for sustainable water management in the future. The chapter will further highlight the challenges and future prospects of their application in this field including their potential impacts on the environment and human health. It will provide insights into the current state of the field, and identify knowledge gaps in this rapidly evolving area of environmental science and engineering.

6.2 CLASSIFICATION OF NANOMATERIALS (NMS)

Contaminated water adversely affects all ecosystems, predominantly aquatic and terrestrial ecosystems, and human health. The availability of pure water is a global challenge for the whole world in the 21st century because, without water, all living beings' existence is impossible. To overcome this severe concern, several techniques were used in ancient times to purify water, such as micro-ultrafiltration, reverse osmosis, sedimentation, and precipitation (Anjum *et al.*, 2019). However, these methods cannot remove minimal amounts of harmful toxins in water. Therefore, there is a need to develop more sensitive and practical techniques for water remediation. Nanotechnology provides advanced methods for completely removing contaminants in water using nanomaterials (Mondal *et al.*, 2023). At the nanoscale, materials exhibit excellent and distinctive characteristics including a strong chemical reactivity, a high surface-to-volume ratio, significant physical/chemical stability, high absorption capacity, and strong charge transfer ability compared to bulk material (Ealia & Saravanakumar, 2017). These properties of nanomaterials are leading to significant improvements in efficiency for sustainable wastewater treatment. Due to their small size, nanomaterials easily penetrate deep and remove contaminants from water.

Nanomaterials also aid in fabricating more efficient and advanced water filtration membranes, allowing for permeability control and fouling resistance. Recently, highly reactive nanomaterials have

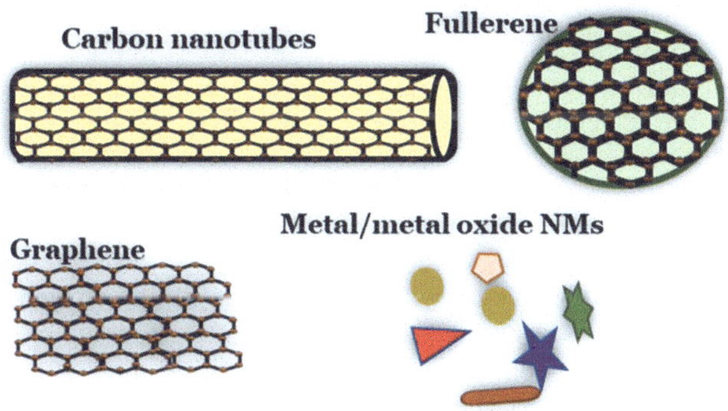

Figure 6.1 Structure of different types of nanomaterials.

been thought to be unique and the best solutions for wastewater treatment, as they offer prospective properties that would make them more efficient in providing pure and toxic-free water (Xu *et al.*, 2023). There are many types of nanomaterials have been synthesized. Nanomaterials can be generally classified into two categories: (i) carbon-based nanomaterials and (ii) metal/metal oxide-based nanomaterials. Figure 6.1 shows different types of nanomaterials.

Carbon has been the primary source of water filtration and purification since antiquity in the form of wood charcoal. Carbon allotropes such as fullerene, CNT, and graphene are attracting interest from researchers in current technology because of their quick mechanisms, such as adsorption, filtered membrane, photocatalytic oxidation, and sensing components (Gill *et al.*, 2023). Carbon remains a popular alternative for water treatment. On the other hand, the possibility of metal and metal oxide-based nanoparticles for wastewater treatment has attracted many academics and technologists. Metal-based nanomaterials have extraordinary qualities such as a high aspect ratio, atomically accurate pores, a wide range of functionality, and high absorption activity (Deng *et al.*, 2023).

6.2.1 Carbon-based nanomaterial

Carbon-based nanoparticles are utilized to detect and remove water contaminants broadly. To improve and develop water treatment, a wide array of carbon-based nanomaterials has been used.

6.2.1.1 Fullerene

Fullerene (C60), known as Buckminsterfullerene, is a carbon-based compound with 60 carbon atoms. sp^2 hybridization holds all carbon atoms together, forming a hollow spherical shape. By adding and attaching functional groups to the surface of fullerene, many kinds of fullerene may be created. Fullerenes can be used as filter membranes, adsorbents, and biofilm-resistant surfaces in water treatment engineering. The use of fullerenes in combination with ultraviolet (UV) irradiation is the most modern and advanced disinfection procedure (ADP) for contaminant elimination (Ealia & Saravanakumar, 2017).

6.2.1.2 Carbon nanotubes

Carbon nanotubes (CNT) are one-dimensional cylindrical structures formed by a honeycomb lattice of carbon atoms. It has lengths ranging from a few micrometers to many centimeters with diameters as small as 0.7 nm. Carbon nanotube ends can be empty or closed with a half-fullerene molecule. There are two varieties of CNTs based on the number of layers: (i) Single Walled Carbon Nanotubes (SWCNTs), which have a single layer of carbon atoms in cylindrical shape, and (ii) Multi Walled Carbon

Nanotubes (MWCNTs), which have more than one or two layers of carbon atoms in cylindrical form. Physical and chemical approaches can be used to create CNT (Parveen *et al.*, 2017), and physical procedures at high temperatures result in large-scale CNTs. Due to the hollow cylindrical geometry of CNTs, it is most widely used to remove contaminants and other organic contaminants from aqueous solution through adsorption.

6.2.1.3 Graphene

Graphene is a two-dimensional planar sheet of hexagonally organized carbon atoms one atom thick. It is a carbon allotropy with good capabilities for eliminating pollutants in an aqueous medium as a selective membrane. Graphene also has promising qualities for wastewater treatment as an adsorbent for the rapid and active removal of heavy metal ions in water. Graphene may be synthesized using ordinary scotch tape and the Hammer method (Lingamdinne *et al.*, 2016). Graphene oxide (GO) is a functionalized graphene created by a graphite sheet's chemical oxidation. The addition of this functional group as oxide improves heavy metal adsorption overall.

6.2.2 Metal/metal oxide-based nanomaterials

Metal and metal oxide nanomaterials are effectively associated with developing a more efficient, eco-friendly, cost-effective, and reliable water filtration process. Membrane technology is also going on to its advanced stage by using metal and metal oxide nanomaterials. Different type of metal nanoparticles has been synthesized for wastewater treatment. Silver, gold, copper, and iron are frequently used in developing filtration membranes, nanosensors, and nano adsorbents to detect and remove water contaminates. In addition, zinc oxide (ZnO), iron oxide, titanium oxide (TiO_2), tin oxide (SnO_2), are examples of metal oxide nanoparticles that are effectively utilized for the photocatalytic decomposition of contaminants in water (Rashid *et al.*, 2014). These metal oxide NPs are considered significant components in developing a more simple, easy, and accurate treatment process, that is, photocatalysis among the advanced wastewater treatment.

6.3 SYNTHESIS AND CHARACTERIZATION OF NANOMATERIALS

Nanomaterials can be developed through a variety of techniques. These techniques can be roughly divided into two categories: top-down and bottom-up. Figure 6.2 shows the different synthesis methods for these two types.

6.3.1 Green synthesis of nanomaterials

The main goal of all technological advancements is to improve human comfort without jeopardizing health. For this innovative cause of safety, an advanced and environmentally friendly development technique is necessary. Sustainable development is also a big component of this since it involves increasing the living system for all by not disrupting our ecosystem. Green nanotechnology contributes to sustainable development by synthesizing and using nanomaterials (Schulte *et al.*, 2013). The synthesis of nanomaterials by the green method is gaining the attention of researchers and technologists due to its non-toxic, cost-effective, and eco-friendly nature. Green technology is very efficient and a major component of sustainable development as it is a requirement of the next generation or the upcoming future.

The most common interpretation of 'green' is the production of nanomaterials using plant-based ingredients. However, the application of non-toxic solvents in synthesizing highly effective nanomaterials is covered by green technology, which is limited and inclusive. Additionally, because they use less mass and more potent ingredients, products made using green processes are anticipated to significantly contribute to preserving the environment and ecosystems (Centi & Perathoner, 2011). An excellent way to eliminate adverse environmental effects is through green nanotechnology. Table 6.2 shows the green synthesis of nanomaterials for different applications.

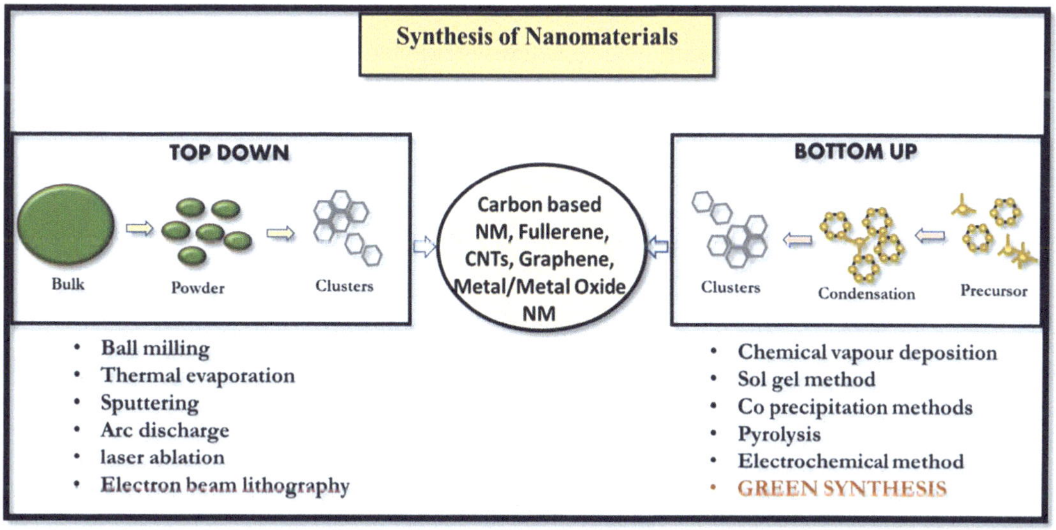

Figure 6.2 Representation of two kinds of approaches followed for the synthesis of nanomaterials.

6.3.2 Characterization of nanomaterials

In the following section, the numerous techniques used to characterize nanomaterials have been discussed extensively. These techniques can be integrated or used solely to conduct research on an exclusive attribute. These techniques can be compared, considering their accessibility, cost, accuracy, non-destructiveness, flexibility of use, and propensity to work with specific compounds or materials.

Despite the several approaches presented here, each one undergoes a comprehensive assessment. Microscopy-based techniques, such as SEM, TEM, and AFM, can be used to determine the size, shape, and crystalline makeup of the nanomaterials. Other methods include UV–visible, X-ray, spectroscopy, and Raman scattering techniques to determine the elemental and compositional elemental analyses and the optical properties of nanomaterials (Parveen *et al.*, 2022a, 2022b). Figure 6.3 is shown the various characterization techniques to get complete information on nanomaterials.

6.4 NANOMATERIALS-BASED APPROACHES OF WASTEWATER TREATMENT (WWT)

With rapid industrialization, urbanization, and population growth, the generation of wastewater has increased significantly, leading to environmental pollution and water scarcity issues. Conventional wastewater treatment methods often fall short of effectively removing various pollutants, including heavy metals, organic contaminants, and pathogens. In recent years, nanotechnology has emerged as a promising approach for wastewater treatment due to the unique properties (physical, chemical, and biological properties compared to their bulk counterparts) of nanomaterials that can address the limitations of traditional treatment methods. These unique properties make nanomaterials highly effective in treating wastewater. Various types of nanomaterials, such as nanoparticles, nanotubes, nanofibers, and nanocomposites, have been explored for their applications in wastewater treatment. When it comes to the mechanism involved in the nanomaterials-based approach for wastewater treatment, several key processes come into play. These processes include adsorption, photocatalysis, membrane filtration, and disinfection.

Table 6.2 Green techniques involved for the synthesis of nanomaterials and their operating parameters.

S. No.	NMs	Synthesis Methods	Synthesis Parameter	References
1	Au	Chitosan as a reducing agent	Trihydrate tetra chloroauric acid ($HAuCl_4.3H_2O$), chitosan, glacial acetic acid, distilled water, temp: 100°C, time:15 min	Barajas et al. (2019)
2	Ag	Leaf extract of *Malachra capitata* (L.)	Silver nitrate ($AgNO_3$), n-hexane, *Malachra capitata* (L.), distilled water, temp: 90°C, time: 24 h	Srirangam and Rao (2017)
3	Cu	*Capparis spinosa* fruit	Copper sulfate solution, *Capparis spinosa*, ethanol, deionized water, time: 24 h 20 min, temp: 60°C	Ebrahimi et al. (2017)
4	ZnO	Ecofriendly wet chemical method	Zn $(NO_3)_2$, sodium hydroxide (NaOH), distilled water, temp: 60°C, time: 2 h 10 min	Yadav and Sisodia (2022)
5	ZnO	Using garlic skin	Garlic skin, zinc chloride, sodium hydroxide (NaOH), distilled water, time: 25 h	Modi and Fulekar (2020)
6	TiO_2	Sol–gel method	Titanium tetra iso prop oxide [Ti $(OCH(CH_3)_2]_4$, isopropanol $[(CH_3)_2CHOH]$, nitric acid (HNO_3), deionized water, temp: 60–80°C, time: 9 h	Sharma et al. (2014)
7	Graphene	Various concentrations of *Tecoma stans* leaves as reducing/capping agents	H_2SO_4, and 1 g of sodium nitrate (ISO-CHEM). time: 1 hr, temperature: 90°C	Mahmoud et al. (2022)
8	CNTs	Using pyrolysis technique	Coconut and olive oils are used as precursors and acetone, nickel chloride (5 wt%.) as a catalyst and argon as a carrier gas. temp: 900°C, time: 15 min	Hamid et al. (2017)

(i) **Adsorption:** Adsorption is a fundamental mechanism utilized in wastewater treatment, and it plays a crucial role in the removal of various contaminants from water. The adsorption process involves the attachment of pollutants onto the surface of a solid material, known as an adsorbent. Nanomaterials, such as activated carbon nanoparticles, metal oxide nanoparticles (e.g., titanium dioxide, iron oxide), and carbon nanotubes, exhibit a high surface area and porosity, allowing them to effectively adsorb and remove pollutants from wastewater. In the context of nanomaterials-based wastewater treatment, nanomaterials with high surface area and porosity are commonly employed as adsorbents (Ahmed *et al.*, 2022).

(ii) The adsorption mechanism relies on the interactions between the adsorbent surface and the contaminants present in the wastewater. Such interactions are largely classified into physical (physisorption) and chemical (chemisorption) adsorption. Physical adsorption occurred due to weak forces between molecules like van der Waals forces, hydrogen bonding, and electrostatic interactions (Fanourakis *et al.*, 2020). Nanomaterials with large surface area-to-volume ratios, such as activated carbon nanoparticles, graphene, and metal oxide nanoparticles, exhibit high physisorption capacity. The contaminants in the wastewater come into contact with the adsorbent surface, and the weak forces between the adsorbent and contaminants allow them to adhere to the surface (Fanourakis *et al.*, 2020). Physisorption is particularly effective

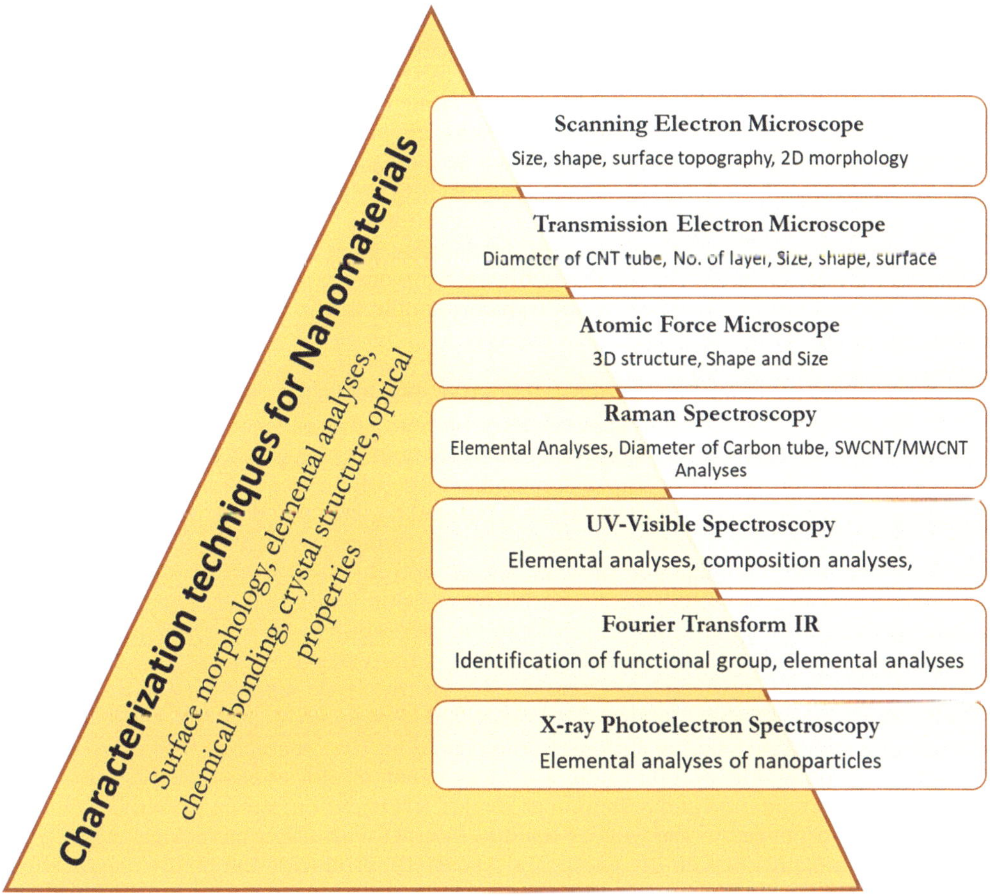

Figure 6.3 Different characterization techniques used for the study of nanomaterials.

at removing organic chemicals such as dyes, pesticides, volatile organic compounds (VOCs) and pharmaceuticals. The porous structure of nanomaterials provides many adsorption sites, enabling them to capture and retain organic molecules effectively. Chemical adsorption involves stronger chemical bonds between the adsorbent surface and the contaminants which usually occurs through covalent or ionic interactions. Nanomaterials with specific functional groups, such as metal oxide nanoparticles and zeolites, can undergo chemisorption with certain contaminants. Chemisorption is especially useful in removing heavy metals and metal ions from wastewater. These nanomaterials possess active sites on their surfaces that can form strong chemical bonds with metal species, effectively immobilizing them. Additionally, ion exchange processes may take place, where the nanomaterials release other ions (such as H^+ or Na^+) into the wastewater while capturing the metal ions. Iron oxides (Fe_3O_4), TiO_2, and zinc oxides (ZnO) are the most widely used metal oxide-based nano-adsorbents for wastewater treatment. Heavy metals like lead, chromium, arsenic, mercury, cadmium, nickel, and copper could be eradicated from wastewater more efficiently by using nano-metal oxides compared to activated carbon. TiO_2's appealing adsorption capability can be indicated by the effective removal of phosphate, nitrate, and methyl blue. Manganese oxide nanoparticles have found extensive application in the elimination of lead (II), arsenic, cadmium (II), and zinc (II) from wastewater due to their substantial surface area (Ahmed *et al.*, 2022). In many cases, adsorbents can be regenerated and reused, reducing the overall treatment cost. Techniques like desorption using appropriate solvents or changing environmental conditions can release the adsorbed contaminants from the nanomaterials, allowing them to be recycled. Also, adsorption can complement other treatment processes, such as biological treatment or membrane filtration, by removing specific pollutants that may not be effectively treated by those methods alone. By harnessing the adsorption capabilities of nanomaterials, wastewater treatment can achieve improved water quality and contribute to sustainable water management practices.

(iii) **Photocatalysis**: It is a process that utilizes light energy and a photocatalyst to initiate chemical reactions that can degrade or transform pollutants in wastewater. The photocatalytic mechanism involves several steps that take place on the photocatalyst's surface (Ahmed *et al.*, 2022). A few nanomaterials have photocatalytic characteristics, especially semiconducting metallic oxides like zinc oxide (ZnO) and titanium dioxide (TiO_2). When exposed to ultraviolet (UV) light, these nanomaterials generate electron-hole pairs, which initiate redox reactions and produce highly reactive oxygen species (ROS) (Fanourakis *et al.*, 2020). These ROS, such as hydroxyl radicals, possess strong oxidative capabilities that can degrade organic contaminants into harmless by-products. Photocatalysis is particularly effective in treating organic pollutants and emerging contaminants, (i.e., personal care goods and pharmaceuticals) present in wastewater. The photocatalytic process begins with the absorption of photons from a light source, typically ultraviolet (UV) or visible light. Upon photon absorption, the photocatalyst undergoes electronic excitation, resulting in the generation of electron-hole pairs. The excited electrons leave behind positive-charged holes in the valence band as they transition from the valence band to the conduction band. This electron-hole separation is crucial in order to perform subsequent photocatalytic processes. The separated electrons (e^-) and holes (h^+) on the photocatalyst surface can engage in redox reactions that interact with the adsorbed species or dissolved pollutants present in the wastewater (Fanourakis *et al.*, 2020). The reactions can be classified as oxidation and reduction processes. (i) Oxidation: strong oxidative abilities allow the h^+ holes produced in the valence band to interact with water molecules or hydroxyl ions (OH^-) that are adsorbed on the photocatalyst surface, producing the highly reactive hydroxyl radicals ($^{\cdot}OH$). These hydroxyl radicals are potent oxidizing agents that can attack and degrade organic contaminants, breaking them down into smaller and less harmful molecules.

(iv) **Reduction:** Simultaneously, the photogenerated electrons (e^-) in the conduction band can reduce certain species, such as dissolved oxygen (O_2), creating superoxide radicals ($\bullet O^{2-}$) or hydrogen peroxide (H_2O_2). These reactive species can participate in further redox reactions, facilitating the degradation or transformation of pollutants. The reactive oxygen species (ROS) generated during the photocatalytic process, comprise highly reactive species such as hydroxyl radicals ($\bullet OH$), superoxide radicals ($\bullet O^{2-}$), and hydrogen peroxide (H_2O_2). These species have strong oxidizing abilities and have the potential to efficiently reduce a broad spectrum of contaminants. These ROS target the chemical bonds of organic compounds, allowing them to break down into smaller, less harmful molecules like carbon dioxide (CO_2) and water (H_2O). The photocatalytic process is continuous, with the regenerated photocatalyst being available for subsequent cycles. The photogenerated electrons and holes recombine, or they can be scavenged by sacrificial agents present in the wastewater, preventing their recombination and maintaining a continuous supply of active species for photocatalytic reactions. TiO_2, ZnO, Fe_2O_3, zinc sulfide (ZnS), zirconium dioxide (ZrO_2), cadmium sulfide (CdS), and tungsten trioxide (WO_3) are the most commonly used photocatalysts. The bandgap energy of TiO_2 has been successfully lowered and the TiO_2 adsorption from the UV light zone has been red-shifted by doping TiO_2 with Cr, Ag, Zn, Al, Mn, Co, Fe, Ni, Pt, Bi, Pd, S, Au, and N. The presence of UV light increases the efficiency of photocatalysts (Ahmed *et al.*, 2022). An ordered mesoporous silica (SBA-15) molecular sieve loaded with TiO_2 and combined with zirconium nanophotocatalyst was reported to minimize reactive red X-3B by 96% in a study by Bai *et al.* (2020). Additionally, photocatalysis is environmentally friendly as it utilizes light energy and does not require the addition of chemicals. However, challenges such as catalyst stability, efficient utilization of solar energy, and optimization of reaction conditions need to be addressed for large-scale implementation of nanomaterials in waste water treatment.

(v) **Membrane filtration**: It is a widely used technique in wastewater treatment for the separation and removal of suspended solids, particulate matter, microorganisms, and dissolved contaminants. Membrane filtration operates on the principle of physical sieving. The wastewater is passed through the membrane, and particles larger than the membrane's pore size are retained and accumulate on the feed side of the membrane. The clean permeate, consisting of purified water and smaller dissolved molecules, passes through the membrane and is collected. Membranes have different pore sizes and selectivity, which determine their filtration capabilities. Nanofiltration and reverse osmosis membranes have even smaller pore sizes (<0.001 μm) and can remove dissolved salts, ions, and small organic molecules. During membrane filtration, fouling (deposition and accumulation of particles, microorganisms, organic matter, and other contaminants on the membrane surface or within its pores) can occur, leading to reduced filtration efficiency. Fouling can result in decreased permeate flow, increased pressure requirements, and reduced membrane lifespan. Nanomaterials are also employed in membrane filtration processes for wastewater treatment. Membranes with nanoscale pores, such as nanofiltration (NF) and reverse osmosis (RO) membranes, can effectively separate and remove dissolved salts, heavy metals, and other contaminants. Nanomaterials, including carbon nanotubes, graphene oxide, and zeolites, are integrated into membranes to enhance their selectivity, permeability, and fouling resistance. These modified membranes offer improved water quality, higher water recovery rates, and prolonged membrane lifespan. Periodic cleaning of the membrane such as backwashing, chemical cleaning, or air scouring is necessary to remove fouling and maintain optimal filtration performance.

(vi) **Disinfection**: Disinfection is an essential step in wastewater treatment to ensure the removal or inactivation of harmful microorganisms, including bacteria, viruses, and protozoa. Disinfection processes often involve the use of strong oxidizing agents, such as chlorine (Cl_2), chlorine dioxide (ClO_2), ozone (O_3), or hydrogen peroxide (H_2O_2) (Ahmed *et al.*, 2022). These oxidants work by reacting with the microorganisms' cellular components, including proteins, enzymes, and nucleic acids. The oxidation disrupts the microorganisms' structure and metabolic processes, leading to their inactivation or death. Oxidizing agents can penetrate the cell walls or membranes of microorganisms. Once inside, they cause damage to the lipid bilayer, resulting in the disruption of the cell's integrity. This leads to the leakage of cellular contents and the loss of essential functions, ultimately leading to the inactivation of the microorganisms (Fanourakis *et al.*, 2020). Oxidizing agents can also directly damage the genetic material (DNA and RNA) of microorganisms. They induce structural modifications and breaks in the DNA or RNA strands, preventing replication and transcription processes. Disinfectants can denature proteins by breaking the hydrogen bonds, disulfide bridges, and other non-covalent interactions that maintain the protein's native structure. This denaturation disrupts the protein's functional properties, leading to the loss of enzymatic activity and vital cellular processes. Without functional proteins, microorganisms are unable to carry out essential metabolic functions, resulting in their inactivation. Some disinfection processes, such as chlorination or ozonation, produce reactive oxygen species (ROS) as by-products. These ROS, including hydroxyl radicals (•OH) and superoxide radicals ($•O^{2-}$), are highly reactive and can damage cellular components, including proteins, lipids, and DNA. The ROS attack the microorganisms' cellular structures and biomolecules, leading to their inactivation. Nanomaterials have shown potential for disinfection purposes in wastewater treatment. Silver nanoparticles (AgNPs) and copper nanoparticles (CuNPs) possess antimicrobial properties, making them effective in inactivating bacteria, viruses, and other microorganisms present in wastewater. The nanoparticles release metal ions that interact with microbial cell membranes, leading to cell damage and death. In addition to MWNTs, *Pseudomonas aeruginosa* PAO1 has been successfully inhibited by AgNP coating (Kim *et al.*, 2018). Similar results were obtained with ZnO nanocrystal-doped 3D nano-customized silicon (Si)-wafers against *E. coli*, one of the most prevalent organic pollutants in wastewater (Rahman *et al.*, 2021). CuONPs impregnated with a mixed matrix of PES and cellulose acetate showed a 75% inhibition in another green synthesis process utilizing copper (Singh *et al.*, 2021). Additionally, nanomaterials like graphene oxide have been explored for their ability to generate reactive oxygen species upon exposure to light, providing disinfection capabilities through a photocatalytic mechanism.

6.5 ADVANCES IN TERMS OF GREEN APPROACH FOR THE LARGE-SCALE USE OF NANOMATERIALS IN WASTEWATER TREATMENT

Nanotechnology has several physical, chemical, and biological processes for cleaning water. These include membrane separation, adsorption, ion exchange, chemical precipitation, photocatalyst splitting and bioremediation using various nanomaterials. In light of the potential for eliminating adsorptive material and subsequent metal(loid) extraction, adsorption-based approaches are acceptable. Carbon-based nanomaterials such as CNTs, graphene and fullerene may be employed for adsorption. The technique of using nanoparticles or nanomaterials to remove environmental contaminants from contaminated areas is known as nano-remediation. Metal/metal oxide-based nanomaterials such as TiO_2, ZnO, SnO_2, and silver (Ag), Gold (Au) and lead (Pb) nanoparticles are successfully used to eliminate metal(loid)s and other contaminants in wastewater treatment (Hairom *et al.*, 2021). Hence, nanomaterials have emerged as the most effective strategy for clean-up due to their increased features, such as a high surface-area-to-volume ratio and strong reactivity.

In the next section, we will focus on four main categories of nanomaterials in wastewater treatment applications as shown in Figure 6.4. These include nano-adsorbents, nano-catalysts, nanofiltration membranes and nanosensors integrating the nanotechnologies as mentioned earlier with advanced processes.

6.5.1 Nanofiltration

Generally, the process of removing contaminated particles from water using a porous membrane is known as filtration. This porous membrane traps the contaminated particles on their surface. If the membrane has the size of pores in the nanoscale range, then nanofiltration process occurs. Nanofiltration refers the excellent combination of uniform nanopores membrane with a unique ability to remove hardness, total dissolved solids and also microorganisms. Based on pore size, there are four main types of separation membranes have been developed, including nanofiltration (NF), microfiltration (MF), ultrafiltration (UF), and reverse osmosis (RO) membranes. The pore sizes of NF membranes are less than 1–2 nm, which are the smallest ones and much smaller than those of MF and UF membranes. Due to its small pore size, it can be easily employed to filter water pollutants very efficiently at the minute level. Hence, nanofiltration is the most effective and widely used technique in water quality treatment because of its advanced filtration mechanism and applicability for removing nano pollutants.

Carbon nanotube-based filtration membranes display excellent properties due to its nanoscale-porosity. Many studies on the CNTs-based filtration process have been reported in the last decades. Vertically aligned (VA) and mixed matrix (MM) CNT membranes are the two types of nanotube membranes that are based on existing manufacturing technologies (Ahn *et al.*, 2012). Fluid can only

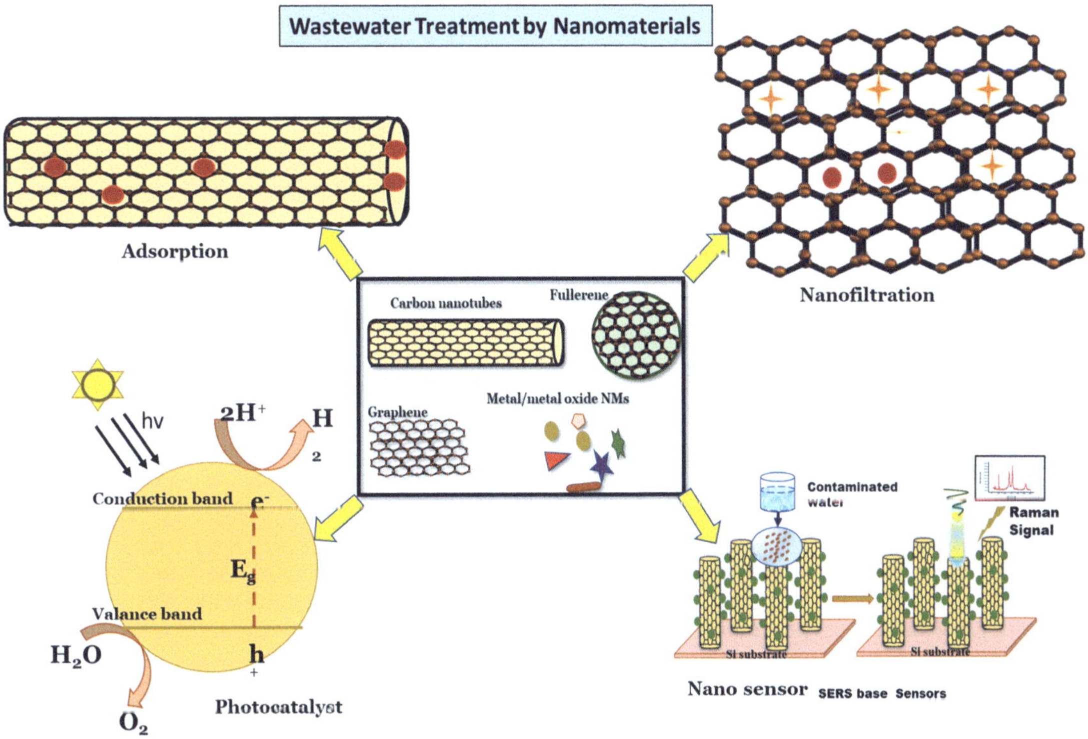

Figure 6.4 Remediation techniques using nanomaterials for the treatment of wastewater.

enter through the hollow interior of the CNT or between the bundles of CNTs because of how the CNTs are organized in the VA-CNT membranes. By matching perpendicular CNTs with a supporting filler material (epoxy, silicon nitride), the VA-CNT membranes may be created between the tubes (Hinds *et al.*, 2004).

On the other hand, a mixed matrix (MM) CNTs membrane is made up of many layers of polymers or another composite material. Due to the frictionless ability of CNT to transfer water through hydrophobic hollow cavity nanotubes, such membranes work with minimal energy consumption. The membrane is highly susceptible to many salts and contaminants, being self-cleaning, reusable, and good water.

Nanocomposite membranes made of PVDF/poly(styrene-butadiene-styrene)/thiocyanate and silver-modified MWCNTs have been created by solution blending (Mehwish *et al.*, 2015). The spray-assisted layer-by-layer method was employed by Liu *et al.* (2013) to construct the PES/functionalized MWCNT (F-MWCNTs) membrane. N, N-dimethylacetamide (DMAc) was employed by Shawky *et al.* (2011) to create MWCNT/aromatic PA nanocomposite membranes using a polymer grafting process. MWCNTs/polyaniline (PANI)/PES membranes were used by in-situ polymerization, according to Lee *et al.* (2016). The absence of reactants in the raw material makes in-situ polymerization stand out from other processes. Instead of taking place on both sides of the interface between the continuous phase and core material, all polymerization occurs in the continuous phase, as in interfacial polymerization. The in-situ colloidal precipitation production of the oxidized MWCNTs (OMWCNTs)/graphene oxide (GO)/PVDF membrane was described by Ho *et al.* (2017).

6.5.2 Nano adsorbents

The nano adsorption process refers to the adsorption of water contaminants on the surface of nanoparticles. Nano sorbents are nanomaterials that absorb water contaminants such as heavy metal ions and organic and inorganic impurities. Nano-sorptions are now regarded as a developing remediation strategy for the elimination of water contaminants in order to produce pure drinking water. Furthermore, these have garnered additional attention due to their unique properties and potential as a viable alternative to conventional adsorbents. Technologists and researchers have documented coagulation's effect on the adsorption activity of carbon-based nanomaterials. More recently, various research studies suggested that nano sorbent materials have excellent and quite promising properties for water purification and remediation, including carbon nanotube, graphene, fullerene, polymers, and metal and metal oxides nanosorbents (Nasrollahzadeh *et al.*, 2021).

Burakov *et al.* (2018) compared the general heavy metal adsorption potential of CNTs with several conventional materials. It has been suggested that plasma-oxidized multiwalled CNTs rather than chemically oxidized multiwalled CNTs might be employed as adsorbents for specific metals. The presence of functional groups containing oxygen on the CNT surface gave plasma-oxidized tubes better adsorption performance than chemically oxidized CNTs. Experimental findings demonstrate that plasma-oxidized CNTs tend to desorb metal ions more readily (Burakov *et al.*, 2018).

6.5.3 Photocatalysis

This photocatalysis remediation significantly benefits eliminating trace contaminants in aqueous solution due to its low cost, environmental friendliness, and universal applicability. Fast-developing technology has moved this procedure to the forefront of the sustainable wastewater treatment strategy. Photocatalysis is a most demanded and well-established technique for water pollutant degradation by UV-oxidation-based water splitting (Lazar *et al.*, 2012). In simple words, Photocatalysis refers to 'the splitting of the water by absorption of light'. It is used to quickly break down various pollutants such as organic materials, pesticides, dyes, and so on. Due to its vast variety of applicability, Photocatalysis with nanoparticles as catalysts is utilized to enhance the water quality by removing the comminates.

In this context, metal oxide nanoparticles are widely used for water purification as practical elements of photocatalysts. TiO_2 and ZnO NPs have been synthesized by green methods and analyzed as photocatalysts for water treatments (Li *et al.*, 2012).

Recent research has demonstrated the excitement of CNTs to TiO_2 nanoparticles. It has been demonstrated that reduced photoluminescence strength caused by a reduced recombination load indicates enhanced photocatalytic oxidation efficiency of CNT/TiO_2 composites to phenol. Single-walled carbon nanotubes (SWCNTs) are more able to increase the photocatalytic activity of TiO_2 than MWCNTs because they have more direct contact with the surface of the TiO_2 nanoparticles. At the interface between SWCNTs and TiO_2, indium tin oxide (ITO) thin films have also been applied. This considerably impacted the photoelectrochemical behavior and reduced the resistance between the two layers (Duong *et al.*, 2011).

6.5.4 Nano sensors

The surface water has been contaminated by household and industrial garbage exposure, which might lead to significant aquatic life diseases. Because of the low concentrations of toxins in water bodies, small-scale monitoring of these dangerous pollutants has become critical. Researchers are becoming interested in nanomaterial-based sensors for detecting water pollutants at low concentrations in this context.

Nanosensors are device that produces fast responses towards any change in the surroundings. Change may be in the form of light, heat, temperature, or change in volume concentration as input signal and are used to convey the information about the behavior as output signal. Nanosensors can sense the analyte at low detection limits, high selectivity, sensitivity, and long-term stability. Due to its low detection limit and fast response, it is widely used for advanced water treatment. These nanosensors are developed using nanomaterials such as carbon nanotubes, graphene, fullerene, metal, and metal oxide Nanomaterials, possessing high surface-area-for fast reactivity and binding affinity to target molecules even at very low concentrations, which produce a fast response. Integrating nanosensors in the conventional water quality sensor may enhance treatment efficiency because nanosensors have contributed excellent performance in removing contaminants in wastewater.

Nanosensors have been designed for water quality monitoring, including optical sensors, surface plasmon resonance, calorimetric, electrochemical sensors, and surface-enhanced Raman spectroscopic sensors (Parveen *et al.*, 2022a, 2022b).

It is generally safe to utilize CNTs as an electrode in biosensor applications. Rarely, the CNT electrode interacts directly with water, and however, some risk measurements may be observed. First off, 2D NMs, particularly graphene, are frequently combined with 1D CNTs for excellent electroconductivity and mechanical strength. These macro frameworks provide a variety of environmental concerns and differ in their physicochemical properties, all of which require careful consideration (Georgakilas *et al.*, 2015). Electrochemical biosensor applications frequently use CNTs functionalized with poly diallyl dimethylammonium chloride, or (PDDA) (Zhang *et al.*, 2011).

Adopting revolutionary advanced water technologies to provide high-quality drinking water with an improved ability to reduce micropollutants is critical. Water treatment technologies that are flexible and adaptive must be used to boost industrial production processes. Compared to traditional water treatment technologies, one of the most significant qualities of nanoparticles is their ability to combine diverse properties, creating multifunctional systems like nanocomposite membrane that is capable of holding particles and removing contaminants. Nanomaterials can also help to improve process efficiency because of their unique features, such as their huge surface area.

6.6 BARRIERS ASSOCIATED AND ENVIRONMENTAL CONCERNS OF NANOTECHNOLOGIES

Although nanotechnology has great implications in wastewater treatment, due to its detrimental impact on humans, the aquatic ecosystem, and the environment as a whole, they are frequently seen as problematic when they are released from the treatment system to the environment after use. There is a lack of understanding and information on the nanomaterials (NMs) fate, behavior, and toxicity upon exposure to humans, aquatic ecosystem, and the environment; therefore, it is critical to gain more knowledge in this area. Limited information about the aggregation and deposition of particular NMs due to the limitation of standard methods and instrumentation for their monitoring and detection poses a major barrier/ challenge for the use of nanotechnologies in wastewater treatment. The same properties such as shape, small size, high reactivity and many more which are useful for treating industrial wastewater become a hazardous problem for the environment. It is necessary to thoroughly study the dose–response impact and the resulting exposure pathways in order to comprehend the risk assessment of NMs. It has been suggested that the effects on lower organisms should be investigated because they are an essential part of the food chain (Patil *et al.*, 2016). Among all the nanomaterials, nanoparticles (NPs) are considered for having the most detrimental effects on the aquatic ecosystem, human health, and the environment. Most NPs do not cause instant death, but they do have an extensive range of harmful effects. For example, silver nanoparticles, titanium dioxide nanoparticles (TiO_2), carbon nanotubes (CNTs), and so on. TiO_2 NPs have been reported to cause the formation of reactive oxygen species (ROS), necrosis, mutagenesis, lipid peroxidation, apoptosis, changes in cell morphology, and mitochondrial dysfunctioning. The physicochemical properties of the NPs (size, morphology, aggregation, reactivity, surface charge, and dissolution) along with the intra and extracellular environment are the key regulators of their toxicity (Thangadurai *et al.*, 2020). In 2010, The American Environmental Protection Agency published the effects of TiO_2 NPs on a variety of aquatic life such as bacteria, algae, plants and fish, and invertebrates. Depending on the concentration and exposure time, various effects were recorded like pathological alterations in the gills, respiratory distress, reduced reproductive output in Daphnia, and behavioral abnormalities in fish (Khan *et al.*, 2021). Zerovalent iron (nZVI) has been proven to be toxic to *Escherichia coli*, *Bacillus subtilis* var. *niger*, and *Pseudomonas fluorescens* and increased the production of ROS in their cells. The nZVI has also been reported to cause phytotoxicity and accumulation, reduction in transpiration and growth of hybrid poplars (*Populous deltoids* × *Populous nigra*) and cattail (*Typha latifolia*) plants (Patil *et al.*, 2016). The CuO NPs cause oxidative stress in fishes even upon short-term exposure. Additionally, they have been found to enhance the activity of different oxidative enzymes including superoxide dismutase (SOD), glutathione S-transferases (GST), and catalase (CAT) in various organs (gills, livers, and kidney) and due to overproduction of ROS and disturbs homeostasis (Thangadurai *et al.*, 2020). Further, in order to better understand the ecotoxicological effects of NMs, there is a need to study not only their environmental effects but also their genotoxic effects and their effects at molecular levels. The nanomaterials are very beneficial to the industries for the wastewater treatment; therefore, it is essential to focus the research on the systemic release and monitor their effects to protect from their ecotoxicity. In this concern, the optimization and design of nanotechnology-based wastewater treatment require a thorough knowledge of the structure–activity of nanomaterials. This will help suggest the material selection, structural integrity, improve durability and process reliability. More green synthesis and eco-friendly approaches will also work to help in this direction

6.7 FUTURE PERSPECTIVES OF NANOMATERIALS IN WASTEWATER TREATMENT (WWT)

Although there are related challenges but breakthroughs in the realm of nanotechnology are associated with the advancement of wastewater treatment technologies. Since the preceding decade, NMs have been widely used in the wastewater treatment industry. Due to the specific physicochemical properties of these materials, including their size, structure, durability, reactivity, mobility, surface-to-volume ratio, and others, they have several advantages over other treatment technologies for emerging pollutants from wastewater, including faster kinetics, higher efficiency, selective affinities for specific pollutants and remarkable antimicrobial activity. There is so much research going on in various nanotechnology areas; for instance, nanomembrane filtration, nano photocatalysts, nano sorbents, and nanomotors to maximize the efficiency of the process. The synthesis of nanocomposites is another potential area to facilitate the dispersion capability and stability of NMs during the reaction. The green synthesis approach for the formation of NMs using biological life forms has attracted a lot of attention in the past few years in order to overcome the toxicity of the NMs. Due to the presence of various structures of the oxides available for metals and the diversity of the biological life forms, a detailed investigation in this research area will prove to be very helpful in overcoming the challenges of the wastewater treatment sector (Hoseinpour & Ghaemi, 2018). However, the development of safe and inexpensive nano-engineered materials opens up many opportunities for novelties in the near future, especially for the decentralized treatment systems, heavily degradable contaminants, and point-of-use devices. Furthermore, cutting-edge analytical and imaging technologies can be beneficial for the analysis and measurement of nanoscale objects, particularly for water treatment processes. The overall cost-effectiveness of the advanced technologies over conventional technologies needs to be understood for large-scale implications of NMs in the treatment of wastewater. Regardless of substantial advancements in the use of NMs for wastewater treatment, "real-time monitoring models" are essential for ensuring the efficacy and effectiveness of the NMs in the treated water while developing a better understanding of the performance of nanotechnology in the treatment process. Additionally, in order to mitigate the health and environmental concerns associated with NMs in this industry, several national and international regulations should be proposed and implemented critically. Only then, they can be adaptable to a large-scale treatment process. All the research communities working in this sector should follow proper guidelines and regulatory standards to reduce the eco-toxicity effects of NMs in the future (Ahmed *et al.*, 2022).

6.8 CONCLUSION

Nanomaterials-based approaches provide innovative solutions for addressing the challenges associated with wastewater treatment. The unique properties of nanomaterials enable enhanced adsorption, photocatalytic degradation, membrane filtration, and disinfection of pollutants in wastewater. It is of critical importance for the development of modified nanomaterials, their oxides, and hybrid nano-based frameworks for tackling waste-water treatment problems. To ensure long-term viability and sustainability, it is crucial to use appropriate and affordable nanomaterials for the waste-water treatment process. Nanomaterials are very effective because of their high reactivity, yet they still have several drawbacks which need to be rectified. One of the major hurdles in implementing nanomaterials is the potential health risks linked with their toxic effects. One promising way to enable acceptable nanomaterial applications is the production of nanocomposite materials consisting of functional polymers and inorganic solids. With further research and development, nanomaterials have the potential to revolutionize the field of wastewater treatment and contribute to sustainable water management practices.

REFERENCES

Ahmed S. F., Mofijur M., Nuzhat S., Chowdhury A. T., Rafa N., Uddin M. A., Inayat A., Mahlia T. M. I., Ong H. C., Chia W. Y. and Show P. L. (2021). Recent developments in physical, biological, chemical, and hybrid treatment techniques for removing emerging contaminants from wastewater. *Journal of Hazardous Materials*, **416**, 125912, https://doi.org/10.1016/j.jhazmat.2021.125912

Ahmed S. F., Mofijur M., Ahmed B., Mehnaz T., Mehejabin F., Maliat D., Hoang A. T. and Shafiullah G. M. (2022). Nanomaterials as a sustainable choice for treating wastewater. *Environmental Research*, **214**, 113807, https://doi.org/10.1016/j.envres.2022.113807

Ahn C. H., Baek Y., Lee C., Kim S. O., Kim S., Lee S., Kim S.-H., Bae S. S., Park J. and Yoon J. (2012). Carbon nanotube-based membranes: fabrication and application to desalination. *Journal of Industrial and Engineering Chemistry*, **18**(5), 1551–1559, https://doi.org/10.1016/j.jiec.2012.04.005

Alvarino T., Suarez S., Lema J. and Omil F. (2018). Understanding the sorption and biotransformation of organic micropollutants in innovative biological wastewater treatment technologies. *Science of the Total Environment*, **615**, 297–306, https://doi.org/10.1016/j.scitotenv.2017.09.278

Anjum M., Miandad R., Waqas M., Gehany F. and Barakat M. A. (2019). Remediation of wastewater using various nano-materials. *Arabian Journal of Chemistry*, **12**(8), 4897–4919, https://doi.org/10.1016/j.arabjc.2016.10.004

Bai L., Minghui W., Enlv H., Dan S., Lumin L., Wanli Y., Xiaojun T. and Baiqi W. (2020). Study on the controlled synthesis of Zr/TiO_2/SBA-15 nanophotocatalyst and its photocatalytic performance for industrial dye reactive red X–3B. *Materials Chemistry and Physics*, **246**, 122825.

Barajas F. J. F., Acevedo Z. C. S. and Pedraza H. P. (2019). Synthesis and characterization of gold nanoparticles in solution using chitosan as reducing agent. *Respuestas*, **24**(2), 49–55, https://doi.org/10.22463/0122820X.1830

Burakov A. E., Galunin E. V., Burakova I. V., Kucherova A. E., Agarwal S., Tkachev A. G. and Gupta V. K. (2018). Adsorption of heavy metals on conventional and nanostructured materials for wastewater treatment purposes: a review. *Ecotoxicology and Environmental Safety*, **148**, 702–712, https://doi.org/10.1016/j.ecoenv.2017.11.034

Centi G. and Perathoner S. (2011). Carbon nanotubes for sustainable energy applications. *ChemSusChem*, **4**(7), 913–925, https://doi.org/10.1002/cssc.201100084

Cheng N., Wang B., Wu P., Lee X., Xing Y., Chen M. and Gao B. (2021). Adsorption of emerging contaminants from water and wastewater by modified biochar: a review. *Environmental Pollution*, **273**, 116448, https://doi.org/10.1016/j.envpol.2021.116448

Deng A., Yin Y., Liu Y., Xu Y., He H., Yang S., Qin Q., Sun D. and Li S. (2023). Unlocking the potential of MOF-derived carbon-based nanomaterials for water purification through advanced oxidation processes: a comprehensive review on the impact of process parameter modulation. *Separation and Purification Technology*, **318**, 123998, https://doi.org/10.1016/j.seppur.2023.123998

Duong T. T., Nguyen Q. D., Hong S. K., Kim D., Yoon S. G. and Pham T. H. (2011). Enhanced photoelectrochemical activity of the TiO_2/ITO nanocomposites grown onto single-walled carbon nanotubes at a low temperature by nanocluster deposition. *Advanced Materials*, **23**(46), 5557–5562, https://doi.org/10.1002/adma.201103030

Ealia S. A. M. and Saravanakumar M. P. (2017, November). A review on the classification, characterisation, synthesis of nanoparticles and their application. In: IOP Conference Series: Materials Science and Engineering (Vol. **263**, No. 3). IOP Publishing, p. 032019.

Ebrahimi K., Shiravand S. and Mahmoudvand H. (2017). Biosynthesis of copper nanoparticles using aqueous extract of *Capparis spinosa* fruit and investigation of its antibacterial activity. *Marmara Pharmaceutical Journal*, **21**(4), 866–871, https://doi.org/10.12991/mpj.2017.31

Fang X., Li J., Li X., Pan S., Zhang X., Sun X., Shen J., Han W. and Wang L. (2017). Internal pore decoration with polydopamine nanoparticle on polymeric ultrafiltration membrane for enhanced heavy metal removal. *Chemical Engineering Journal*, **314**, 38–49, https://doi.org/10.1016/j.cej.2016.12.125

Fanourakis S. K., Peña-Bahamonde J., Bandara P. C. and Rodrigues D. F. (2020). Nano-based adsorbent and photocatalyst use for pharmaceutical contaminant removal during indirect potable water reuse. *NPJ Clean Water*, **3**(1), 1–, https://doi.org/10.1038/s41545-019-0048-8

Georgakilas V., Perman J. A., Tucek J. and Zboril R. (2015). Broad family of carbon nanoallotropes: classification, chemistry, and applications of fullerenes, carbon dots, nanotubes, graphene, nanodiamonds, and combined superstructures. *Chemical Reviews*, **115**(11), 4744–4822, https://doi.org/10.1021/cr500304f

Gill S. S., Goyal T., Goswami M., Patel P., Das Gupta G. and Verma S. K. (2023). Remediation of environmental toxicants using carbonaceous materials: opportunity and challenges. *Environmental Science and Pollution Research*, **30**(27), 69727–69750, https://doi.org/10.1007/s11356-023-27364-9

Hairom N. H. H., Soon C. F., Mohamed R. M. S. R., Morsin M., Zainal N., Nayan N., Zulkifli C. Z. and Harun N. H. (2021). A review of nanotechnological applications to detect and control surface water pollution. *Environmental Technology & Innovation*, **24**, 102032, https://doi.org/10.1016/j.eti.2021.102032

Hamid Z. A., Azim A. A., Mouez F. A. and Rehim S. A. (2017). Challenges on synthesis of carbon nanotubes from environmentally friendly green oil using pyrolysis technique. *Journal of Analytical and Applied Pyrolysis*, **126**, 218–229, https://doi.org/10.1016/j.jaap.2017.06.005

Hinds B. J., Chopra N., Rantell T., Andrews R., Gavalas V. and Bachas L. G. (2004). Aligned multiwalled carbon nanotube membranes. *Science*, **303**(5654), 62–65, https://doi.org/10.1126/science.1092048

Ho K. C., Teow Y. H., Ang W. L. and Mohammad A. W. (2017). Novel GO/OMWCNTs mixed-matrix membrane with enhanced antifouling property for palm oil mill effluent treatment. *Separation and Purification Technology*, **177**, 337–349, https://doi.org/10.1016/j.seppur.2017.01.014

Hodges B. C., Cates E. L. and Kim J. H. (2018). Challenges and prospects of advanced oxidation water treatment processes using catalytic nanomaterials. *Nature Nanotechnology*, **13**(8), 642–650, https://doi.org/10.1038/s41565-018-0216-x

Hoseinpour V. and Ghaemi N. (2018). Green synthesis of manganese nanoparticles: applications and future perspective – a review. *Journal of Photochemistry and Photobiology B: Biology*, **189**, 234–243, https://doi.org/10.1016/j.jphotobiol.2018.10.022

Karpińska J. and Kotowska U. (2021). New aspects of occurrence and removal of emerging pollutants. *Water*, **13**(17), 2418, https://doi.org/10.3390/w13172418

Karthigadevi G., Manikandan S., Karmegam N., Subbaiya R., Chozhavendhan S., Ravindran B., Chang S. W. and Awasthi M. K. (2021). Chemico-nanotreatment methods for the removal of persistent organic pollutants and xenobiotics in water – A review. *Bioresource Technology*, **324**, 124678, https://doi.org/10.1016/j.biortech.2021.124678

Khan S., Naushad M., Al-Gheethi A. and Iqbal J. (2021). Engineered nanoparticles for removal of pollutants from wastewater: current status and future prospects of nanotechnology for remediation strategies. *Journal of Environmental Chemical Engineering*, **9**(5), 106160, https://doi.org/10.1016/j.jece.2021.106160

Kim Y., Choudhry Q. N., Chatterjee N. and Choi J. (2018). Immune and xenobiotic response crosstalk to chemical exposure by PA01 infection in the nematode *Caenorhabditis elegans*. *Chemosphere*, **210**, 1082–1090, https://doi.org/10.1016/j.chemosphere.2018.07.031

Kollarahithlu S. C. and Balakrishnan R. M. (2021). Adsorption of pharmaceuticals pollutants, ibuprofen, acetaminophen, and streptomycin from the aqueous phase using amine functionalized superparamagnetic silica nanocomposite. *Journal of Cleaner Production*, **294**, 126155, https://doi.org/10.1016/j.jclepro.2021.126155

Kurniawan T. A., Haider A., Ahmad H. M., Mohyuddin A., Umer Aslam H. M., Nadeem S., Javed M., Othman M. H. D., Goh H. H. and Chew K. W. (2023). Source, occurrence, distribution, fate, and implications of microplastic pollutants in freshwater on environment: a critical review and way forward. *Chemosphere*, **325**, 138367, https://doi.org/10.1016/j.chemosphere.2023.138367

Lazar M. A., Varghese S. and Nair S. S. (2012). Photocatalytic water treatment by titanium dioxide: recent updates. *Catalysts*, **2**(4), 572–601, https://doi.org/10.3390/catal2040572

Lee J., Ye Y., Ward A. J., Zhou C., Chen V., Minett A. I., Sanghyup L., Zongwen L., So-Ryong C. and Shi J. (2016). High flux and high selectivity carbon nanotube composite membranes for natural organic matter removal. *Separation and Purification Technology*, **163**, 109–119, https://doi.org/10.1016/j.seppur.2016.02.032

Li W., Guo C., Su B. and Xu J. (2012). Photodegradation of four fluoroquinolone compounds by titanium dioxide under simulated solar light irradiation. *Journal of Chemical Technology and Biotechnology*, **87**(5), 643–650, https://doi.org/10.1002/JCTB.2759

Lingamdinne L. P., Koduru J. R., Roh H., Choi Y. L., Chang Y. Y. and Yang J. K. (2016). Adsorption removal of Co (II) from waste-water using graphene oxide. *Hydrometallurgy*, **165**, 90–96, https://doi.org/10.1016/j.hydromet.2015.10.021

Liu L., Son M., Chakraborty S., Bhattacharjee C. and Choi H. (2013). Fabrication of ultra-thin polyelectrolyte/carbon nanotube membrane by spray-assisted layer-by-layer technique: characterization and its anti-protein fouling properties for water treatment. *Desalination and Water Treatment*, **51**(31–33), 6194–6200, https://doi.org/10.1080/19443994.2013.780767

Mahmoud A. E. D., Hosny M., El-Maghrabi N. and Fawzy M. (2022). Facile synthesis of reduced graphene oxide by *Tecoma stans* extracts for efficient removal of Ni (II) from water: batch experiments and response surface methodology. *Sustainable Environment Research*, **32**(1), 1–16, https://doi.org/10.1186/S42834-022-00131-0

Mehwish N., Kausar A. and Siddiq M. (2015). High-performance polyvinylidene fluoride/poly(styrene–butadiene–styrene)/functionalized MWCNTs-SCN-Ag nanocomposite membranes. *Iranian Polymer Journal (English Edition)*, **24**(7), 549–559, https://doi.org/10.1007/S13726-015-0346-Z

Mirzaei A., Chen Z., Haghighat F. and Yerushalmi L. (2017). Removal of pharmaceuticals from water by homo/heterogonous Fenton-type processes – a review. *Chemosphere*, **174**, 665–688, https://doi.org/10.1016/j.chemosphere.2017.02.019

Modi S. and Fulekar M. H. (2020). Green synthesis of zinc oxide nanoparticles using garlic skin extract and its characterization. *Journal of Nanostructures*, **10**(1), 20–27.

Mondal P., Nandan A., Ajithkumar S., Siddiqui N. A., Raja S., Kola A. K. and Balakrishnan D. (2023). Sustainable application of nanoparticles in wastewater treatment: fate, current trend & paradigm shift. *Environmental Research*, **232**, 116071, https://doi.org/10.1016/j.envres.2023.116071

Morin-Crini N., Lichtfouse E., Fourmentin M., Ribeiro A. R. L., Noutsopoulos C., Mapelli F., Fenyvesi É, Vieira M. G. A., Picos-Corrales L. A., Moreno-Piraján J. C., Giraldo L., Sohajda T., Huq M. M., Soltan J., Torri G., Magureanu M., Bradu C. and Crini G. (2022). Removal of emerging contaminants from wastewater using advanced treatments. A review. *Environmental Chemistry Letters*, **20**(2), 1333–1375, https://doi.org/10.1007/S10311-021-01379-5

Nasrollahzadeh M., Sajjadi M., Iravani S. and Varma R. S. (2021). Carbon-based sustainable nanomaterials for water treatment: state-of-art and future perspectives. *Chemosphere*, **263**, 128005, https://doi.org/10.1016/j.chemosphere.2020.128005

Ollier R. P., Villanueva M. E., Copello G. J., Alvarez V. A. and Sanchez L. M. (2021). Engineered nanomaterials for emerging contaminant removal from wastewater. In: Kharissova, O. V., Torres-Martínez, L. M. and Kharisov, B. I. (eds), Handbook of Nanomaterials and Nanocomposites for Energy and Environmental Applications. Springer, Cham. https://doi.org/10.1007/978-3-030-36268-3_63

Parida V. K., Saidulu D., Majumder A., Srivastava A., Gupta B. and Gupta A. K. (2021). Emerging contaminants in wastewater: a critical review on occurrence, existing legislations, risk assessment, and sustainable treatment alternatives. *Journal of Environmental Chemical Engineering*, **9**(5), 105966, https://doi.org/10.1016/j.jece.2021.105966

Parveen S., Kumar A., Husain S. and Husain M. (2017). Fowler Nordheim theory of carbon nanotube-based field emitters. *Physica B: Condensed Matter*, **505**, 1–8, https://doi.org/10.1016/j.physb.2016.10.031

Parveen S., Husain S., Husain M. and Zulfequar M. (2022a). Tailoring SWCNTs surface morphology using PEI for highly selective and stable detection of Cu^{2+} heavy metal ion: a nanosensing platform. International Journal of Environmental Analytical Chemistry, 1–14, https://doi.org/10.1080/03067319.2022.2052867

Parveen S., Saifi S., Akram S., Husain M. and Zulfequar M. (2022b). ZnO nanoparticles functionalized SWCNTs as highly sensitive SERS substrate for heavy metal ions detection. *Materials Science in Semiconductor Processing*, **149**, 106852, https://doi.org/10.1016/j.mssp.2022.106852

Patil S. S., Shedbalkar U. U., Truskewycz A., Chopade B. A. and Ball A. S. (2016). Nanoparticles for environmental clean-up: a review of potential risks and emerging solutions. *Environmental Technology & Innovation*, **5**, 10–21, https://doi.org/10.1016/j.eti.2015.11.001

Rahman A., Tan A. L., Harunsani M. H., Ahmad N., Hojamberdiev M. and Khan M. M. (2021). Visible light induced antibacterial and antioxidant studies of ZnO and Cu-doped ZnO fabricated using aqueous leaf extract of *Ziziphus mauritiana* Lam. *Journal of Environmental Chemical Engineering*, **9**(4), 105481, https://doi.org/10.1016/j.jece.2021.105481

Rashid J., Barakat M., Salah N., Barakat M. A. and Habib S. S. (2014, 4, pp. 56892–56899). Ag/ZnO nanoparticles thin films as visible light photocatalysts. RSC Advances, **4**, 56892–56899, https://doi.org/10.1039/C4RA12990C

Rathi B. S., Kumar P. S. and Show P. L. (2021). A review on effective removal of emerging contaminants from aquatic systems: current trends and scope for further research. *Journal of Hazardous Materials*, **409**, 124413, https://doi.org/10.1016/j.jhazmat.2020.124413

Rout P. R., Zhang T. C., Bhunia P. and Surampalli R. Y. (2021). Treatment technologies for emerging contaminants in wastewater treatment plants: a review. *Science of the Total Environment*, **753**, 141990, https://doi.org/10.1016/j.scitotenv.2020.141990

Schulte P. A., McKernan L. T., Heidel D. S., Okun A. H., Dotson G. S., Lentz T. J., Geraci C. L., Heckel P. E. and Branche C. M. (2013). Occupational safety and health, green chemistry, and sustainability: a review of areas of convergence. *Environmental Health: A Global Access Science Source*, **12**(1), 1–9, https://doi.org/10.1186/1476-069X-12-31

Sharma A., Karn R. K. and Pandiyan S. K. (2014). Synthesis of TiO_2 nanoparticles by sol-gel method and their characterization. *Journal of Basic and Applied Engineering Research*, **1**(9), 1–5.

Shawky H. A., Chae S. R., Lin S. and Wiesner M. R. (2011). Synthesis and characterization of a carbon nanotube/polymer nanocomposite membrane for water treatment. *Desalination*, **272**(1–3), 46–50, https://doi.org/10.1016/j.desal.2010.12.051

Singh H., Bamrah A., Bhardwaj S. K., Deep A., Khatri M., Kim K. H. and Bhardwaj N. (2021). Nanomaterial-based fluorescent sensors for the detection of lead ions. *Journal of Hazardous Materials*, **407**, 124379, https://doi.org/10.1016/j.jhazmat.2020.124379

Srirangam G. M. and Rao K. P. (2017). Synthesis and charcterization of silver nanoparticles from the leaf extract of *Malachra capitata* (L.). *Rasayan Journal of Chemistry*, **10**(1), 46–53.

Sufiani O., Sahini M. G. and Elisadiki J. (2023). Towards attaining SDG 6: the opportunities available for capacitive deionization technology to provide clean water to the African population. *Environmental Research*, **216**, 114671, https://doi.org/10.1016/j.envres.2022.114671

Thangadurai, D., Sangeetha, J. and Prasad, R. (Eds.). (2020). Nanotechnology for Food, Agriculture, and Environment. Springer, Berlin/Heidelberg, Germany.

Wen D., Fu R. and Li Q. (2021). Removal of inorganic contaminants in soil by electrokinetic remediation technologies: a review. *Journal of Hazardous Materials*, **401**, 123345, https://doi.org/10.1016/j.jhazmat.2020.123345

Wu Y., Pang H., Liu Y., Wang X., Yu S., Fu D., Chen J. and Wang X. (2019). Environmental remediation of heavy metal ions by novel-nanomaterials: a review. *Environmental Pollution*, **246**, 608–620, https://doi.org/10.1016/j.envpol.2018.12.076

Xu Z., Zhao D., Lu J., Liu J., Dao G., Chen B., Huang B. and Pan X. (2023). Multiple roles of nanomaterials along with their based nanotechnologies in the elimination and dissemination of antibiotic resistance. *Chemical Engineering Journal*, **455**, 140927, https://doi.org/10.1016/j.cej.2022.140927

Yadav S. and Sisodia V. (2022). Synthesis and characterization of zinc oxide nanoparticles for antifungal and antibacterial activity, https://doi.org/10.21203/rs.3.rs-1538022/v1

Yaqoob A. A. and Ibrahim M. N. M. (2019). A review article of nanoparticles; synthetic approaches and wastewater treatment methods. *International Research Journal of Engineering and Technology*, **6**(1–7), 2395.

Yaqoob A. A., Parveen T., Umar K., Nasir M. and Ibrahim M. (2020). Role of nanomaterials in the treatment of wastewater: A review. Water 2020, **12**(2), 495, https://doi.org/10.3390/w12020495

Zhang X., Teng Y., Fu Y., Zhang S., Wang T., Wang C., Jin L. and Zhang W. (2011). Lectin-based electrochemical biosensor constructed by functionalized carbon nanotubes for the competitive assay of glycan expression on living cancer cells. *Chemical Science*, **2**(12), 2353–2360, https://doi.org/10.1039/C1SC00562F

doi: 10.2166/9781789063752_0113

Chapter 7

Treatment approaches for emerging contaminants in sludge and wastewater

Rayane Kunert Langbehn, Felipe Matheus Müller, Elisângela Edila Schneider, Camila Pereira Senna, Eric Sanches-Simões, Júlia Pedó Gutkoski, Maikon Kelbert, Camila Michels and Hugo Moreira Soares*

Department of Chemical Engineering and Food Engineering, Federal University of Santa Catarina, Florianópolis, Santa Catarina 88040-900, Brazil
**Corresponding author: rayane.kunert@posgrad.ufsc.br*

ABSTRACT

The presence of emerging contaminants (ECs) in superficial and drinking water is a reality worldwide. Among them, we can highlight compounds such as personal care products and pharmaceuticals used to treat and prevent human or animal diseases. ECs end up in the environment mainly because of inefficient wastewater or sludge treatment. It is well known that conventional treatment does not completely remove these substances. In addition, ECs pose a potential risk if released into the environment because of their toxicity, recalcitrance, and biouptake in animals and plants. This chapter will cover biological and physicochemical treatments to remove EC from wastewater and sludge. We will address the most recent advances for each process, focusing on their main parameters, operation conditions, and applications. Moreover, we will compare the advantages and disadvantages of each process. This chapter provides a comprehensive understanding of the role of biological and physicochemical processes, applied individually or combined, in treating wastewater and sludge containing EC. Lastly, we will present future perspectives to improve the treatment of ECs in wastewater treatment plants.

Keywords: biological processes, degradation, micropollutants, operating conditions, physicochemical processes

7.1 INTRODUCTION

Urbanization and industrialization in the last decades have increased the demand and use of chemicals in the agriculture, health, and technology sectors. These development processes have promoted several advances, including increasing life expectations and quality. However, this paradigm shift has also introduced several anthropogenic chemical substances into the environment. Recently, the presence of emerging contaminants (ECs) – also known as emerging micropollutants, organic micropollutants, emerging pollutants, and contaminants of emerging concern – in the environment has attracted attention and raised concern (Besha *et al.*, 2017; Dharupaneedi *et al.*, 2019). These xenobiotic compounds are recognized as pollutants, in their majority, because of their potentially harmful effect on human health and the environment (Alvarino *et al.*, 2018; Dubey *et al.*, 2021).

ECs are found in the environment and wastewater treatment plants (WWTPs) at concentrations ranging from nanograms per liter to micrograms per liter (Couto *et al.*, 2019). They are classified according to their source and characteristics (Peña-Guzmán *et al.*, 2019). Some common groups include: (i) pharmaceuticals (e.g., antibiotics, analgesics, and antidepressants); (ii) personal care products (PCPs) (e.g., shampoos, soaps, perfumes, cosmetics, and oral care products); (iii) hormones and similar molecules; (iv) hydrocarbons (combustion of molecules which release polycyclic aromatic hydrocarbon); (v) food additives (synthetic molecules such as sweeteners and antioxidants); (vi) transformation products (TPs) of pharmaceuticals by the human or animal body; and (vii) pesticides.

The first mention of ECs dates back to the last decades of the 20th century. In the 1970s, Hines (1979) reported the presence of pharmaceutical substances in the environment. Ever since, the spread of ECs and concerns around this matter have arisen. All ECs present a substantial risk to the environment; however, their long-term effect is still unknown. Moreover, once they reach the environment, EC might persist for long periods, bioaccumulate in the food chain, and affect biodiversity (Khan *et al.*, 2020; Rempel *et al.*, 2021).

Several examples of the rising EC bioaccumulation in different environments and its influence on the food chain can be found in the literature. Guillette *et al.* (1994) reported that juvenile alligators living in contaminated lakes in Florida exhibited abnormal levels of hormones because of exposure to dicofol and DDT pesticides. These ECs deregulated the endocrine system, affecting the reproductive system and causing the population to decline. Later on, during the 2000s, the vulture population in Pakistan decreased by over 95%. Their death was caused by renal failure because of the anti-inflammatory drug diclofenac, ingested through the food source of dead domestic livestock (Oaks *et al.*, 2004). Moreover, despite the banishment of polychlorinated biphenyls, Jepson *et al.* (2016) demonstrated that European cetacean species continue to have high concentrations of these ECs. According to this study, polychlorinated biphenyls induced reproductive toxicity, causing the long-term population 0 and suppressing cetaceans' population recovery.

Similar to other living beings, the effect of ECs on human health is also a concern because of their direct and indirect risks. Indirectly, exposure to some ECs (e.g., antibiotics) might contribute to the occurrence of antimicrobial resistance genes in bacteria, generating superbugs (Wang *et al.*, 2020). Furthermore, human exposure to EC might also have a harmful direct effect on human health, but the knowledge of long-term exposure to EC in low concentrations is still unknown. Studies indicate that ECs, such as endocrine disruptors, might be related to hormonal disruptions and increased cancer risk in humans (Starling *et al.*, 2018). A major exposure route cited by authorities and the academic community that can lead to human contamination is the presence of ECs in drinking water. Therefore, facing this imminent risk requires that both remediation practices for already-contaminated environments and new alternative substitutes to replace synthetic recalcitrant chemicals become the focal points of future studies.

Regarding remediation practices, it is crucial to implement effective measures to remove ECs from contaminated carriers, such as wastewater and sludge, before their discharge. These implementations require studies on the fate of ECs in WWTPs and parameter optimization for efficient removal. This chapter will cover the removal of ECs through several biological and physicochemical processes for sludge and wastewater treatment (Figure 7.1). For each process, its pros and cons, principal operating conditions, and applications will be presented. Lastly, we will cover the perspectives for EC removal in WWTP.

7.2 BIOLOGICAL PROCESSES

Biological processes rely on microbial activity to remove conventional contaminants and ECs in WWTP. These processes have been used for decades to treat diverse types of waste, for example, municipal, industrial, and agricultural. This versatility of bioprocesses stands out because they usually

Figure 7.1 Biological and physicochemical treatment approaches for ECs in sludge and wastewater.

require mild operating conditions (e.g., temperature, pressure, and pH), implying less energy and chemical inputs. Moreover, some bioprocesses are applied for resource recovery, generating fuels, fertilizers, and other valuable substances from renewable waste.

The primary goal of bioprocesses in WWTP is to remove organic matter and nutrients. Although ECs have a complex structure, they might also be removed by bioprocesses in WWTP. However, the removal efficiency will depend on the ECs physicochemical properties, operating conditions, and the microbial community (Langbehn *et al.*, 2021). Microorganisms may use several mechanisms to remove ECs from wastewater and sludge, as depicted in Figure 7.2. Depending on the microorganism community and the EC, these mechanisms can occur together or separately.

Biodegradation is the most common removal mechanism observed in bioprocesses; it is mediated by enzymes and may occur in the extra or intracellular medium. External biodegradation occurs through enzymes secreted by microorganisms (Figure 7.2f). Conversely, internal biodegradation (Figure 7.2e) uses intracellular enzymes and requires an additional step of bioadsorption (Figure 7.2b) to transfer the EC to the intracellular medium. In the intracellular medium, TPs can either follow the bioaccumulation or biouptake processes on their original structure or after their biodegradation. TPs can also return to the extracellular medium through vesicles. Sometimes, side mechanisms, like photolysis and oxidative degradation, assist biodegradation by partially degrading ECs (Figures 7.2a and 2c) (Gondi *et al.*, 2022; Oberoi *et al.*, 2019).

EC biodegradation occurs via metabolic and co-metabolic pathways, which are determined by EC concentration (Alvarino *et al.*, 2018). The metabolic pathway uses EC as the sole carbon and energy source for microorganisms. Conversely, in the co-metabolic pathway, EC biodegradation is dependent on primary substrate degradation. The primary substrate degradation allows microbial growth,

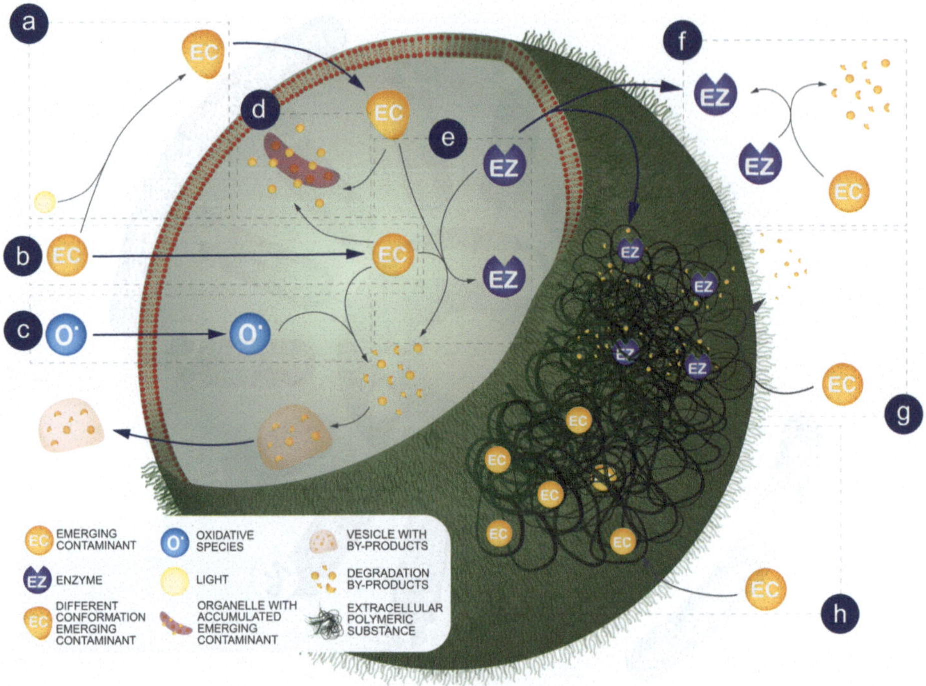

Figure 7.2 General EC removal mechanisms in a generic cellular structure: (a) photolysis, (b) bioadsorption, (c) oxidative degradation, (d) biouptake/bioaccumulation, (e) internal biodegradation, (f) external biodegradation, (g) adsorption and biodegradation, and (h) adsorption.

induces enzyme production, and acts as a source of electron donors for EC biodegradation (Alvarino *et al.*, 2018; James & Vijayanandan, 2023). However, the low concentrations of ECs in wastewater do not provide the required energy for its use as a primary substrate; therefore, their biodegradation occurs mainly via the co-metabolic pathway (Alvarino *et al.*, 2018).

Adsorption is a frequently reported pathway for EC removal in bioprocesses (Figure 7.2h). It occurs through physicochemical interactions between EC and microorganisms or extracellular polymeric substances (EPS). After adsorption in EPS, biodegradation may occur through the action of extracellular enzymes (Figure 7.2g) (Oberoi *et al.*, 2019).

During biological processes, the role of ECs in WWTPs will depend on their capability to be sorbed (absorption or adsorption) for both sludge and wastewater treatment (Besha *et al.*, 2017). When the ECs are ionizable in the medium, adsorption occurs by electrostatic attraction to other surfaces, as indicated by the acid dissociation constant (pK_a) (e.g., positively charged groups interact by electrostatic forces with microorganism surface cells that are negatively charged) (Mai *et al.*, 2018). Conversely, the absorption of ECs happens through hydrophobic interactions. Hydrophobic functional groups interact with the lipid fractions in the medium. In these cases, the octanol–water partition coefficient (K_{OW}) value is a crucial parameter (Zhang *et al.*, 2013).

By understanding the removal mechanisms of EC during biological wastewater treatment, it is possible to evaluate and propose new approaches for effective EC removal from solid and liquid wastes in WWTP. The following sections will elucidate the pros and cons of the most common bioprocesses in a non-exhaustive way. First, we describe the features of conventional processes – technologies already available for industrial implementation and used in WWTP; then, we present non-conventional processes – technologies that are not yet well-established for WWTP (Figure 7.3).

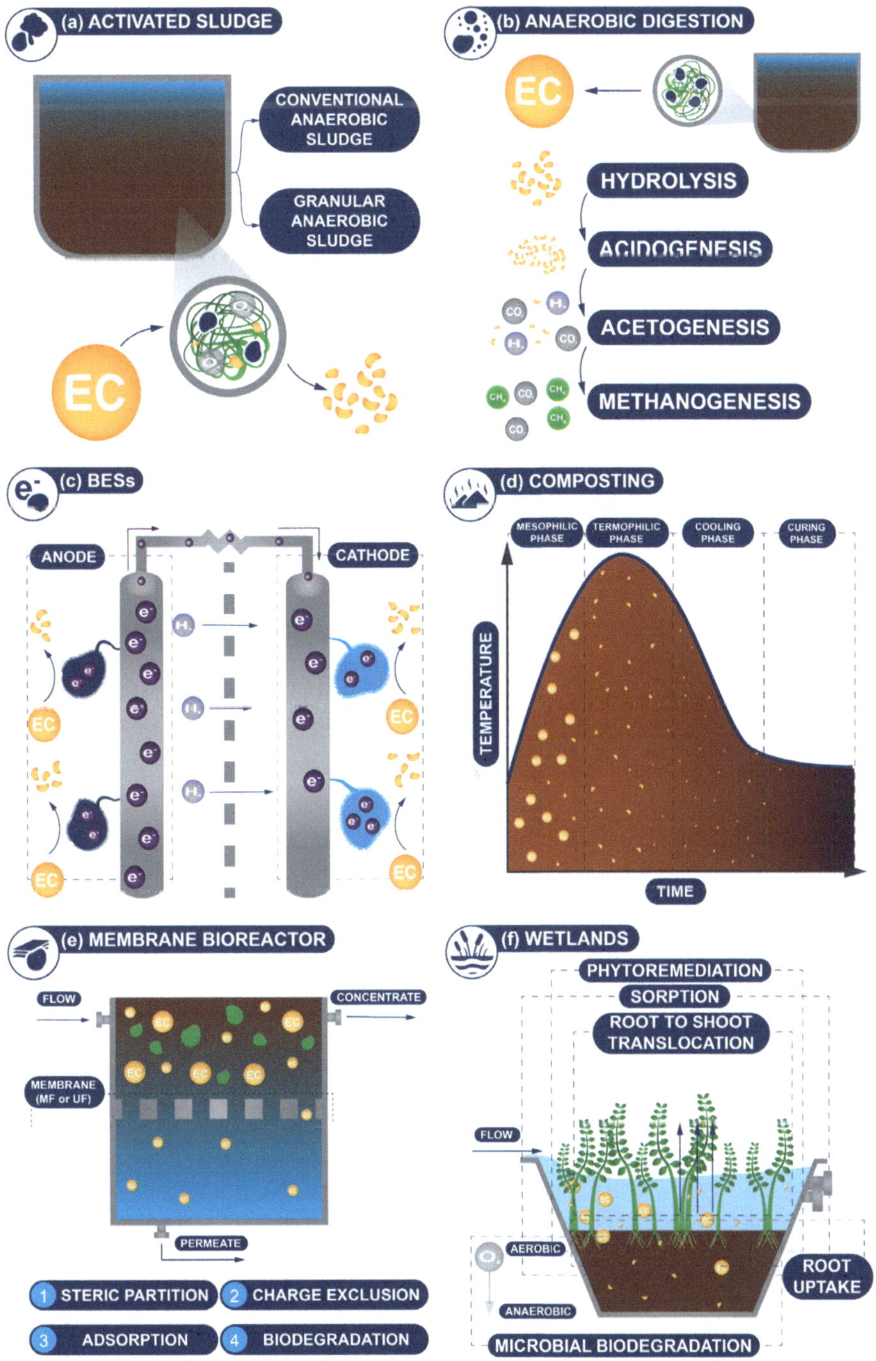

Figure 7.3 Conventional and non-conventional bioprocesses used to treat wastewater and sludge: (a) activated sludge, (b) anaerobic digestion, (c) bioelectrochemical systems, (d) composting, (e) membrane bioreactor, and (f) wetlands.

7.2.1 Conventional

7.2.1.1 Activated sludge

Activated sludge processes, both conventional activated sludge (CAS) and granular activated sludge (GAS), are key components of WWTP (Figure 7.3a). In both CAS and GAS processes, a diverse microbial community works to degrade organic matter. While these biological processes can contribute to reducing any compound classified as an EC, their efficiency in degrading ECs is not optimal and can vary significantly (Burzio *et al.*, 2022). This variation is influenced by factors such as the nature of the contaminants and the operating conditions (i.e., aeration, temperature, pH, inhibitors, and disturbances). For instance, ECs have been shown to persist through physical, chemical, and biological treatment of wastewater (Castaño-Trias *et al.*, 2021; Petrie *et al.*, 2014).

The GAS process is a more recent development. In GAS systems, the microbial community forms compact granules, which can enhance the degradation of ECs and improve process efficiency (Winkler *et al.*, 2018). The granular structure provides a favorable environment for the growth of specialized microorganisms, which can degrade specific ECs resistant to conventional treatment processes (Wilén *et al.*, 2018).

However, while both CAS and GAS processes can reduce the concentration of ECs in wastewater, they may not completely eliminate them. Specific recalcitrant molecules may resist degradation and persist in the treated wastewater. The degradation pathways may also generate TPs that can also be total or partially not degraded during the process (Wang and Wang, 2018). Therefore, further research is needed to enhance the efficiency of these processes and to understand the fate of ECs during treatment.

In the context of sludge treatment, both CAS and GAS are operationally applicable. However, the fate of these contaminants in the sludge is a critical consideration. The EPS present in the sludge can adsorb ECs, potentially retaining them in the sludge. This highlights the need for careful handling and further investigation into the potential environmental and public health risks associated with sludge disposal (Burzio *et al.*, 2022; Petrie *et al.*, 2014).

7.2.1.2 Membrane bioreactor

Membrane bioreactor (MBR) is a hybrid technology used in WWTP that combines mostly CAS and membrane filtration for biodegradation and solid–liquid separation (Figure 7.3e). In MBRs, microfiltration or ultrafiltration membranes are used to substitute the secondary clarifier of CAS. Commercial MBR modules use flat sheet or hollow-fiber membranes in two configurations: (i) submerged membranes (immersed into the biological tank) or (ii) external circulation (side-stream) (Alvarino *et al.*, 2018).

MBR is an alternative to CAS for final effluent quality improvement. A higher sludge concentration can be employed during MBR operation and, consequently, a longer solids retention time (SRT), resulting in a lower space requirement and reducing biomass production (Liu *et al.*, 2022). Besides that, MBR's permeate presents high quality, and there is a complete decoupling of hydraulic retention time (HRT) and SRT (Grandclément *et al.*, 2017). However, MBR requires high energy consumption for its operations (e.g., aeration used to minimize membrane fouling and pressurization for pressure-driven membranes), and ECs are usually not fully removed (Besha *et al.*, 2017). Recently, MBR has gained attention because of the development of new processes that aim to reduce membrane cost and energy consumption (Goswami *et al.*, 2018).

Despite all these advantages, ECs are usually not fully removed in MBR. ECs removal mechanisms by MBR have been widely investigated, which include size exclusion, volatilization, biodegradation, and sorption (Goswami *et al.*, 2018). Depending on the characteristics of ECs, different removal mechanisms can occur: sorption in sludge for apolar and hydrophobic ECs, and biodegradation for polar and hydrophilic ECs (Besha *et al.*, 2017). Overall, the main removal mechanism couples the sequence of those two (sorption followed by biodegradation), being directly dependent on operational conditions (SRT, HRT, temperature, pH, redox conditions, and sludge concentration) (Xiao *et al.*,

2019). Apart from these, secondary mechanisms such as sorption in the membrane and volatilization have minor significance in the removal of ECs by MBR. However, EC removal can also happen by retention in a second layer formed by particles retained in the membrane (Goswami *et al.*, 2018).

Long SRT (>15 days) used in MBR promotes the enrichment of slowly growing microorganisms, which improves the removal of ECs (Khan *et al.*, 2020). Furthermore, other strategies can help increase EC removal efficiency: (i) high HRT for longer EC retention and sorption improvement (Liu *et al.*, 2022); (ii) combination of diverse redox conditions (aerobic, anaerobic, and anoxic) to enhance pollutant biodegradation (Dharupaneedi *et al.*, 2019); and (iii) application of high-retention membranes (i.e., forward osmosis and nanofiltration) to improve EC removal by size exclusion (Liu *et al.*, 2022).

7.2.1.3 Anaerobic digestion

The anaerobic digestion (AD) process involves a consortium of bacteria and archaea living in syntrophism, applicable for both sludge and wastewater treatment (Figure 7.3b). Overall, organic matter is enzymatically hydrolyzed by fermentative bacteria (hydrolysis step). Hydrolysis products go through acidogenesis and acetogenic steps, generating less complex compounds. These compounds are then converted to methane (CH_4) and carbon dioxide (CO_2) (Cremonez *et al.*, 2021). Its application depends on the characteristics of the wastewater and sludge, which directly influence the reactor configuration. Wastewater treatment takes place in AD reactors such as the upflow anaerobic sludge blanket reactor (USAB), and the anaerobic membrane bioreactor (AnMBR). Meanwhile, sludge treatment occurs in digestors, a type of AD reactor normally applied to stabilize waste-activated sludge from municipal WWTPs (Gonzalez-Gil *et al.*, 2020; Oberoi *et al.*, 2019). AD reactors require a smaller installation area and can be employed in decentralized WWTPs to treat sludge and wastewater generated in hospitals, industries, agriculture, and urban areas (Khan *et al.*, 2020).

The affinity of AD for EC removal has been an object of study in the last few years. AD efficiency for EC removal usually follows other biological processes' tendencies toward lower efficiency because of the molecular complexity and low biodegradability. Some ECs might negatively impact AD, reducing the efficiency of organic matter removal and the generation of possible value-added products, such as biogas production (Hube & Wu, 2021). However, its efficiency appears to be higher than that of other processes, such as CAS and GAS. Therefore, ECs can be easily biodegraded or endured after AD. It is suggested that reductive dehalogenation and cleavage of ether bond reactions are predominant, and the presence of electron-withdrawing or electron-donating groups plays a role in the fate of ECs in AD (Akpasi *et al.*, 2023; Dubey *et al.*, 2021; Gonzalez-Gil *et al.*, 2020). This ease of biodegradability is shown in the literature, where higher removal (~80%) was found in ECs with ether bonds and electron-donating groups in their molecules (e.g., acetaminophen, sulfamethoxazole, trimethoprim, and naproxen) when compared to the removal (~30%) of more complex molecules, such as those with esther and multiple cyclic and multiple bonds (e.g., carbamazepine, diclofenac, ibuprofen, and terbutryn). It is important to highlight that recalcitrant ECs and some TPs can remain sorbed in the sludge, which might negatively affect the microbial community (Dubey *et al.*, 2021; Gonzalez-Gil *et al.*, 2020).

In summary, biodegradation of ECs by AD requires compatibility between their chemical structure and the active site of the enzymes produced in one of the four steps of AD (Gonzalez-Gil *et al.*, 2017). Operational strategies to improve AD efficiency to remove ECs are: correct choice of inoculum source to reduce the acclimatation period; careful acclimation period to allow the microbial community adaptation to the sludge or wastewater characteristics; and adjustment of the operational parameters (e.g., temperature, micro-aeration, pH, SRT, and organic loading rate) (Gonzalez-Salgado *et al.*, 2020; Nascimento *et al.*, 2021; Panigrahi & Dubey, 2019; Venegas *et al.*, 2021).

7.2.1.4 Nitrogen removal

Nitrogen removal at WWTP is essential for environmental protection. CAS and GAS can partially remove the nitrogen fraction from wastewater; however, highly concentrated streams require a complementary treatment to meet treated wastewater standards. The conventional treatment for nitrogen removal is the nitrification and denitrification process, which takes place under distinct conditions. Nitrification converts ammonia into nitrate by the action of two autotrophic and aerobic bacteria groups: ammonia-oxidizing bacteria and nitrite-oxidizing bacteria. Then, denitrification occurs by converting nitrate into nitrogen through denitrifying bacteria, which requires heterotrophic and anoxic conditions.

Both nitrification and denitrification contribute to EC removal through biodegradation and adsorption mechanisms (Figure 7.2e, g, and h). However, mechanistic studies have focused on understanding the role of nitrifying bacteria rather than the denitrifying group. EC removal during nitrification has been associated with biodegradation via co-metabolism (James & Vijayanandan, 2023). Several studies suggest that the enzyme ammonium monooxygenase (AMO) acts as a catalyst for EC biodegradation in the presence of ammonia (Alvarino *et al.*, 2018). Nevertheless, AMO might be selective for some ECs owing to its physicochemical properties (e.g., polarity, size, and functional groups) and ability to diffuse across the cell membranes (Dawas-Massalha *et al.*, 2014). Conversely, the role of denitrification in EC removal still needs to be clarified. Some studies found no relevant contribution from denitrifying bacteria on EC removal (Alvarino *et al.*, 2018), while others observed a correlation between denitrifying activity and antibiotic biodegradation (Langbehn *et al.*, 2021; Oberoi *et al.*, 2019).

Aside from bacterial activity, modifications in the nitrification and denitrification process could also impact EC removal. Simultaneous nitrification and denitrification (SND) has been considered a promising treatment for the simultaneous removal of nitrogen and EC (James & Vijayanandan, 2023). SND has economic advantages over conventional nitrification and denitrification, requiring a small implementation area and fewer carbon sources and oxygen inputs. In an SND system, nitrification and denitrification occur in the same reactor, allowing the growth of a wide diversity of microorganisms, developing different redox conditions, and, consequently, novel metabolic pathways for nitrogen removal (James & Vijayanandan, 2023). Likewise, those modifications could influence EC removal in SND (Liu *et al.*, 2017; Sun *et al.*, 2019). Nevertheless, further studies are necessary to understand the removal mechanisms and evaluate if SND improves the EC removal efficiency compared to the conventional nitrification and denitrification process.

7.2.2 Non-conventional

7.2.2.1 Constructed wetlands

Constructed wetlands (CW) is a wastewater treatment with low operating costs and energy input (Xiong *et al.*, 2023). These systems are composed of water, substrate, microorganisms, and plants, which combine physical, chemical, and biological processes to remove organic and inorganic compounds during secondary or tertiary treatment (Figure 7.3f). Different mechanisms, such as volatilization, sorption, plant uptake, photodegradation, and biodegradation can co-occur. These mechanisms can be identified according to the targeted EC through controlling operational conditions (e.g., pH, temperature, HRT, batch or continuous operation mode, and plant species) (Gorito *et al.*, 2017).

CWs can be configured by employing different locations for the water matrices and directions of flow. Some configurations are surface flow, horizontal subsurface flow, and vertical subsurface flow (Gorito *et al.*, 2017), apart from specific configurations such as hydroponic gravel bed configuration (allowing growth without soil) and restoration wetland (employing wastewater or WWTP discharges as matrix). The EC removal in these systems is challenging to predict since multiple factors, such as operational and environmental conditions, vegetation, EC chemical properties, and insolation, can influence it (Verlicchi & Zambello, 2014).

According to Verlicchi and Zambello (2014), high removal percentages for multiple ECs were observed in the literature, that is, acetaminophen, caffeine, ibuprofen, naproxen (>99%), and triclosan (98%), in surface flow CW acting as a primary treatment. Venditti *et al.* (2022) observed a >90% removal efficiency for 27 EC spiked (1–5 μg/ L) in synthetic wastewater in a vertical flow CW that operated for 357 days. CW enhances EC removal mechanisms because of its anoxic–aerobic–anaerobic microregions; as a consequence, CW can remove a broad spectrum of EC.

Despite the good EC removal obtained by CW, more studies are needed to verify the feasibility and applicability of WWTPs with optimized operational conditions to enhance the removal of critical compounds. Looking at it from a more realistic point of view, CWs are suitable for small urban or rural communities because of their low implementation and maintenance costs. Also, it could be utilized as a polishing step in specific locations, such as hospital facilities (Verlicchi & Zambello, 2014).

7.2.2.2 Composting

Composting is a recognized method for managing organic waste, including sludge from WWTP. This process involves the biological decomposition of organic matter under controlled conditions, resulting in a safe soil amendment product (Congilosi & Aga, 2021).

In the context of ECs, composting can play a significant role in their reduction. The microbial communities involved in composting can degrade various ECs, reducing their concentrations in the final compost product (Iranzo *et al.*, 2018). However, the efficiency of this process can vary significantly depending on the nature of the contaminants and the composting conditions.

The composting process can also contribute to reducing ECs in the sludge. However, the fate of these contaminants in the compost and the potential environmental and public health risks associated with compost require further investigation (Kakimoto & Onoda, 2019).

As depicted in Figure 7.3d, the composting process involves several stages, each characterized by different microbial activities and environmental conditions. These stages can influence the degradation of ECs and the overall efficiency of composting (Xia *et al.*, 2005).

Nonetheless, it is important to note that while composting can effectively reduce the concentration of ECs in sludge, it may not completely eliminate all contaminants. Certain ECs may exhibit resistance to degradation and persist in the compost. Therefore, further research is needed to enhance the efficiency of the composting process and to understand the fate of ECs during composting. One of the concerns regarding this residual EC concentration remains the further use of the composting product, such as in agriculture, where the leftover concentration can lead to accumulation in soil and related environments in the long term (Lillenberg *et al.*, 2010).

In summary, composting is a valuable tool for managing sludge from WWTP. It can contribute to the reduction of ECs and the production of a valuable soil amendment. Careful management of the composting process is necessary to ensure the safe use of the resulting compost and to minimize potential environmental and public health risks.

7.2.2.3 Microalgae-mediated processes

Microalgae-mediated processes are considered an alternative to treating wastewater contaminated with EC because of their resistance to harsh environmental conditions, high biomass growth rate, and capacity to adsorb amino and nitro groups (typically present in EC molecules). Treatment costs can also be lower than for other biological processes since the wastewater can already present the nutrients needed for growth and oxygen supply is not required (Gondi *et al.*, 2022; Rempel *et al.*, 2021). In this view, the application of microalgae-mediated processes to remove EC has been reported as a circular economy strategy since CO_2 is fixated during the process, generating a biomass that can produce value-added products (Chhandama *et al.*, 2023).

EC removal by microalgae can occur through five mechanisms: (i) bioadsorption, (ii) bioaccumulation, (iii) biouptake, (iv) biodegradation, and (v) photodegradation (according to Figure 7.2a, b, d, and e). Briefly, (i) EC can be adsorbed into the algal cell wall or the EPS; (ii) it can bind to the intracellular proteins in the non-living cell, and (iii) in the living cell; (iv) complex EC are broken into simpler compounds, a process that is catalyzed by enzymes such as cytochrome P450 and glutathione; if EC could not be bioremediated by the above, it can still be biotransformed through (v) photodegradation, either by the EC reaction with oxidants (photooxidative degradation) or by the absorption of light (photolysis) (Gondi *et al.*, 2022). Over these five mechanisms, biodegradation is reported as the main mechanism for EC removal. Ouada *et al.* (2019) observed that, while analyzing diclofenac removal by *Picocystis* sp., bioadsorption and bioaccumulation contributed only 3.87%, while biodegradation accounted for 69% of the removal efficiency.

Although the microalgal-mediated process presents some cost advantages, some challenges are involved. Contamination with zooplankton and herbivorous protozoa in open tanks can reduce algal biomass production by up to 90% (Gondi *et al.*, 2022). Biomass harvesting after the treatment and its downstream processing are key factors that hinder the scale-up since they account for 20–30% of the total operational cost on the pilot scale. Also, further research about integrated approaches with other treatments for removing EC from wastewater is necessary. More investigations are needed to evaluate the actual applicability of microalgae in this context, such as optimized strains and operational conditions that can maximize EC removal efficiency and biomass production (Gondi *et al.*, 2022).

7.2.2.4 Mycoremediation

Mycoremediation uses fungi to mitigate the impact of pollutants on the environment. A fungus can remove ECs from wastewater by different pathways. It can work as an adsorbent, a phase-transfer process, adsorbing the molecule on its surface, and biotransforming using enzymes, as depicted in Figure 7.2e, f, g, and h (Akpasi *et al.*, 2023). Fungal metabolism works through enzyme production with intra and extracellular mechanisms (Vaksmaa *et al.*, 2023). In both ways, enzymes degrade the compound into different and less harmful TP (Pereira *et al.*, 2020). Among the fungi that can be employed in EC degradation, the white-rot fungi are the most used, being the group responsible for degrading lignin and allowing wood decomposition in nature (Gao *et al.*, 2010).

Researchers demonstrated the *Trametes versicolor* potential to degrade some pesticides, reaching 100% malathion degradation (Hu *et al.*, 2022). In enzymatic exploration, they showed that the intracellular enzyme cytochrome P450 was substantially involved in the process. *Aspergillus terreus* GS28 could adsorb/degrade 98% of azo dye from the sludge of the carpet industry within seven days, showing its potential to decolorize textile wastewater (Hu *et al.*, 2022). Moreover, Agrawal and Shahi (2017) degraded 96.1% of the polycyclic aromatic hydrocarbon pyrene with *Coriolopsis byrsina* APC5 at pH 6 and 25°C.

Cruz-Morató *et al.* (2014) operated a 10-liter pilot-scale reactor filled with *Trametes versicolor* pellets to assess the removal of pharmaceuticals and endocrine disruptors from non-sterile real hospital wastewater. They obtained promising results in the degradation of antibiotics, with removal percentages above 77%. Additionally, the process reached a removal efficiency of 80% and 100% for venlafaxine and diclofenac, respectively. However, other ECs remained resistant to the mycoremediation. For instance, caffeine had a removal rate of 38%.

By combining adsorption, biotransformation, and low substrate specificity, mycoremediation allows the degradation of a wide range of ECs (Vaksmaa *et al.*, 2023). However, its use on an industrial scale is still challenging because of the slow growth of biomass, substrate competition with bacteria, and the need for additional nutrients since wastewater does not provide enough for fungal growth. Furthermore, mycoremediation requires longer HRTs that can last for days, and the fungal growth should be as pellets or on carriers to avoid operational problems associated with fungal dispersion in the liquid medium (Mir-Tutusaus *et al.*, 2018).

7.2.2.5 Enzymatic processes

Enzymes are biocatalysts that can be employed to degrade a broad spectrum of ECs, such as dyes, pesticides, and pharmaceuticals. Direct application of enzymes arose in the last few years as a way to amplify the enzymatic activity that naturally occurs in other biological processes (e.g., CAS/GAS and AD). Its sources are mainly bacteria and fungi (Rao *et al.*, 2014). Within enzymatic processes, oxireductases attracted interest because of the possibility of extracellular secretion, high biocatalyst activity, operational flexibility, and lower substrate selectivity (Zdarta *et al.*, 2018). A research review showed its potential to degrade various pharmaceutical compounds (e.g., non-steroidal hormones, antibiotics, and anticancer drugs) (Pereira *et al.*, 2020).

Studies showed that enzymes can biotransform hazardous substances into less harmful compounds, mitigating their impact on human health and the environment (Ahsan *et al.*, 2021; Pereira *et al.*, 2023). Tyrosinase from *Agaricus sbisporus* completely biotransformed phenol in an aqueous solution after 1 h at 25°C (Shesterenko *et al.*, 2012). Laccase from *Myceliophthora thermophila* could efficiently degrade a mix of pharmaceuticals (anti-inflammatories and estrogen hormones) at trace concentrations within less than 24 h (Lloret *et al.*, 2010). Laccase from *Trametes versicolor* could also efficiently degrade the anticancer drugs etoposide and doxorubicin into less harmful TP (Kelbert *et al.*, 2021; Pereira *et al.*, 2023).

Although enzymes could perform reactions that even other catalysts could not achieve, their use on a large or industrial scale can be tricky. Problems like enzymatic activity loss, biocatalyst reuse, high enzyme cost, and low availability are the main drawbacks to the process's scalability. Therefore, researchers have worked on immobilization techniques to improve the enzymatic process and choose the best support for each application (Daronch *et al.*, 2020). Enzymatic immobilization allows its reuse, increases enzymatic stability, and facilitates its separation from the final product, enabling its use on an industrial scale to mitigate EC in wastewater (Zdarta *et al.*, 2018). Despite several studies aiming to make enzymatic processes feasible, the cost barrier has yet to be overcome; thus, this is not yet a widely applied process. Thereby, even with decades of research, it can be considered an emerging technology for WWTP.

7.2.2.6 Bioelectrochemical systems

Bioelectrochemical systems (BES) are technological platforms based on the ability of some microorganisms to generate an electrical current from the biodegradation of organic matter (Figure 7.3c). This electric current can be employed for several purposes, for example, power generation, synthesis of value-added products, and remediation (Langbehn *et al.*, 2021). The versatility of BES includes the reactor design (single, double, or multiple channels), the process or bioprocess incorporated in the system, and the electrode materials (carbon, metal, or modified).

The types of BES most explored for EC removal are microbial fuel cells (MFC) and microbial electrolysis cells (MEC). Other BES arrangements have also been explored in the EC field, specifically the MFC hybridization with other processes like CW, microalgae, and Fenton (Gupta *et al.*, 2022; Sathe *et al.*, 2022). Sathe *et al.* (2022) analyzed the published literature about MFC coupled with Fenton oxidation (MFC-Fenton) and its potential to remove ECs. This process has been demonstrated to be effective for removing several ECs, for example, dyes, pesticides, pharmaceuticals, and heavy metals. The degradation of most of the EC occurred very rapidly; however, the complete mineralization might require a prolonged reaction time. Moreover, MFC-Fenton requires fewer chemicals and energy inputs to degrade ECs than Fenton oxidation.

As in other bioprocesses aforementioned, biodegradation and adsorption are the main mechanisms involved in EC removal. However, EC adsorption can occur in the biofilm or electrode material (Gupta *et al.*, 2022; Syed *et al.*, 2021). Moreover, BES can exhibit different redox conditions that may play a role in EC removal, improving its performance in contrast to other processes. The increase in current with lower external resistance in MFC contributes to improving the removal efficiency of conventional pollutants. However, the available literature indicates that the best MFC performance to remove ECs might occur with external resistance values close to the MFC internal resistance (Fernando *et al.*, 2014; Jiang *et al.*, 2021).

In the context of WWTP, BES faces physical, energetic, and biological challenges to achieve its commercial implementation (Gupta *et al.*, 2022). Physical challenges comprise scalability and capital costs. Electrodes and membranes commonly used in lab-scale BES are unaffordable for the economic viability of large scale. Moreover, large-scale reactors rarely achieve treatment and energetic efficiencies comparable to lab-scale ones. The energetic challenges include high internal resistance, limited electrode conductivity, and sub-optimal contact between the electrodes.

In summary, BES shows promising results at a lab-scale level for simultaneous EC removal and power generation. However, the mechanisms involved in EC removal, the behavior of mixed-culture bioelectrodes, and the electron transfer mechanism still need to be well elucidated. Likewise, process optimization regarding reactor design, electrode and membrane materials, and energy storage methods is necessary for the large-scale application of BES.

7.3 PHYSICOCHEMICAL PROCESSES

The physicochemical processes for EC removal in sludge and wastewater are presented in this topic. Some conventional and emergent technologies are discussed: advanced oxidation processes (AOPs), adsorption, membrane filtration, and pyrolysis. ECs are converted chemically in AOPs and pyrolysis technologies, while in adsorption and membrane filtration, there is only physical retention, moving the contaminant to another matrix. Figure 7.4 summarizes the mechanisms involved in EC removal/degradation by the physicochemical processes described in this topic.

The physicochemical processes studied for EC removal are generally used as a tertiary treatment in WWTP after a biological treatment. Using these technologies as the primary treatment could significantly increase operational costs. If AOPs were applied to raw wastewater, the free radicals produced would attack not only ECs but all other substances present in the waste (Khan *et al.*, 2020). A pre-treatment should be employed to prevent scaling and fouling in membrane filtration processes, reducing the frequency of cleaning cycles and membrane deterioration (Dharupaneedi *et al.*, 2019). For the adsorption process, saturation of the active sites will happen fast with a higher organic load in the waste, reducing the life-cycle of columns and increasing process costs (regeneration and replacement of the adsorbent). In contrast, the pyrolysis application is more common in sludge treatment and is an effective option for ECs and other substance degradation.

7.3.1 Advanced oxidation processes

AOPs are a set of technologies designed to treat wastewater and sludge by generating highly reactive radicals capable of degrading a wide range of ECs. Overall, these processes are based on the production of different types of free radicals that exert specific oxidation effects on the treated matrix, as depicted in Figure 7.4a.

The most common methods include ozonation, Fenton's process, photoprocesses, catalytic processes, photocatalysis, electrochemical oxidation, persulfate technology, and plasma technology (Oturan & Aaron, 2014; Wang *et al.*, 2023). Each AOP has unique characteristics and operational parameters that influence its effectiveness in treating wastewater and sludge. For instance, while ozonation can degrade a wide range of ECs, it requires a high energy input and can produce harmful by-products (Dubey *et al.*, 2021). On the other hand, Fenton's process is highly effective but can be limited by the need for acidic conditions and sludge production (Gomes *et al.*, 2017). Photocatalysis and electrochemical oxidation are versatile and effective but have limitations, such as the need for ultraviolet (UV) light and a power source, respectively (Lin *et al.*, 2022; Oturan & Aaron, 2014). Despite their effectiveness, the efficiency of AOPs can be influenced by various factors, including the nature and concentration of the ECs, the presence of other substances in the wastewater or sludge, and the operating conditions of the process (Ribeiro *et al.*, 2019).

While AOPs can significantly reduce the concentration of ECs in wastewater and sludge, they may not mineralize all contaminants. Certain ECs may resist degradation and persist in the treated wastewater or sludge, as well as resistant TPs, posing potential risks to the environment and public health (Wang *et al.*, 2023). According to Lin *et al.* (2022), the TPs generated via various pathways (e.g., hydroxylation, dehydrogenation, and ring-opening reactions) during AOPs may be more toxic than the parent compounds. Therefore, while AOPs can significantly reduce the concentration of ECs in wastewater and sludge, the potential formation of toxic TPs and the persistence of certain ECs in the treated wastewater or sludge should be taken into consideration (Lin *et al.*, 2022; Wang *et al.*, 2023).

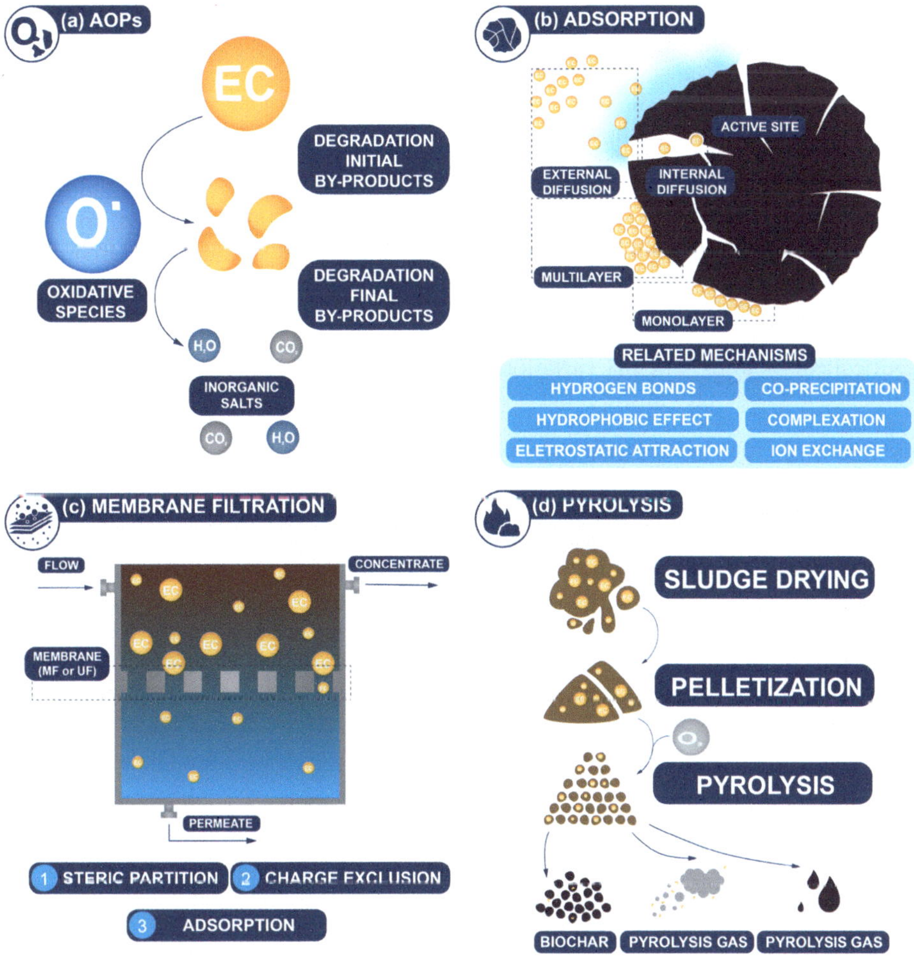

Figure 7.4 Physicochemical processes used to treat wastewater and sludge: (a) advanced oxidation processes, (b) adsorption, (c) membrane filtration, and (d) pyrolysis.

Overall, AOPs offer a promising approach to treating wastewater and sludge-containing ECs. However, further research is needed to optimize these processes, overcome their limitations, and ensure their effective and sustainable application in real-world settings.

7.3.2 Adsorption

Adsorption is a separation process that consists of the accumulation of the adsorbate onto the adsorbent surface through physical and chemical interactions (Bonilla-Petriciolet *et al.*, 2019; Rempel *et al.*, 2021). The efficiency of this process depends on the adsorbent, adsorbate, and solution characteristics, such as surface area, porosity, point of zero charge, chemical structure, pKa, solubility, temperature, and pH. The most crucial choice is which adsorbent will be used since it impacts adsorption capacity and overall treatment cost (Dotto & McKay, 2020).

There are various types of adsorbents, and they can be classified as conventional and unconventional. Conventional adsorbents include activated carbon, zeolites, silica gel, and activated alumina. In contrast, examples of unconventional adsorbents are new materials from agro-industrial wastes and other renewable sources developed mainly to reduce costs (Crini *et al.*, 2019). The process can operate in continuous or batch mode. The continuous process is applied to larger volumes with small physical space in a fixed-bed column, while the batch process optimizes operational conditions (Patel, 2022). One challenge in scaling up is increasing adsorption efficiency by developing new column designs that are feasible for WWTPs (Dotto & McKay, 2020).

Adsorption occurs in three main steps: (i) external diffusion, (ii) intraparticle diffusion, and (iii) surface reaction (Figure 7.4b). Usually, the third step is not the limiting step, so external or intraparticle diffusion controls the total adsorption rate. Studies should be carried out to fully evaluate the adsorption process, such as the effects of pH, kinetics, isotherms, thermodynamics, desorption, and regeneration. In column studies, it should also include the breakthrough curve. Various authors mention the relevance of matrix pH and the adsorbent point of zero charge in EC adsorption, mainly for pharmaceutical compounds. The pH variation can influence pharmaceuticals' functional groups' protonation and promote changes in the superficial adsorbent charge (Quesada *et al.*, 2019).

The adsorption process can be applied as a complementary treatment for removing EC, as it presents high removal efficiencies and a broad spectrum of low-cost adsorbents from renewable sources. However, persistent issues, such as the adsorbents' lifespan, must be addressed to scale up the process. The saturation of adsorbents and prior regeneration can be time- and money-consuming. Also, few papers address the disposal of adsorbents that can no longer be regenerated, which generates another EC-contaminated waste. When treating a complex matrix containing organic loads and other substances, such as wastewater, competition between the organic and inorganic compounds could impair the overall EC removal efficiency. In the literature, articles that perform EC adsorption in synthetic water/wastewater are more common, which denotes an 'unrealistic' scenario (Quesada *et al.*, 2019). Considering these challenges, a careful study should verify the feasibility of applying adsorption as a step in WWTP.

7.3.3 Membrane filtration

Membrane filtration is a physical treatment that uses synthetic membranes (semi-permeable or porous) to remove pollutants. Membrane processes based on pressure-driven operation are classified according to their pore size as microfiltration (MF), ultrafiltration (UF), nanofiltration (NF), and reverse osmosis (RO). ECs removal by membrane filtration in wastewater treatment has been widely investigated in the last two decades (Coccia & Bontempi, 2023). The removal efficiency is dependent on interactions between membrane properties (hydrophobicity, surface charge, pore size, and morphology) and EC characteristics (hydrophobicity, pK_a, size, and charge) (Dharupaneedi *et al.*, 2019).

According to membrane type and contaminants, some removal mechanisms can be involved: size exclusion, sorption, and electrostatic repulsion (Figure 7.4c) (Khanzada *et al.*, 2020). Size or steric exclusion is related to membrane pore size and ECs molecular weight. Sorption is significant only at the beginning of porous membrane system operation and irrelevant after membrane sites' saturation. Most of the driven-pressure membranes used in WWTPs are polymeric, which presents a negatively charged surface that rejects ECs with the same charge.

It is difficult to delineate a single rule for ECs removal by membrane processes since their removal efficiencies depend on the wide range of compounds, their respective characteristics, and the diverse operational conditions (Khanzada *et al.*, 2020). However, a general tendency observed was cited by Kim *et al.* (2018), with an increasing order of ECs removal efficiencies: MF $<$ UF $<$ NF $<$ RO.

In MF and UF processes, EC removal occurs predominantly by sorption mechanisms because the membrane pore size is larger than the overall pollutants' molecular sizes (Khanzada *et al.*, 2020). In those cases, membrane treatment is ineffective in removing more polar and less hydrophobic ECs (Kim *et al.*, 2018). NF and RO systems are the most effective for ECs removal, reaching 70 to $>$90% rejection for the majority of ECs mainly because of the size exclusion mechanism (Dharupaneedi *et al.*,

2019). However, low removal efficiencies were observed for compounds with smaller molecular sizes, opposite charges to the membrane, and elevated log K_{OW} values (Khanzada *et al.*, 2020). High energy consumption is the more relevant disadvantage of these systems, increasing the final operational cost of wastewater treatment.

Despite membrane technology efficacy, there is a gap in ECs removal evaluation in full-scale systems with real wastewater and the presence of multiple compounds (Kim *et al.*, 2018). It will be possible to better define membrane filtration operational conditions for this additional purpose for ECs removal only in this case (Kumar *et al.*, 2023). Furthermore, using membrane technologies for ECs removal does not eliminate the problem because contaminants remain in the concentrated stream and need to be discarded appropriately.

7.3.4 Pyrolysis

Pyrolysis is a thermochemical reductive process that occurs at high temperatures (100–1300°C) and in an oxygen-free atmosphere to transform the organic matter of biosolids into stable and low-molecular substances. The main products of pyrolysis are (i) biochar, a carbon-rich solid that may also retain several nutrients; (ii) syngas, containing mainly CO, CO_2, CH_4, and H_2; and (iii) bio-oil, a dark and viscous liquid mixture of organic compounds and water (Figure 7.4d) (Racek *et al.*, 2020).

Pyrolysis can be applied to the sludge treatment from WWTP, and its potential to remove EC sorbed on biosolids has been investigated in recent years. The main mechanisms involved in EC removal during pyrolysis are volatilization, transformation into other compounds, and mineralization (Buss, 2021). Ross *et al.* (2016) observed that ECs with higher vapor pressures tend to be removed primarily through volatilization. Conversely, ECs with lower vapor pressures are exposed to pyrolytic conditions for a longer time. This extended exposure allows ECs to undergo thermochemical degradation, gradually breaking down into TP until mineralization.

So far, researchers have demonstrated that pyrolysis removes a broad variety of EC with removal efficiencies greater than 95%. Buss (2021) reviewed the degradation conditions of several compounds during the pyrolysis treatment of sludge. The author found strong evidence that an average temperature of 500°C is enough to degrade several classes of EC in sewage sludge, for example, pharmaceuticals, polycyclic aromatic hydrocarbons, hexachlorobenzenes, hormone-like substances, microplastics, and dioxins. Moreover, those studies suggest an ideal residence time of 30–60 min for the degradation of EC during pyrolysis. However, some EC may be completely removed from biosolids with a retention time of 5 min (Ross *et al.*, 2016). Besides the high removal efficiency and the short residence time, another advantage of pyrolysis is that the oxygen-free atmosphere reduces the regeneration of dioxins and polychlorinated biphenyls during the cooling of biochar when compared to other thermochemical processes (Buss, 2021).

The main drawback to treating sludge by pyrolysis is its high energy consumption. The sludge dewatering step is essential for pyrolysis and has high energy consumption. This energy consumption can be up to five times higher than the pyrolysis itself and needs an external energy supply, turning the energy balance negative.

As mentioned before, a fraction of EC and its metabolites may only be shifted to the vapor phase; therefore, a post-treatment of syngas is necessary to remove the volatilized fraction of EC and conclude the degradation. However, this post-treatment may occur in the postcombustors installed in commercial pyrolysis units to oxidize gaseous emissions generated during the sludge treatment (Buss, 2021).

7.4 TREATMENT TRENDS FOR ECS REMOVAL

Conventional WWTP rarely reaches complete EC removal, and, as a result, residual EC and TP are still present in the treated wastewater and sludge. As discussed in Sections 7.2 and 7.3, no single technology can effectively remove and/or degrade all ECs found in wastewater and sludge. Nevertheless, the available literature points out perspectives and trends to enhance EC removal efficiencies with these technologies.

Conventional biological processes in WWTPs could improve EC removal by increasing EC biodegradation. This feature can be reached with the following strategies:

(a) Combining different redox environments: prioritize processes that allow the coexistence of aerobic, anaerobic, and anoxic conditions, like GAS and SND.
(b) Operating parameter optimization: SRT strongly correlates with increased EC biodegradation.
(c) Waste pre-treatment: include a pre-treatment step, for example, hydrothermal, has been demonstrated to enhance EC biodegradation in AD (Díaz *et al.*, 2020).
(d) Employing additives: iron-based additives can regulate reaction conditions, electron transfer mode, and microbial function in AD (He *et al.*, 2022).

Some non-conventional bioprocesses have shown huge potential to improve EC removal and can be considered a trend for WWTP. Enzymatic processes show high EC efficiency removal; however, immobilizing enzymes on appropriate supports is critical to their use in WWTP. Therefore, creating supports and developing reactors that offer more stability and allow effective enzyme reuse is crucial. BES has also stood out due to its ability to remove EC while generating in-situ energy and other value-added products. Moreover, the BES concept can be applied to bioprocesses (e.g., AD, CW, and nitrogen removal) to amend microbial communities and enhance their efficiency. However, both technologies need further studies to understand EC removal in actual scenarios, that is, multiple EC removal, real wastewater, and scale-up.

The application of physicochemical processes to remove EC in WWTP can be favored by developing new and advanced materials, for example, catalyzers, adsorbents, and membranes. Membranes stand out for their versatility in WWTP, being applied in physicochemical and biological treatments. They contribute to EC removal by size-exclusion mechanisms and also reduce the area required for the treatment. Trends in membrane development are focused on increasing membrane rejection and lifespan. Studies suggest that using high-retention membranes and developing advanced membranes with mixed-based matrices and active layer functionalization enhance membrane rejection (Coccia & Bontempi, 2023; Dharupaneedi *et al.*, 2019; Khanzada *et al.*, 2020). Meanwhile, forward osmosis and membrane distillation have the potential to reduce membrane fouling and increase EC removal (Goswami *et al.*, 2018).

Despite all the previously mentioned strategies to improve EC removal in each process, there is a consensus that a single technology will not solve the problem of EC in WWTP, and the best approach would be to combine different processes (Besha *et al.*, 2017; Langbehn *et al.*, 2021). This approach enables the application of complementary removal mechanisms (e.g., sorption, biodegradation, chemical oxidation, and size exclusion) to remove and mineralize EC and improve the final effluent quality. Recently, promising results were obtained with the following configurations: biological treatment + membrane filtration (NF or RO) (Besha *et al.*, 2017); MBR + AOPs (Liu *et al.*, 2022); AOPs + adsorption (Kumar *et al.*, 2023); anaerobic–anoxic–oxic systems (AAO) (Ashfaq *et al.*, 2017); and A2/O-MBR + UV/chlorine (Ren *et al.*, 2022).

In summary, several processes can potentially mitigate EC contamination in wastewater and sludge, and the removal efficiency is enhanced with optimized configurations, operational conditions, and advanced materials. Nevertheless, it is worth noting that most of the studies published so far explore the removal of pollutants in lab-scale and ideal conditions (i.e., one or similar ECs and synthetic wastewater). While these studies provide insight into the fundamentals of removing EC, their findings cannot be applied to real-life scenarios, which are far more complex than those simulated in these studies. Therefore, it is also necessary to encourage further research to understand EC removal in real-life scenarios to advance this topic.

REFERENCES

Agrawal N. and Shahi S. K. (2017). Degradation of polycyclic aromatic hydrocarbon (pyrene) using novel fungal strain Coriolopsis byrsina strain APC5. *International Biodeterioration & Biodegradation*, **122**, 69–81, https://doi.org/10.1016/j.ibiod.2017.04.024

Ahsan Z., Kalsoom U., Bhatti H. N., Aftab K., Khalid N. and Bilal M. (2021). Enzyme-assisted bioremediation approach for synthetic dyes and polycyclic aromatic hydrocarbons degradation. *Journal of Basic Microbiology*, **61**(11), 960–981, https://doi.org/10.1002/jobm.202100218

Akpasi S. O., Anekwe I. M. S., Tetteh E. K., Amune U. O., Shoyiga H. O., Mahlangu T. P. and Kiambi S. L. (2023). Mycoremediation as a potentially promising technology: current status and prospects – a review. *Applied Sciences*, **13**(8), 4978, https://doi.org/10.3390/app13084978

Alvarino T., Suarez S., Lema J. and Omil F. (2018). Understanding the sorption and biotransformation of organic micropollutants in innovative biological wastewater treatment technologies. *Science of the Total Environment*, **615**, 297–306, https://doi.org/10.1016/j.scitotenv.2017.09.278

Ashfaq M., Li Y., Wang Y., Chen W., Wang H., Chen X., Wu W., Huang Z., Yu C. P. and Sun Q. (2017). Occurrence, fate, and mass balance of different classes of pharmaceuticals and personal care products in an anaerobic-anoxic-oxic wastewater treatment plant in Xiamen, China. *Water Research*, **123**, 655–667, https://doi.org/10.1016/j.watres.2017.07.014

Besha A. T., Gebreyohannes A. Y., Tufa R. A., Bekele D. N., Curcio E. and Giorno L. (2017). Removal of emerging micropollutants by activated sludge process and membrane bioreactors and the effects of micropollutants on membrane fouling: a review. *Journal of Environmental Chemical Engineering*, **5**(3), 2395–2414, https://doi.org/10.1016/j.jece.2017.04.027

Bonilla-Petriciolet A., Mendoza-Castillo D. I., Dotto G. L. and Duran-Valle C. J. (2019). Adsorption in water treatment. Reference Module in Chemistry, Molecular Sciences and Chemical Engineering. Cent Conf. https://doi.org/10.1016/b978-0-12-409547-2.14390-2 (accessed on May 26, 2023)

Burzio C., Ekholm J., Modin O., Falås P., Svahn O., Persson F., van Erp T., Gustavsson D. J. I. and Wilén B. M. (2022). Removal of organic micropollutants from municipal wastewater by aerobic granular sludge and conventional activated sludge. *Journal of Hazardous Materials*, **438**(June), 129528, https://doi.org/10.1016/j.jhazmat.2022.129528

Buss W. (2021). Pyrolysis solves the issue of organic contaminants in sewage sludge while retaining carbon – making the case for sewage sludge treatment via pyrolysis. *ACS Sustainable Chemistry and Engineering*, **9**(30), 10048–10053, https://doi.org/10.1021/acssuschemeng.1c03651

Castaño-Trias M., Brienza M., Tomei M. C. and Buttiglieri G. (2021). Fate and removal of pharmaceuticals in CAS for water and sewage sludge reuse. *Handbook of Environmental Chemistry*, **108**, 23–51.

Chhandama M. V. L., Rai P. K. and Lalawmpuii. (2023). Coupling bioremediation and biorefinery prospects of microalgae for circular economy. *Bioresource Technology Reports*, **22**, 101479, https://doi.org/10.1016/j.biteb.2023.101479

Coccia M. and Bontempi E. (2023). New trajectories of technologies for the removal of pollutants and emerging contaminants in the environment. *Environmental Research*, **229**, 115938, https://doi.org/10.1016/j.envres.2023.115938

Congilosi J. L. and Aga D. S. (2021). Review on the fate of antimicrobials, antimicrobial resistance genes, and other micropollutants in manure during enhanced anaerobic digestion and composting. *Journal of Hazardous Materials*, **405**(July 2020), 123634, https://doi.org/10.1016/j.jhazmat.2020.123634

Couto C. F., Lange L. C. and Amaral M. C. S. (2019). Occurrence, fate and removal of pharmaceutically active compounds (PhACs) in water and wastewater treatment plants – a review. *Journal of Water Process Engineering*, **32**, 100927, https://doi.org/10.1016/j.jwpe.2019.100927

Cremonez P. A., Teleken J. G., Weiser Meier T. R. and Alves H. J. (2021). Two-stage anaerobic digestion in agroindustrial waste treatment: a review. *Journal of Environmental Management*, **281**, 111854, https://doi.org/10.1016/j.jenvman.2020.111854

Crini G., Lichtfouse E., Wilson L. D. and Morin-Crini N. (2019). Conventional and non-conventional adsorbents for wastewater treatment. *Environmental Chemistry Letters*, **17**(1), 195–213, https://doi.org/10.1007/s10311-018-0786-8

Cruz-Morató C., Lucas D., Llorca M., Rodriguez-Mozaz S., Gorga M., Petrovic M., Barceló D., Vicent T., Sarrà M. and Marco-Urrea E. (2014). Hospital wastewater treatment by fungal bioreactor: removal efficiency for pharmaceuticals and endocrine disruptor compounds. *Science of the Total Environment*, **493**, 365–376, https://doi.org/10.1016/j.scitotenv.2014.05.117

Daronch N. A., Kelbert M., Pereira C. S., de Araújo P. H. H. and de Oliveira D. (2020). Elucidating the choice for a precise matrix for laccase immobilization: a review. *Chemical Engineering Journal*, **397**, 125506, https://doi.org/10.1016/j.cej.2020.125506

Dawas-Massalha A., Gur-Reznik S., Lerman S., Sabbah I. and Dosoretz C. G. (2014). Co-metabolic oxidation of pharmaceutical compounds by a nitrifying bacterial enrichment. *Bioresource Technology*, **167**, 336–342, https://doi.org/10.1016/J.BIORTECH.2014.06.003

Dharupaneedi S. P., Nataraj S. K., Nadagouda M., Reddy K. R., Shukla S. S. and Aminabhavi T. M. (2019). Membrane-based separation of potential emerging pollutants. *Separation and Purification Technology*, **210**, 850–866, https://doi.org/10.1016/j.seppur.2018.09.003

Díaz I., Díaz-Curbelo A., Pérez-Lemus N., Fdz-Polanco F. and Pérez-Elvira S. I. (2020). Traceability of organic contaminants in the sludge line of wastewater treatment plants: a comparison study among schemes incorporating thermal hydrolysis treatment and the conventional anaerobic digestion. *Bioresource Technology*, **305**, 123028, https://doi.org/10.1016/j.biortech.2020.123028

Dotto G. L. and McKay G. (2020). Current scenario and challenges in adsorption for water treatment. *Journal of Environmental Chemical Engineering*, **8**(4), 103988, https://doi.org/10.1016/j.jece.2020.103988

Dubey M., Mohapatra S., Tyagi V. K., Suthar S. and Kazmi A. A. (2021). Occurrence, fate, and persistence of emerging micropollutants in sewage sludge treatment. *Environmental Pollution*, **273**, 116515, https://doi.org/10.1016/j.envpol.2021.116515

Fernando E., Keshavarz T. and Kyazze G. (2014). External resistance as a potential tool for influencing azo dye reductive decolourisation kinetics in microbial fuel cells. *International Biodeterioration & Biodegradation*, **89**, 7–14, https://doi.org/10.1016/j.ibiod.2013.12.011

Gao D., Du L., Yang J., Wu W. M. and Liang H. (2010). A critical review of the application of white rot fungus to environmental pollution control. *Critical Reviews in Biotechnology*, **30**(1), 70–77, https://doi.org/10.3109/07388550903427272

Gomes J., Costa R., Quinta-Ferreira R. M. and Martins R. C. (2017). Application of ozonation for pharmaceuticals and personal care products removal from water. *Science of the Total Environment*, **586**, 265–283, https://doi.org/10.1016/j.scitotenv.2017.01.216

Gondi R., Kavitha S., Yukesh Kannah R., Parthiba Karthikeyan O., Kumar G., Kumar Tyagi V. and Rajesh Banu J. (2022). Algal-based system for removal of emerging pollutants from wastewater: a review. *Bioresource Technology*, **344**, 126245, https://doi.org/10.1016/j.biortech.2021.126245

Gonzalez-Gil L., Carballa M. and Lema J. M. (2017). Cometabolic enzymatic transformation of organic micropollutants under methanogenic conditions. *Environmental Science and Technology*, **51**(5), 2963–2971, https://doi.org/10.1021/acs.est.6b05549

Gonzalez-Gil L., Carballa M. and Lema J. M. (2020). Removal of organic micro-pollutants by anaerobic microbes and enzymes. In: Current Developments in Biotechnology and Bioengineering. Elsevier, Amsterdam, pp. 397–426, https://doi.org/10.1016/b978-0-12-819594-9.00016-4 (accessed on June 15, 2023)

Gonzalez-Salgado I., Cavaillé L., Dubos S., Mengelle E., Kim C., Bounouba M., Paul E., Pommier S. and Bessiere Y. (2020). Combining thermophilic aerobic reactor (TAR) with mesophilic anaerobic digestion (MAD) improves the degradation of pharmaceutical compounds. *Water Research*, **182**, 116033, https://doi.org/10.1016/j.watres.2020.116033

Gorito A. M., Ribeiro A. R., Almeida C. M. R. and Silva A. M. T. (2017). A review on the application of constructed wetlands for the removal of priority substances and contaminants of emerging concern listed in recently launched EU legislation. *Environmental Pollution*, **227**, 428–443, https://doi.org/10.1016/j.envpol.2017.04.060

Goswami L., Vinoth Kumar R., Borah S. N., Arul Manikandan N., Pakshirajan K. and Pugazhenthi G. (2018). Membrane bioreactor and integrated membrane bioreactor systems for micropollutant removal from wastewater: a review. *Journal of Water Process Engineering*, **26**, 314–328, https://doi.org/10.1016/j.jwpe.2018.10.024

Grandclément C., Seyssiecq I., Piram A., Wong-Wah-Chung P., Vanot G., Tiliacos N., Roche N. and Doumenq P. (2017). From the conventional biological wastewater treatment to hybrid processes, the evaluation of organic micropollutant removal: a review. *Water Research*, **111**, 297–317, https://doi.org/10.1016/j.watres.2017.01.005

Guillette L. J., Gross T. S., Masson G. R., Matter J. M., Percival H. F. and Woodward A. R. (1994). Developmental abnormalities of the gonad and abnormal sex hormone concentrations in juvenile alligators from contaminated and control lakes in Florida. *Environmental Health Perspectives*, **102**(8), 680–688, https://doi.org/10.1289/ehp.94102680

Gupta S. K., Rachna Singh B., Mungray A. K., Bharti R., Nema A. K., Pant K. K. and Mulla S. I. (2022). Bioelectrochemical technologies for removal of xenobiotics from wastewater. *Sustainable Energy Technologies and Assessments*, **49**, 101652, https://doi.org/10.1016/j.seta.2021.101652

He Z. W., Zou Z. S., Ren Y. X., Tang C. C., Zhou A. J., Liu W., Wang L., Li Z. and Wang A. (2022). Roles of zero-valent iron in anaerobic digestion: mechanisms, advances and perspectives. *Science of the Total Environment*, **852**, 158420, https://doi.org/10.1016/j.scitotenv.2022.158420

Hines T. C. (1979). The past and present status of the alligator in Florida. *Proceedings of the Annual Conference Southeast. Association of Fish & Wildlife Agencies*, **33**, 224–232.

Hu K., Barbieri M. V., López-García E., Postigo C., López de Alda M., Caminal G. and Sarrà M. (2022). Fungal degradation of selected medium to highly polar pesticides by *Trametes versicolor*: kinetics, biodegradation pathways, and ecotoxicity of treated waters. *Analytical and Bioanalytical Chemistry*, **414**(1), 439–449, https://doi.org/10.1007/s00216-021-03267-x

Hube S. and Wu B. (2021). Mitigation of emerging pollutants and pathogens in decentralized wastewater treatment processes: a review. *Science of the Total Environment*, **779**, 146545, https://doi.org/10.1016/j.scitotenv.2021.146545

Iranzo M., Gamón M., Boluda R. and Mormeneo S. (2018). Analysis of pharmaceutical biodegradation of WWTP sludge using composting and identification of certain microorganisms involved in the process. *Science of the Total Environment*, **640–641**, 840–848, https://doi.org/10.1016/j.scitotenv.2018.05.366

James S. N. and Vijayanandan A. (2023). Recent advances in simultaneous nitrification and denitrification for nitrogen and micropollutant removal: a review. *Biodegradation*, **34**(2), 103–123, https://doi.org/10.1007/s10532-023-10015-8

Jepson P. D., Deaville R., Barber J. L., Aguilar À., Borrell A., Murphy S., Barry J., Brownlow A., Barnett J., Berrow S., Cunningham A. A., Davison N. J., Ten Doeschate M., Esteban R., Ferreira M., Foote A. D., Genov T., Giménez J., Loveridge J., Llavona À., Martin V., Maxwell D. L., Papachlimitzou A., Penrose R. and Law R. J. (2016). PCB pollution continues to impact populations of orcas and other dolphins in European waters. *Scientific Reports 2016 6:1*, **6**(1), 1–17, https://doi.org/10.1038/srep18573

Jiang J., Wang H., Zhang S., Li S., Zeng W. and Li F. (2021). The influence of external resistance on the performance of microbial fuel cell and the removal of sulfamethoxazole wastewater. *Bioresource Technology*, **336**, 125308, https://doi.org/10.1016/j.biortech.2021.125308

Kakimoto T. and Onoda Y. (2019). Fate of pharmaceuticals in composting process. In: Resource-Oriented Agro-Sanitation Systems, N. Funamizu (ed.), Springer, Tokyo, pp. 79–96, https://doi.org/10.1007/978-4-431-56835-3_6 (accessed on June 13, 2023)

Kelbert M., Pereira C. S., Daronch N. A., Cesca K., Michels C., de Oliveira D. and Soares H. M. (2021). Laccase as an efficacious approach to remove anticancer drugs: a study of doxorubicin degradation, kinetic parameters, and toxicity assessment. *Journal of Hazardous Materials*, **409**, 124520, https://doi.org/10.1016/j.jhazmat.2020.124520

Khan N. A., Khan S. U., Ahmed S., Farooqi I. H., Yousefi M., Mohammadi A. A. and Changani F. (2020). Recent trends in disposal and treatment technologies of emerging-pollutants – a critical review. *Trends in Analytical Chemistry*, **122**, 115744, https://doi.org/10.1016/j.trac.2019.115744

Khanzada N. K., Farid M. U., Kharraz J. A., Choi J., Tang C. Y., Nghiem L. D., Jang A. and An A. K. (2020). Removal of organic micropollutants using advanced membrane-based water and wastewater treatment: a review. *Journal of Membrane Science*, **598**, 117672, https://doi.org/10.1016/j.memsci.2019.117672

Kim S., Chu K. H., Al-Hamadani Y. A. J., Park C. M., Jang M., Kim D. H., Yu M., Heo J. and Yoon Y. (2018). Removal of contaminants of emerging concern by membranes in water and wastewater: a review. *Chemical Engineering Journal*, **335**, 896–914, https://doi.org/10.1016/j.cej.2017.11.044

Kumar M., Sridharan S., Sawarkar A. D., Shakeel A., Anerao P., Mannina G., Sharma P. and Pandey A. (2023). Current research trends on emerging contaminants pharmaceutical and personal care products (PPCPs): a comprehensive review. *Science of the Total Environment*, **859**, 160031, https://doi.org/10.1016/j.scitotenv.2022.160031

Langbehn R. K., Michels C. and Soares H. M. (2021). Antibiotics in wastewater: from its occurrence to the biological removal by environmentally conscious technologies. *Environmental Pollution*, **275**, 116603, https://doi.org/10.1016/j.envpol.2021.116603

Lillenberg M., Yurchenko S., Kipper K., Herodes K., Pihl V., Lõhmus R., Ivask M., Kuu A., Kutti S., Litvin S. V. and Nei L. (2010). Presence of fluoroquinolones and sulfonamides in urban sewage sludge and their degradation as a result of composting. *International Journal of Environmental Science and Technology*, **7**(2), 307–312, https://doi.org/10.1007/BF03326140

Lin W., Liu X., Ding A., Ngo H. H., Zhang R., Nan J., Ma J. and Li G. (2022). Advanced oxidation processes (AOPs)-based sludge conditioning for enhanced sludge dewatering and micropollutants removal: a critical review. *Journal of Water Process Engineering*, **45**, 102468, https://doi.org/10.1016/j.jwpe.2021.102468

Liu J., Wang J., Zhao C., Liu J., Xie H., Wang S., Zhang J. and Hu Z. (2017). Performance and mechanism of triclosan removal in simultaneous nitrification and denitrification (SND) process under low-oxygen condition. *Applied Microbiology and Biotechnology*, **101**(4), 1653–1660, https://doi.org/10.1007/s00253-016-7952-3

Liu W., Song X., Na Z., Li G. and Luo W. (2022). Strategies to enhance micropollutant removal from wastewater by membrane bioreactors: recent advances and future perspectives. *Bioresource Technology*, **344**(Part B), 126322, https://doi.org/10.1016/j.biortech.2021.126322

Lloret L., Eibes G., Lú-Chau T. A., Moreira M. T., Feijoo G. and Lema J. M. (2010). Laccase-catalyzed degradation of anti-inflammatories and estrogens. *Biochemical Engineering Journal*, **51**(3), 124–131, https://doi.org/10:1016/j.bej.2010.06.005

Mai D. T., Stuckey D. C. and Oh S. (2018). Effect of ciprofloxacin on methane production and anaerobic microbial community. *Bioresource Technology*, **261**, 240–248, https://doi.org/10.1016/j.biortech.2018.04.009

Mir-Tutusaus J. A., Baccar R., Caminal G. and Sarrà M. (2018). Can white-rot fungi be a real wastewater treatment alternative for organic micropollutants removal? A review. *Water Research*, **138**, 137–151, https://doi.org/10.1016/j.watres.2018.02.056

Nascimento J. G. da S. do, Silva E. V. A., dos Santos A. B., da Silva M. E. R. and Firmino P. I. M. (2021). Microaeration improves the removal/biotransformation of organic micropollutants in anaerobic wastewater treatment systems. *Environmental Research*, **198**, 111313, https://doi.org/10.1016/j.envres.2021.111313

Oaks J. L., Gilbert M., Virani M. Z., Watson R. T., Meteyer C. U., Rideout B. A., Shivaprasad H. L., Ahmed S., Chaudhry M. J. I., Arshad M., Mahmood S., Ali A. and Khan A. A. (2004). Diclofenac residues as the cause of vulture population decline in Pakistan. *Nature*, **427**(6975), 630–633, https://doi.org/10.1038/nature02317

Oberoi A. S., Jia Y., Zhang H., Khanal S. K. and Lu H. (2019). Insights into the fate and removal of antibiotics in engineered biological treatment systems: a critical review. *Environmental Science and Technology*, **53**(13), 7234–7264, https://doi.org/10.1021/acs.est.9b01131

Oturan M. A. and Aaron J. J. (2014). Advanced oxidation processes in water/wastewater treatment: principles and applications. A review. *Critical Reviews in Environmental Science and Technology*, **44**(23), 2577–2641, https://doi.org/10.1080/10643389.2013.829765

Ouada S. Ben, Ali R. Ben, Cimetiere N., Leboulanger C., Ouada H. Ben and Sayadi S. (2019). Biodegradation of diclofenac by two green microalgae: *Picocystis* sp. and *Graesiella* sp. *Ecotoxicology and Environmental Safety*, **186**, 109769, https://doi.org/10.1016/j.ecoenv.2019.109769

Panigrahi S. and Dubey B. K. (2019). A critical review on operating parameters and strategies to improve the biogas yield from anaerobic digestion of organic fraction of municipal solid waste. *Renewable Energy*, **143**, 779–797, https://doi.org/10.1016/j.renene.2019.05.040

Patel H. (2022). Comparison of batch and fixed bed column adsorption: a critical review. *International Journal of Environmental Science and Technology*, **19**(10), 10409–10426, https://doi.org/10.1007/s13762-021-03492-y

Peña-Guzmán C., Ulloa-Sánchez S., Mora K., Helena-Bustos R., Lopez-Barrera E., Alvarez J. and Rodriguez-Pinzón M. (2019). Emerging pollutants in the urban water cycle in Latin America: a review of the current literature. *Journal of Environmental Management*, **237**, 408–423, https://doi.org/10.1016/j.jenvman.2019.02.100

Pereira C. S., Kelbert M., Daronch N. A., Michels C., de Oliveira D. and Soares H. M. (2020). Potential of enzymatic process as an innovative technology to remove anticancer drugs in wastewater. *Applied Microbiology and Biotechnology*, **104**(1), 23–31, https://doi.org/10.1007/s00253-019-10229-y

Pereira C. S., Kelbert M., Daronch N. A., Cordeiro A. P., Cesca K., Michels C., de Oliveira D. and Soares H. M. (2023). Laccase-assisted degradation of anticancer drug etoposide: by-products and cytotoxicity. *Bioenergy Research*, **1**, 1–10.

Petrie B., McAdam E. J., Lester J. N. and Cartmell E. (2014). Assessing potential modifications to the activated sludge process to improve simultaneous removal of a diverse range of micropollutants. *Water Research*, **62**, 180–192, https://doi.org/10.1016/j.watres.2014.05.036

Quesada H. B., Baptista A. T. A., Cusioli L. F., Seibert D., de Oliveira Bezerra C. and Bergamasco R. (2019). Surface water pollution by pharmaceuticals and an alternative of removal by low-cost adsorbents: a review. *Chemosphere*, **222**, 766–780, https://doi.org/10.1016/j.chemosphere.2019.02.009

Racek J., Sevcik J., Chorazy T., Kucerik J. and Hlavinek P. (2020). Biochar – recovery material from pyrolysis of sewage sludge: a review. *Waste and Biomass Valorization*, **11**(7), 3677–3709, https://doi.org/10.1007/s12649-019-00679-w

Rao M. A., Scelza R., Acevedo F., Diez M. C. and Gianfreda L. (2014). Enzymes as useful tools for environmental purposes. *Chemosphere*, **107**, 145–162, https://doi.org/10.1016/j.chemosphere.2013.12.059

Rempel A., Gutkoski J. P., Nazari M. T., Biolchi G. N., Cavanhi V. A. F., Treichel H. and Colla L. M. (2021). Current advances in microalgae-based bioremediation and other technologies for emerging contaminants treatment. *Science of the Total Environment*, **772**, 144918, https://doi.org/10.1016/j.scitotenv.2020.144918

Ren X., Wang Q., Chen H., Dai X. and He Q. (2022). Removal effect and mechanism of typical pharmaceuticals and personal care products by AAO-MBR and UV/chlorine in black water. *Journal of Cleaner Production*, **346**, 131104, https://doi.org/10.1016/j.jclepro.2022.131104

Ribeiro A. R. L., Moreira N. F. F., Li Puma G. and Silva A. M. T. (2019). Impact of water matrix on the removal of micropollutants by advanced oxidation technologies. *Chemical Engineering Journal*, **363**, 155–173, https://doi.org/10.1016/j.cej.2019.01.080

Ross J. J., Zitomer D. H., Miller T. R., Weirich C. A. and McNamara P. J. (2016). Emerging investigators series: pyrolysis removes common microconstituents triclocarban, triclosan, and nonylphenol from biosolids. *Environmental Science: Water Research & Technology*, **2**(2), 282–289, https://doi.org/10.1039/C5EW00229J

Sathe S. M., Chakraborty I., Dubey B. K. and Ghangrekar M. M. (2022). Microbial fuel cell coupled Fenton oxidation for the cathodic degradation of emerging contaminants from wastewater: applications and challenges. *Environmental Research*, **204**, 112135, https://doi.org/10.1016/j.envres.2021.112135

Shesterenko Y. A., Sevast'yanov O. V. and Romanovskaya I. I. (2012). Removal of phenols from aqueous solutions using tyrosinase immobilized on polymer carriers and inorganic coagulants. *Journal of Water Chemistry and Technology*, **34**(2), 107–111, https://doi.org/10.3103/S1063455X12020063

Starling M. C. V. M., Amorim C. C. and Leão M. M. D. (2018). Occurrence, control and fate of contaminants of emerging concern in environmental compartments in Brazil. *Journal of Hazardous Materials*, **372**, 17–36, https://doi.org/10.1016/j.jhazmat.2018.04.043

Sun H., Wang T., Yang Z., Yu C. and Wu W. (2019). Simultaneous removal of nitrogen and pharmaceutical and personal care products from the effluent of waste water treatment plants using aerated solid-phase denitrification system. *Bioresource Technology*, **287**, 121389, https://doi.org/10.1016/j.biortech.2019.121389

Syed Z., Sogani M., Dongre A., Kumar A., Sonu K., Sharma G. and Gupta A. B. (2021). Bioelectrochemical systems for environmental remediation of estrogens: a review and way forward. *Science of the Total Environment*, **780**, 146544, https://doi.org/10.1016/j.scitotenv.2021.146544

Vaksmaa A., Guerrero-Cruz S., Ghosh P., Zeghal E., Hernando-Morales V. and Niemann H. (2023). Role of fungi in bioremediation of emerging pollutants. *Frontiers in Marine Science*, **10**, 235, https://doi.org/10.3389/fmars.2023.1070905

Venditti S., Brunhoferova H. and Hansen J. (2022). Behaviour of 27 selected emerging contaminants in vertical flow constructed wetlands as post-treatment for municipal wastewater. *Science of the Total Environment*, **819**, 153234, https://doi.org/10.1016/j.scitotenv.2022.153234

Venegas M., Leiva A. M., Reyes-Contreras C., Neumann P., Piña B. and Vidal G. (2021). Presence and fate of micropollutants during anaerobic digestion of sewage and their implications for the circular economy: a short review. *Journal of Environmental Chemical Engineering*, **9**(1), 104931, https://doi.org/10.1016/j.jece.2020.104931

Verlicchi P. and Zambello E. (2014). How efficient are constructed wetlands in removing pharmaceuticals from untreated and treated urban wastewaters? A review. *Science of the Total Environment*, **470–471**, 1281–1306, https://doi.org/10.1016/j.scitotenv.2013.10.085

Wang S. and Wang J. (2018). Degradation of emerging contaminants by acclimated activated sludge. *Environmental Technology*, **39**(15), 1985–1993, https://doi.org/10.1080/09593330.2017.1345989

Wang J., Chu L., Wojnárovits L. and Takács E. (2020). Occurrence and fate of antibiotics, antibiotic resistant genes (ARGs) and antibiotic resistant bacteria (ARB) in municipal wastewater treatment plant: an overview. *Science of the Total Environment*, **744**, 140997, https://doi.org/10.1016/j.scitotenv.2020.140997

Wang H., Wang Y. and Dionysiou D. D. (2023). Advanced oxidation processes for removal of emerging contaminants in water. *Water*, **15**(3), 398, https://doi.org/10.3390/w15030398

Wilén B. M., Liébana R., Persson F., Modin O. and Hermansson M. (2018). The mechanisms of granulation of activated sludge in wastewater treatment, its optimization, and impact on effluent quality. *Applied Microbiology and Biotechnology*, **102**(12), 5005–5020, https://doi.org/10.1007/s00253-018-8990-9

Winkler M. K. H., Meunier C., Henriet O., Mahillon J., Suárez-Ojeda M. E., Del Moro G., De Sanctis M., Di Iaconi C. and Weissbrodt D. G. (2018). An integrative review of granular sludge for the biological removal of nutrients and recalcitrant organic matter from wastewater. *Chemical Engineering Journal*, **336**, 489–502, https://doi.org/10.1016/j.cej.2017.12.026

Xia K., Bhandari A., Das K. and Pillar G. (2005). Occurrence and fate of pharmaceuticals and personal care products (PPCPs) in biosolids. *Journal of Environmental Quality*, **34**(1), 91–104, https://doi.org/10.2134/jeq2005.0091

Xiao K., Liang S., Wang X., Chen C. and Huang X. (2019). Current state and challenges of full-scale membrane bioreactor applications: a critical review. *Bioresource Technology*, **271**, 473–481, https://doi.org/10.1016/j.biortech.2018.09.061

Xiong R., Li Y., Gao X., Li N., Lou R., Saeed L. and Huang J. (2023). Effects of a long-term operation wetland for wastewater treatment on the spatial pattern and function of microbial communities in groundwater. *Environmental Research*, **228**, 115929, https://doi.org/10.1016/j.envres.2023.115929

Zdarta J., Meyer A. S., Jesionowski T. and Pinelo M. (2018). Developments in support materials for immobilization of oxidoreductases: a comprehensive review. *Advances in Colloid and Interface Science*, **258**, 1–20, https://doi.org/10.1016/j.cis.2018.07.004

Zhang J., Chang V. W. C., Giannis A. and Wang J. Y. (2013). Removal of cytostatic drugs from aquatic environment: a review. *Science of the Total Environment*, **445–446**, 281–298, https://doi.org/10.1016/j.scitotenv.2012.12.061

doi: 10.2166/9781789063752_0135

Chapter 8

Novel approaches for removing emerging contaminants from sludge using fungal-mediated processes

Lamia Yakkou[1], Sofia Houida[2]*, Maryam Chelkha[1], Imane Sarroukh[3], Sartaj Ahmad Bhat[4], Rabha Abdelwahd[5], Mohammed Ibriz[3], Mohammed Raouane[1], Souad Amghar[1] and Abdellatif El Harti[1]

[1]Laboratoire LBVRN, Faculté des Sciences d'Agadir, Université Ibn Zohr, BP 8106, 80000, Agadir, Morocco/Faculty of Applied Sciences- Ait Melloul, University Ibn Zohr, BP 8106, 80000, Agadir, Morocco
[2]Pasteur Institute of Morocco, Casablanca, Morocco
[3]Laboratory of Biology, nutrition, health and environment, Department of Biology, Faculty of Science, Ibn Tofail University, B.P. 133, Kenitra 14000, Morocco
[4]River Basin Research Center, Gifu University, 1-1 Yanagido, Gifu 501-1193, Japan
[5]Biotechnology Research Unit, Institut National de la Recherche Agronomique (INRA), B.P. 415, Rabat, Morocco
*Corresponding author: sofia.houida@pasteur.ma

ABSTRACT

Emerging contaminants (ECs) such as pharmaceuticals, personal care products, and industrial chemicals pose an increasing threat to both the environment and human health, with their presence being detected more frequently in wastewater treatment plants and sludge. In response, fungal-mediated processes have emerged as a promising bioremediation technology, offering the unique ability to degrade a wide range of pollutants present in sludge. This chapter delves into the fungal species utilized for this purpose and recent advances in fungal-mediated processes, including genetically modified and immobilized fungi and combinations with other treatment methods. Furthermore, the mechanisms by which fungi remove ECs from sludge, such as biosorption, biodegradation, and enzyme production, are comprehensively discussed. Factors affecting the efficiency of fungal-mediated processes, including pH, temperature, fungal species, nutrient availability, and reactor design, are also examined. Finally, the chapter outlines the challenges encountered when using fungal-mediated processes to remove ECs from sludge and potential real-world applications of these processes in wastewater treatment scenarios.

Keywords: biodegradation, bioremediation, biosorption, emerging contaminants, fungal-mediated processes, sludge

8.1 INTRODUCTION

Sludge is a by-product generated while treating industrial wastewater combined with domestic swag at an industrial facility, often containing various organic and inorganic contaminants that can pose risks to human health and the environment (Turovskiy & Mathai, 2006). Traditional sludge treatment methods may not effectively remove these emerging contaminants (ECs), leading to their persistence in the environment (Vicent *et al.*, 2013).

Many fungal species have gained significant attention in recent years for their potential to remove ECs from sludge (Vaksmaa *et al.*, 2023). Using fungi to remediate emerging pollutants has many advantages over other physical and chemical mechanisms, such as its high effectiveness, low cost, and environmentally friendly nature (Tomasini & León-Santiesteban, 2019). In addition, fungus bioremediation has advantages over bacteria due to the diversity of processes, degrading enzymatic capacities, and ability to function under broad pH conditions (Tomasini & León-Santiesteban, 2019; Wani *et al.*, 2017).

Through their mycelium, which is a network of filaments, fungi can absorb the ECs that are present in the sludge. Hence, the contaminants may adhere to the mycelium's surface or penetrate the interior of the cells. The concentration of contaminants is then reduced as a result of this sorption and absorption (Malik *et al.*, 2023). Certain fungi use specific metabolic pathways to convert ECs into less toxic molecules. These metabolites may be more easily degraded or less persistent in the environment, aiding in the decontamination of waterways (Maqsood *et al.*, 2023). Some mushrooms can form complexes with ECs. These complexes may alter the contaminants' solubility or reactivity, making subsequent processes like precipitation or filtration easier to eliminate (Maqsood *et al.*, 2023). Furthermore, fungi develop symbiotic relationships with other microorganisms like bacteria. By fostering interactions between various organisms, these may promote the degradation of ECs and increase the overall effectiveness of the bioremediation process (Nguyen *et al.*, 2013).

Novel approaches to removing ECs from sludge by fungal-mediated processes have been the subject of active research. Some notable strategies and techniques have been investigated, including co-cultivation of fungi, fungal bioaugmentation, genetic modification of fungi, fungal-assisted phytoextraction, and fungal-based nanomaterials (Malik *et al.*, 2022).

Nevertheless, these novel approaches' effectiveness and practical applicability may vary depending on the specific contaminants, fungal species, and environmental conditions. Further research and development are needed to optimize these techniques and evaluate their scalability for large-scale sludge treatment (Marshall *et al.*, 2020).

In this chapter, we delve into the fungal species utilized for this purpose and recent advances in fungal-mediated processes, including genetically modified and immobilized fungi and combinations with other treatment methods. Furthermore, we highlight the mechanisms by which fungi remove ECs from sludge, such as biosorption, biodegradation, and enzyme production are comprehensively discussed. In addition, the factors affecting the efficiency of fungal-mediated processes, including pH, temperature, fungal species, nutrient availability, and reactor design, are also examined. Finally, the chapter outlines the challenges encountered when using fungal-mediated processes to remove ECs from sludge and the potential real-world applications of these processes in wastewater treatment scenarios.

8.2 FUNGAL SPECIES USED FOR THE REMOVAL OF ECS

8.2.1 Fungal species used for the removal of ECs from sludge

Several fungal species have been studied and employed to remove ECs from sludge. Here we present some commonly investigated fungal species in this context (Table 8.1). As shown in Table 8.1, the most commonly used fungi species for removing ECs are *Trametes versicolor* and *Pleurotus ostreatus*. *T. versicolor* is a white-rot fungus known for its versatile enzymatic system, which allows it to degrade a wide range of organic compounds, including ECs like pharmaceuticals and personal care products. It also exhibits metal-binding properties, making it helpful in removing heavy metals from sludge. *T. versicolor* removed >99.9% and 40% of diclofenac and ketoprofen, respectively, within 14 days (Dalecka *et al.*, 2020). Another study using *T. versicolor* for diclofenac degradation reported >87% removal in just 14 days (Hu *et al.*, 2020). *T. versicolor* has also demonstrated its ability to break down certain pesticides, including malathion and chlorpyrifos, by using enzymes such as peroxydase and

Table 8.1 Fungal species used in ECs removal.

Species	Emerging Contaminant	Implicated Enzymes	Duration of Degradation	Removal Efficiency (%)	References
Aspergillus sydowii	Chlorpyrifos Methyl parathion Profenofos	Phosphoesterase Methylesterase	10 days	52–80	Soares *et al.* (2021)
Bjerkandera adjusta	Atrazine	Cytochrome P450	–	92	Dhiman *et al.* (2020)
Ganoderma lucidum	Diuron	Laccase	15 days	>50	da Coelho-Moreira *et al.* (2018)
Phanerochaete chysosporium	Sulfamethoxazole	Laccase	10 days	74	Guo *et al.* (2014)
Pleurotus ostreatus	Diclofenac Ketoprofen	Laccase	5 days	100 70	Palli *et al.* (2017)
Pleurotus ostreatus	Atenolol	Laccase	20 days	50	Palli *et al.* (2017)
Pleurotus ostreatus	Carbamazepine	Cytochrome P450	7 days	58	Buchicchio *et al.* (2016)
Trametes versicolor	Chlorpyrifos	Laccase	7–14 days	94.7	Hu *et al.* (2020)
Trametes versicolor	Gemfibrozil	–	30 days	77–82	Alamo *et al.* (2021)
Trametes versicolor	Hydrochlorothiazide	–	30 days	80–95	del Álamo *et al.* (2018)
Trametes versicolor	Diazepam	Laccase	15 days	66	Badia-Fabregat *et al.* (2016)
Trametes versicolor	Bisphenol Nonylphenol Parabens Phthalates	Laccase	2–7 days	45–60	Pezzella *et al.* (2017)
Trametes versicolor	17α-ethinyl-estradiol	Laccase	24 hours	83	Becker *et al.* (2016)
Trametes versicolor	Carbamazepine Trimethoprim	–	14 days	88.6–89.8	Tormo-Budowski *et al.* (2021)

(*Continued*)

Table 8.1 Fungal species used in ECs removal (*Continued*).

Species	Emerging Contaminant	Implicated Enzymes	Duration of Degradation	Removal Efficiency (%)	References
Trametes versicolor	Sulfamethoxazole	Laccase	48 hours	34–82	Alharbi *et al.* (2019)
Trametes versicolor	Diclofenac	Laccase	2–7 days	39–95	Stenholm *et al.* (2019)
Trametes versicolor	Dicofol	–	14	87.9	Hu *et al.* (2020)
Trametes versicolor	Cypermethrin	–	14	93.1	Hu *et al.* (2020)
Trametes versicolor	Diuron Bentazon	Laccase	27 days	93	Beltrán-Flores *et al.* (2021)
Trametes versicolor	Sulfamethoxazole				del Álamo *et al.* (2018)
Trametes versicolor	Ibuprofen Ketoprofen Naproxen	Laccase	49 days	60–90	Torán *et al.* (2017)
Trametes versicolor	Bezafibrate Gemfibrozil Ketoprofen Ibuprofen Naproxen	Laccase	14–21 days	>80	Mir-Tutusaus *et al.* (2018)
Trichoderma pubescens	Amoxicillin	–	24 hours	98	Cai *et al.* (2023) Badia-Fabregat *et al.* (2016)
T. versicolor	Diclofenac Ketoprofen	–	14 days	>99.9 40	Dalecka *et al.* (2020)

laccase (Ma & Ruan, 2015). The ability of *T. versicolor* to use peroxydases and other enzymes to break down polycyclic aromatic hydrocarbons (PAHs), including naphtalene and pyrene, has also been investigated (Baldrian *et al.*, 2000).

Another edible mushroom species, *P. ostreatus*, has been studied for its ability to degrade organic pollutants, such as PAHs and textile dyes, in sludge. *P. ostreatus* has a high capacity for lignin degradation and can break down complex organic compounds (Purnomo *et al.*, 2010). *P. ostreatus* has proven its capacity to effectively decolorize and break down a variety of synthetic dyes, including azoic and triphenylmethane dyes, by using its ligninolytic enzymes (peroxydases and laccases) (Levin *et al.*, 2004). Additionally, *P. ostreatus* has been studied for its capacity to degrade organic pollutants, such as pesticides and herbicides, in contaminated soil and water (Purnomo *et al.*, 2010). *P. ostreatus* may also degrade PAHs found in soil and sediments, assisting in the degradation of these pollutants (Pozdnyakova *et al.*, 2018).

Known as a potent white-rot fungus, *Phanerochaete chrysosporium* has been extensively studied for its ability to degrade recalcitrant organic pollutants, including pesticides, herbicides, and industrial chemicals. Its unique enzymatic system allows it to break down complex organic compounds and transform them into more straightforward, less harmful forms. *Aspergillus niger* is a filamentous fungus with a robust enzymatic system capable of degrading various organic compounds. It has also been investigated for its potential to remove pharmaceuticals and other organic pollutants from sludge (Arun *et al.*, 2023).

The filamentous fungal species of *Aspergillus* has been reported to be able to degrade PAHs, chlorophenols, and aliphatic hydrocarbons (Malik *et al.*, 2022). *A. niger* can produce extracellular enzymes, such as ligninases and cellulases, which aid in the break down of complex organic molecules. Moreover, several species of *Penicillium*, including *Penicillium chrysogenum* and *Penicillium purpurogenum*, have shown promise in removing ECs from sludge. These fungi possess diverse enzymatic capabilities and can degrade various organic compounds, including pharmaceuticals and industrial pollutants (Leitão, 2009).

It is important to note that selecting a specific fungal species for sludge remediation depends on the targeted contaminants and environmental conditions (Somu *et al.*, 2022). Different fungi may exhibit varying degrees of efficiency and substrate specificity. Additionally, research is ongoing to explore the potential of other fungal species and optimize their performance to remove ECs from sludge (Rafeeq *et al.*, 2023).

The primary agents of ligninous material biodegradation in nature are white-rot fungi (WRF). The majority of studies have shown that WRF, such as *P. chysosporium*, *T. versicolor*, *Bjerkandera adjusta*, and *Pleurotus* sp., can bioremediate through the production of various ligninolytic enzymes like laccases and peroxidases. *T. versicolor* (García-Galán *et al.*, 2011; Rodríguez-Rodríguez *et al.*, 2012b), *P. chrysosporium* (Huang *et al.*, 2017), and *Phlebia tremellosa* (Kum *et al.*, 2011) are some of the most common WRF when it comes to bioremediation of contaminants (Marco-Urrea *et al.*, 2010; Rodríguez-Rodríguez *et al.*, 2010). A similar role has also been reported for *Penicillium* spp. (Li *et al.*, 2020). Other examples of pollutants are anthracene, which is removed by the fungi *Irpexlacteus* and *P. ostreatus* (Drevinskas *et al.*, 2016); pyrene by *Ganoderma lucidum* (Agrawal *et al.*, 2018); and chrysene by *Polyporus* spp. (Hadibarata *et al.*, 2009). These fungal groups have also been able to degrade toxic metals with high bioremediation efficiency. For example, *P. ostreatus* (white-rot fungus) is known to aid in bioremediation by degradation of crude oil and toxic metals, with bioremediation efficiency within the range of 28.2–75.9% (Anacletus *et al.*, 2017).

8.2.2 Mechanisms by which fungi can remove ECs from sludge

Fungal species offer several advantages for the remediation of sludge contaminated with ECs. Firstly, fungi can degrade a wide range of organic compounds due to their diverse enzymatic systems. This versatility allows them to break down complex organic pollutants, including pharmaceuticals, personal care products, pesticides, and industrial chemicals present in sludge (Rodríguez-Rodríguez *et al.*, 2013). Moreover, some fungal species have the capability to transform or bind with heavy metals and metalloids, thereby reducing their toxicity and mobility. This is particularly beneficial when sludge contains elevated levels of heavy metals, such as lead, mercury, or arsenic, which can harm ecosystems and human health (Abd Elnabi *et al.*, 2023; Priyadarshini *et al.*, 2021). In addition to their degradation and binding capacities, fungi can also assist in the removal of nutrients from sludge. Excessive levels of nutrients, such as nitrogen and phosphorus, can contribute to eutrophication when sludge is applied as fertilizer. Certain fungal species, like WRF, can efficiently break down organic forms of nitrogen and phosphorus, reducing their availability and minimizing the risk of nutrient pollution (Bhambri *et al.*, 2021). While fungal species used for sludge remediation show promise, further research and development are needed to fully understand their potential, optimize their performance, and scale up their application in practical wastewater treatment systems. Nevertheless, using fungal species holds excellent potential for removing ECs from sludge, offering a sustainable and environmentally friendly approach to sludge management (Aydin, 2016; Rodríguez-Rodríguez *et al.*, 2013).

Fungi can employ various mechanisms to remove ECs from sludge. Here are some key mechanisms by which fungi contribute to the degradation, transformation, or sequestration of contaminants (Figure 8.1):

Enzymatic degradation: One of the key characteristics of fungi involved in the elimination of ECs is their production of specific enzymes. Fungi possess a diverse array of enzymes that play a crucial role in the degradation of organic contaminants. Extracellular specialized enzymes that can degrade ECs include the peroxydases, hydrolase, oxidoreductase, dehalogenase, oxygenase, transferase, laccases, and ligninases, which have the ability to break down a wide range of chemical compounds, including recalcitrant pollutants like PAHs and pharmaceuticals (Kothawale *et al.*, 2023; Levin *et al.*, 2004; Naghdi *et al.*, 2018).

Pozdnyakova *et al.* (2018) investigated the biodegradation potential of the PAH compounds, phenanthrene and anthracene using the fungi *Agaricus bisporus* and *P. ostreatus* (Pozdnyakova *et al.*, 2018). Evidence showed that the laccases produced by *A. bisporus* turned the PAH compounds into their quinone analogues. On the other hand, *P. ostreatus* has also evolved flexible peroxidase enzymes.

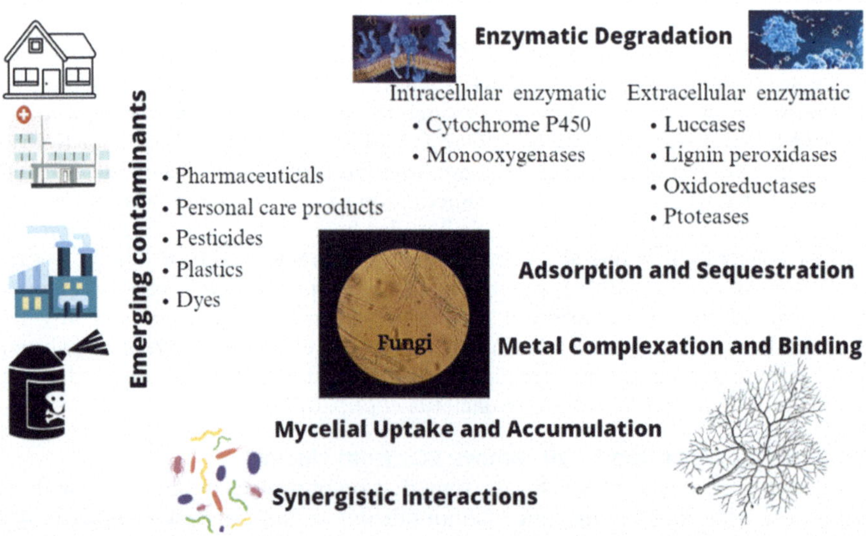

Figure 8.1 Various mechanisms employed by fungi to remove ECs.

On another study, the HPLC analysis of fungal metabolites confirmed that a novel strain of the fungus *Coriolopsis byrsina* could split the aromatic rings of pyrene and degrade it into pyrene trans-4,5-dihydrodiol by producing laccase enzymes. In addition, manganese peroxidase converted anthracene, phenanthrene, pyrene, and fluoranthene into quinones by demineralization and oxidation (Baborová *et al.*, 2006). In the field of environmental biotechnology, these discoveries hold great potential for the remediation of PAH-contaminated sites. Understanding the enzymatic degradation of PAH by fungi can pave the way for developing sustainable and eco-friendly solutions to tackle the persistent problem of polycyclic aromatic hydrocarbon pollution.

The emergence of exogenous redox mediators, which act as transport molecules in the process, may aid in degradation. Azoreductase is a flavodoxin protein that breaks the azo bonds in dyes. This made it possible for the dyes to be changed into the right aromatic acids (Maqsood *et al.*, 2023).

Furthermore, laccases were found to be involved in the break down processes of dicofol and chlorpyrifos in *T. versicolor* (Table 8.1). The first step is dechlorination, which converts dicofol to 2,2-dichloro-1,1-bis(4-chlorophenyl)-ethanol. These oxidative cleavages and dechlorinations result in 4,4′-dichlorobenzophenone. Benzaldehyde is made when enzymes that break down lignin break the ring structure of lignin (Purnomo *et al.*, 2010).

Recently, more and more studies have shown that the enzyme cytochrome P450 is involved in removing and breaking down organic pollutants. The CYP450s are a diverse superfamily of enzymes with specialized folding properties. The hydroxylation, N-, O-, S-dealkylation, sulfurization, epoxidation, deamination, desulfurization, dehalogenation, peroxidation, and N-oxide reduction reactions are just a few examples of the various reactions that these enzymes, which are either located in the cell membrane or the cytoplasm, contribute to the transport, metabolism, and catabolism of organic substrates (Roccatano, 2015). In order to oxidize an inert substrate, for instance, CYP450 first breaks the strong bond between C and H before creating a stronger bond between O and H (Dacco *et al.*, 2020). Therefore, CYP450s can catalyze ECs of various sizes and shapes (Deshmukh *et al.*, 2016).

Co-metabolism: Fungi can utilize contaminants as co-substrates or co-metabolites for their metabolic processes. In some cases, fungi may not fully mineralize the contaminants but transform them into less toxic or less persistent metabolites. This co-metabolic activity can lead to the degradation or detoxification of ECs in sludge (Deshmukh *et al.*, 2016).

Adsorption and sequestration: Fungi can bind or adsorb contaminants onto their cell walls or extracellular matrices. The fungal cell wall is composed of polysaccharides, proteins, and other biomolecules that can act as sorbents for various contaminants. This adsorption mechanism can lead to the immobilization or sequestration of contaminants, reducing their bioavailability and potential for environmental impact (Crini *et al.*, 2018).

Metal complexation and binding: Some fungi can sequester and bind heavy metals in sludge. They can produce metal-binding molecules, such as organic acids, peptides, and extracellular polymeric substances that form complexes with metal ions. This metal complexation can reduce the mobility and bioavailability of heavy metals, mitigating their potential toxic effects (Qian *et al.*, 2017).

Mycelial uptake and accumulation: Fungal mycelium, consisting of interconnected branching hyphae, can act as a physical network for the uptake and accumulation of contaminants. Mycelium can penetrate through sludge particles, creating a vast surface area for contact with contaminants. Fungi can absorb contaminants into their mycelium, effectively concentrating and sequestering them within their biomass (Qian *et al.*, 2017).

Synergistic interactions: Fungi can establish synergistic interactions with other microorganisms, such as bacteria, to enhance contaminant removal. Some bacteria can produce enzymes or metabolites that complement the degradation capabilities of fungi. Co-cultivation or biofilm formation involving fungi and bacteria can create a cooperative network that leads to improved degradation efficiency (Angeles-de Paz *et al.*, 2023; Espinosa-Ortiz *et al.*, 2022; Purnomo *et al.*, 2020).

It is important to note that the mechanisms fungi employ for contaminant removal can vary depending on the specific fungal species, contaminants present, environmental conditions, and the interplay with other microorganisms (Kumar *et al.*, 2021).

8.3 RECENT ADVANCES IN FUNGAL-MEDIATED PROCESSES FOR EC REMOVAL

Recent years have witnessed significant advances in fungal-mediated processes for the removal of ECs (Figure 8.2). Moreover, as mentioned in the previous section of this chapter, a considerable research effort on fungi capacities demonstrates that their biochemical and physiological characteristics can be necessary for a wide range of biotechnological applications to be used for degrading organic contaminants. As a result of their great promise, this chapter highlights recent advances and novel approaches for using these microorganisms to remove ECs.

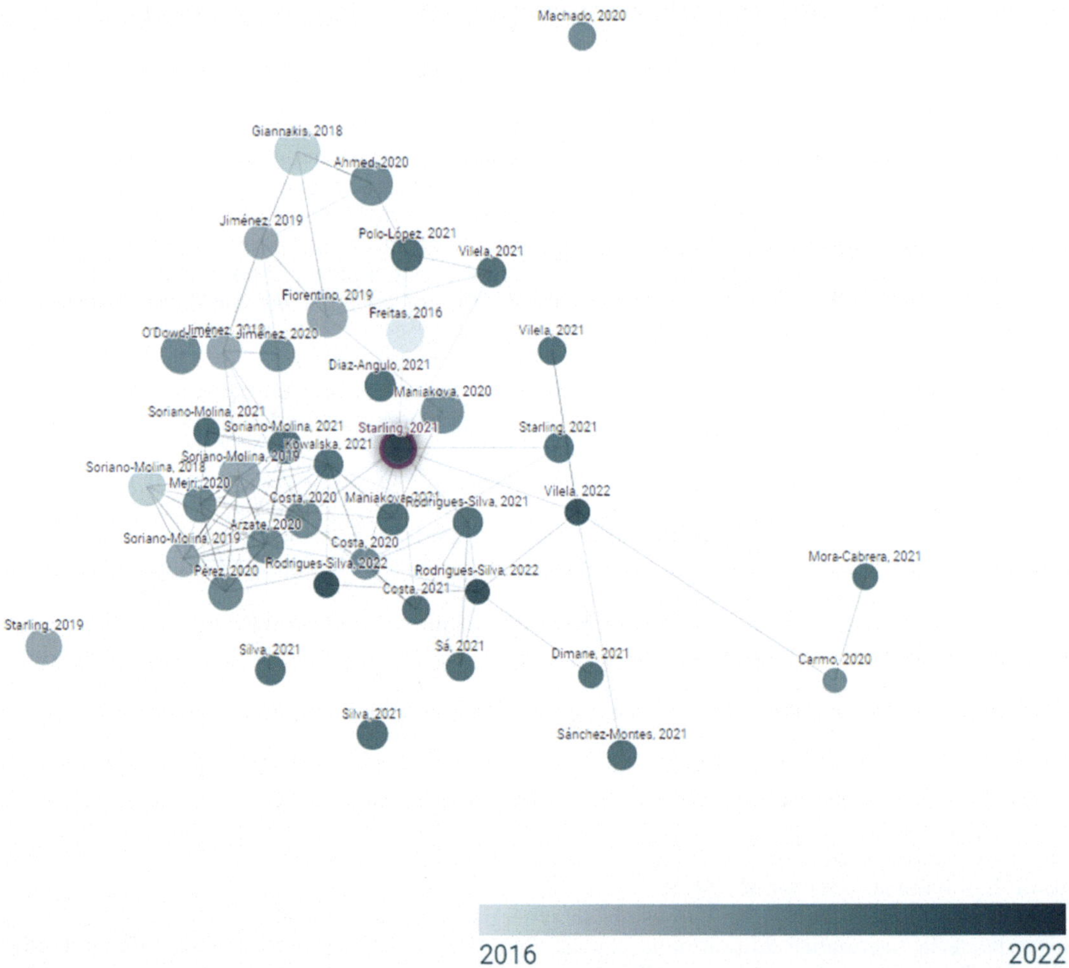

Figure 8.2 Papers related to the new technologies using fungi to remove ECs (the graph was created using Connected Papers website).

Researchers have been exploring the fungi's enzymatic systems and metabolic pathways to better understand their potential for EC removal. Studies have focused on identifying novel enzymes involved in degrading specific contaminants and optimizing their activity. Additionally, transcriptomic and proteomic analyses have provided insights into fungi's gene expression and protein profiles during the degradation process (Malik *et al.*, 2022). Furthermore, the advances in genomic and metagenomic studies have facilitated the discovery and characterization of fungal species and their functional genes involved in EC degradation. High-throughput sequencing technologies have made it possible to identify different fungal communities in sludge and assess their potential for pollutant removal. Metagenomic approaches have also shed light on fungal populations' genetic diversity and metabolic potential in sludge ecosystems (Bala *et al.*, 2022).

The application of molecular tools, such as quantitative polymerase chain reaction (qPCR) and high-throughput sequencing, has facilitated the monitoring and assessment of fungal-mediated processes. These tools allow for quantifying fungal populations, tracking the expression of key genes involved in degradation, and assessing changes in fungal community structure and dynamics during the remediation process (Akerman-Sanchez & Rojas-Jimenez, 2021; Kour *et al.*, 2021; Somu *et al.*, 2022). These recent advances demonstrate the growing understanding of fungal-mediated processes for EC removal and highlight the potential for more efficient and targeted strategies. Continued research and technological development in this field hold promise for the implementation of sustainable and effective fungal-based approaches to sludge remediation (Kour *et al.*, 2021).

8.3.1 Fungal reactors

The development of bioreactor systems for fungal-mediated contaminant removal has gained attention. Bioreactors provide controlled environments that optimize fungal growth and activity for efficient degradation. Various configurations, such as submerged, solid-state, and biofilm reactors, have been explored to enhance fungal performance in sludge treatment. Integration of bioreactors with advanced monitoring and control systems allows for real-time optimization of process conditions (Tormo-Budowski *et al.*, 2021).

In contrast to ex situ bioremediation techniques, using a bioreactor to treat emerging pollutants has a number of advantages as of late. The time needed for bioremediation can be greatly decreased by using an effective bioremediation process based on bioreactors that can precisely control pH, agitation, temperature, aeration, substrate concentration, and inoculum concentration. When the bioreactor can be managed and controlled, biological reactions can occur. Bioreactor designs are flexible enough to maximize microbial degradation while minimizing abiotic losses (Bala *et al.*, 2022). They are therefore a viable option for cleaning contaminated areas. To get rid of various classes of emerging pollutants, a biotechnological method that used the fungus *T. versicolor* in a sludge-bioslurry reactor was evaluated. For 24 pharmaceuticals that were detected but either removed or completely degraded, this technique demonstrated efficiencies over 50% (Rodríguez-Rodríguez *et al.*, 2012a).

In a membrane bioreactor supplemented with the fungus *T. versicolor* and activated sludge, Nguyen *et al.* (2013)'s research examined the break down of 30 trace organic pollutants in synthetic wastewater. In comparison to a typical membrane bioreactor containing simply activated sludge, the fungus, along with bacterial strains, had a better removal rate of trace organic pollutants (80%). Moreover, the fungus-bacteria enhanced bioreactor eliminated fenoprop, clofibric acid, pentachlorophenol, ketoprofen, diclofenac, and naproxen (Pathak *et al.*, 2020). In these experiments, the synergistic break down by the bacteria and fungus was attributed to the improved elimination of organic pollutants. In fact, recent research indicates that using more than one living organism can increase bioremediation's effectiveness and results while allowing for a wider variety of microorganisms.

8.3.2 Coculture-based approach

Researchers have started investigating the potential synergistic interactions between fungi and bacteria for enhanced contaminant removal. Fungi can provide a conducive environment for

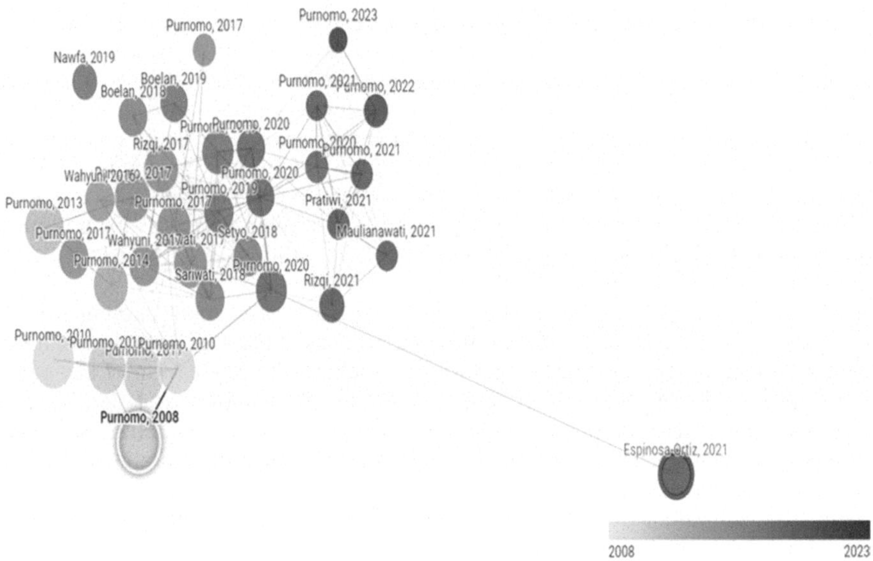

Figure 8.3 Articles related to the use of fungal–bacterial co-cultures to remove ECs (the graph was created using Connected Papers website).

bacterial growth, while bacteria can assist in the break down of complex contaminants and provide additional enzymatic activities. Co-cultivation or biofilm formation involving fungi and bacteria has shown promise in improving degradation efficiency (Espinosa-Ortiz *et al.*, 2022). In fact, a microbial consortia's individual species interact with one another, which results in the development of various traits like stability and the specialization of microbial partners to carry out particular functions like degradation (Feng *et al.*, 2019). However, it is challenging to isolate a sustainable microbial combination that could be effective on a large scale (Gupta *et al.*, 2022).

A published paper recently proposed an innovative method for creating a multi-domain co-culture that can degrade many medicinal chemicals at once. In order to improve their degrading performance, seven previously isolated microorganisms (fungi and bacteria) from sewage sludge were examined. The strains were factorially combined, and they were then utilized to put together several artificial co-cultures. The most effective co-cultures were tested with three distinct pharmaceutical substances to gauge the rate of break down of developing pollutants. The minimum active microbial consortia, which included co-existing bacteria as well as *Cladosporium cladosporoides* and *Penicillium* spp., had the best performance (>80% destruction). It was emphasized that these consortiums converted the pharmaceutical active chemicals by hydroxylation. Therefore, high-throughput detection of co-cultures can be a fast, reliable, and efficient method to reduce co-cultures' degradation suitable for ECs and avoid toxic by-products. However, the fact that microbial consortia are highly stable *in vitro* does not ensure that *in situ* efficiency might not be negatively affected by external factors (Angeles-de Paz *et al.*, 2023). On the other hand, a socially stable consortium with significant degrading capacity should be identified in order to assure the effective implementation of the artificial co-culture construction, and more research is required to understand how it competes with native microbiota. Thus, future metatranscriptomics and metaproteomics investigations will help in understanding the processes of adaptability and the stability over generations of these artificial consortia (Angeles-de Paz *et al.*, 2023).

Mixed microbial communities have distinctive interactions that can lead to more effective systems for separating and removing organic contaminants. Adopting polymicrobial culture methods has made it possible to get a deeper knowledge of microbial dynamics in mixed microbial communities of fungus and bacteria. Compared to pure cultures of either fungus or bacteria, prior studies have found that combining fungal–bacterial treatments results in greater biodegradation of certain contaminants (Espinosa-Ortiz *et al.*, 2022). The interaction of bacterial and fungal systems may result in synergistic partnerships that can speed up the break down of organic contaminants. Graphic 1 summarizes the most recent publications in the literature that used a variety of fungal–bacterial co-cultures to remove organic contaminants.

When bacteria and fungi are co-cultivated, the production of enzymes can increase, helping to fully mineralize contaminants, or secondary metabolites can be produced, which can help with pollutant break down. In co-cultures, interactions that are either synergistic or antagonistic can lead to an increase in the production of degradative enzymes, such as lignin peroxidases, laccases, and manganese peroxidases (Espinosa-Ortiz *et al.*, 2022). Competition for resources and space as well as oxidative stress is other effects of fungi and bacteria co-cultivating together (Wan *et al.*, 2015). Oxidative stress can speed up the transition of fungi to secondary metabolism, which leads to the creation of oxidative enzymes. In fact, oxidative enzymes called fungal laccases efficiently break down persistent pollutants (Wan *et al.*, 2015).

It is common knowledge that the presence of ligninolytic enzymes in fungi facilitates the degradation of pesticides. Most of the time, only a small portion of pesticides are degraded by bacteria. Nonetheless, it has been documented that both aqueous and porous media may be used to co-culture fungal and bacterial species to break down organochlorine and organophosphate insecticides (Jain *et al.*, 2017).

In recent years, dichloro-diphenyl-trichloroethane has been the pesticide most frequently utilized in research on fungal degradation (Vaksmaa *et al.*, 2023). The fungal–bacterial co-cultures showed 86% dichloro-diphenyl-trichloroethane break down after 712 days of incubation. It was proposed that the capacity of the used bacteria (*Pseudomonas aeruginosa*) to create rhamnolipid biosurfactants, which can improve the solubility of dichloro-diphenyl-trichloroethane, was the cause of the increased break down by the fungal–bacterial co-cultures (Deshmukh *et al.*, 2016). The same pesticide degradation process was applied in different research using the fungus *Pleurotus eryngii* and the bacteria *Ralstonia pickettii*, which produce biosurfactants (Purnomo *et al.*, 2020). After seven days of incubation, the co-culture degraded the pesticide by 78%, whereas the single fungal culture had only degraded it by 43%. The authors noted that the bacteria's presence stimulated the fungus's development. Li *et al.* (2016) used a co-culture of the fungus *Pycnoporus sanguineus* and the bacteria *Alcaligenes faecalis* to study the break down of sulfamethoxazole. The co-culture degraded sulfamethoxazole to 73% after 48 hours of incubation. The increased elimination effectiveness was attributable to *P. sanguineus*'s increased laccase activity when *A. faecalis* was present (Li *et al.*, 2016).

8.3.3 Enzymes application-based approach

Many yeasts and filamentous fungi, which are well recognized for their environmental durability and resistance to degradation, have been used to decolorize synthetic dyes, mostly employed in the textile sector (Deshmukh *et al.*, 2016; Levin *et al.*, 2004). Fungal extracellular enzymes, including lignin peroxidases, laccases, and manganese peroxidases, can degrade synthetic dyes without producing toxic aromatic amines (Akerman-Sanchez & Rojas-Jimenez, 2021). However, because fungi grow best at low pH levels, using them in continuously operating reactors typically necessitates lengthy hydraulic retention times for decolorization, which could be problematic given that the majority of effluents containing synthetic dyes have an alkaline pH (Akerman-Sanchez & Rojas-Jimenez, 2021). One potential solution to overcome the pH limitation of fungal extracellular enzymes in decolorizing synthetic dyes is immobilized enzymes (Somu *et al.*, 2022). Immobilization techniques can help maintain enzyme activity at higher pH levels, allowing for more efficient decolorization processes.

Additionally, immobilized enzymes in continuously operating reactors can reduce the hydraulic retention times required for effective dye degradation (Somu *et al.*, 2022).

Tran *et al.* (2010) suggested the combination of enzymes isolated from fungi (laccases) and bacteria (oxygenases) as a plausible strategy to improve the degradation of emerging trace organic contaminants. For instance, bacterial oxygenase can oxidize aromatic pollutants, resulting in smaller phenolic compounds (Tran *et al.*, 2010). These phenols can be toxic to bacteria. However, fungal laccases can promote the degradation of these phenolic compounds. Therefore, the combination of fungal and bacterial enzymes can result in a synergistic break down of organic contaminants. In addition, microbial secondary metabolites like pigments, antibiotics, alkaloids, and carotenoids can facilitate external interactions among partners in microbial communities (Espinosa-Ortiz *et al.*, 2022).

Using the microbial enzyme cytochrome P450 to convert hydrocarbons into less hazardous chemicals is a novel bioremediation technique. P450s are naturally able to break down xenobiotics through a variety of bioremediation-related chemical processes, including aliphatic hydroxylations and epoxidations, dealkylations, dehalogenation, and different mechanism-based inactivations (Bhandari *et al.*, 2021).

Protease is another enzyme from the hydrolase family that catalyzes the peptide bonds in proteins. It was isolated from a fungus like *Aspergillus* sp. Due to their low cost, prolific manufacturing, and effective action, fungi-derived proteases are of utmost significance. They are extensively employed in sectors of the economy such as wastewater treatment, the food sector, and the leather sector. Protease can be utilized in bioremediation for the break down of polymers since it can break down ß-ester linkages, ß-ester bonds made by poly(hydroxybutyrate)depolymerase, and c-bonds made by lipase (de Souza *et al.*, 2015).

8.3.4 Genetically modified fungi application-based approach

Although fungi are well known for their role in the removal of pollutants, the capacity of local species to digest these contaminants is limited, and the process is time-consuming. Hence, the break down process can be sped up by genetically engineered organisms, whose altered metabolic pathways stimulate the oversecretion of a range of proteins helpful to the bioremediation process (Maqsood *et al.*, 2023).

Currently, research is being done on the widespread application of genetically altered microorganisms that can aid in the removal of ECs from the environment, such as petroleum, naphthalene, toluene, benzene, and other xenobiotic chemicals (Rafeeq *et al.*, 2023). However, most studies focus on engineered bacterial species. More research should be done to explore the enzymatic mechanisms of fungal species.

Fungal strains' genomic and metagenomic study has given researchers fresh insight into how to develop and exploit them as biotechnological tools. These omics analyses have assisted in identifying several taxa of fungi in various environments that are being studied for their potential to bioremediate (Malik *et al.*, 2022). To break down phenolic substances, the yeast laccase gene from the fungus *Yarrowia lipolytica* was inserted into the *Pichia pastoris* genome (Kalyani *et al.*, 2015). The yeast laccase gene (YlLac) was extracted from the isolated *Yarrowia lipolytica* using a modified thermal asymmetric interlaced polymerase chain reaction. Compared to other known laccases, the utilization of the engineered yeast demonstrated better catalytic effectiveness toward 2,2-azino-bis(3-ethylbenzothiazoline-6-sulfonate) and 2,6-dimethoxyphenol. Because of this, the yeast recombinant laccase was considered a promising candidate for industry use (Kalyani *et al.*, 2015).

Similarly, the expression of the *Phanerochaete flavido*-alba laccase gene in *Aspergillus niger* as well as the physical and biochemical characteristics of the recombinant enzyme (rLac-LPFA) was studied in order to evaluate it for the biotransformation of synthetic dyes (Benghazi *et al.*, 2014). High amounts of an active recombinant enzyme (30 mg L^{-1}) was produced by *A. niger.* Interestingly, the recombinant enzyme performed better in tests with organic solvents and pH ranges of 2 to 9. Moreover, compared to the native enzyme, synthetic textile dyes showed a greater proportion of decolorization and biotransformation (Benghazi *et al.*, 2014). These findings suggest that the expression of the *Phanerochaete flavido-alba* laccase gene in *A. niger* has the potential to be a promising tool for the biotransformation of synthetic dyes. Further research could focus on optimizing the production and application of the recombinant enzyme for industrial-scale dye degradation processes.

Investigations were done into the introduction of the yeast strain *Kluyveromyces lactis*' Lcc1 laccase in the fungus *Trametes trogii*. Compared to the native enzyme, the secreted recombinant laccase had improved enzyme properties. As a result, they were able to draw the conclusion that particular *K. lactis* strains with advantageous physiological characteristics and transcription regulation of the heterologous gene would be the best hosts for the manufacture of laccase isoenzyme (Ranieri *et al.*, 2009).

The three major categories of gene editing methods are ZFN (Zinc Finger Nucleases), TALEN (Transcription Activator-Like Effector Nucleases), and CRISPR/Cas (Clustered Regularly Interspaced Short Palindromic Repeats and Cas9 Protein) (Maqsood *et al.*, 2023). The area of bioremediation is increasing as a result of scientific concepts on CRISPR toolkits and creating gRNA for the expression of function-specific genes in non-model organisms. Since they can endure and retain a variety of hazardous, resistant, and non-degradable xenobiotics, fungi, like bacteria, make attractive candidates for metabolic engineering and gene editing. The current route is modified using metabolic engineering to optimize the efficiency of the bioremediation procedure. CRISPR/Cas is a superior gene editing technique to ZFNs and TALENs due to its high throughput programming capability. In addition, CRISPR/Cas9 genome editing technology has revolutionized the field of genetic modification by providing a precise and efficient tool for altering an organism's DNA (Maqsood *et al.*, 2023). This technology allowed researchers to target specific genes and make precise changes, such as modifying the genes that code for hydrocarbon break down or generating hybrid strains. For instance, 20 amino acid molecules were changed to enhance the dioxygenase's capability to recognize polychlorinated biphenyls. Later, it was discovered that switching threonine for aspartic acid might improve the dioxygenase activity for the oxidation of other substrates (Suenaga *et al.*, 2002). With CRISPR/Cas9, scientists can now explore new possibilities for enhancing the capabilities of enzymes like dioxygenase, opening up exciting avenues for improving bioremediation strategies.

At its pinnacle, microbiome engineering using the CRISPR/Cas9 tool for emerging pollutant biodegradation would make it important to encourage keystone species for workable advancement. Understanding the functions of microbial genes is based on the application of experimental genetics to cultured microorganisms. However, most fungi remain uncultivated, preventing the application of traditional genetic methods to these organisms (Maqsood *et al.*, 2023). A generalizable strategy for genome editing of specific organisms in microbial communities has been explored through the application of environmental transformation sequencing (ET-seq), in which untargeted transposon insertions are mapped and quantified after delivery to a microbial community (Rubin *et al.*, 2022). Next, DNA-editing all-in-one RNA-guided CRISPR-Cas transposase (DART) systems for targeted DNA insertion into organisms identified as tractable by ET-seq are used to enable organism- and locus-specific genetic manipulation in a community context. By testing the fitness of genes and doing site-specific editing on a variety of non-model bacteria, these methods can be used to increase and improve the preferred population in the soil, including the fungal population.

In contrast to bacteria, biotechnological innovations' role in fungal bioremediation is comparatively less well documented. In addition, bacteria and fungi show different mechanisms for the bioremediation of emerging pollutants. Significant progress in molecular biology related to fungi has been achieved, especially in the extraction of genetic material, gene cloning, and genetic engineering of fungi (Jafari *et al.*, 2013). A great future lies in successful genetic engineering intending to construct new potential fungi that can utilize ECs as the sole carbon source and remove them from nature.

8.4 FACTORS AFFECTING FUNGAL-MEDIATED PROCESSES

Fungal-mediated processes have emerged as promising approaches for the removal of ECs from sludge and wastewater. However, achieving efficient and effective remediation requires a comprehensive understanding of the various factors that influence these processes. Several key factors significantly impact the efficacy of fungal-mediated removal.

The composition and characteristics of the contaminants themselves play a critical role in determining the success of fungal degradation. The chemical structure, concentration, and stability of the contaminants can influence the availability and accessibility of the target compounds to the fungal enzymes (Ashe *et al.*, 2016; Batista-García *et al.*, 2017; Gros *et al.*, 2014; Singh *et al.*, 2021). Understanding the nature of the contaminants is essential for selecting appropriate fungal species or optimizing cultivation conditions. During the treatment process, the partition of ECs from wastewater to sludge is contingent upon various physicochemical properties of the compounds, including molecular weight, acid dissociation constant (pK_a), octanol–water partition coefficient (K_{OW}), solubility, and biodegradability (Mohapatra *et al.*, 2021). To illustrate, the sorption of hydrophobic compounds, like PAH, exhibits an increase with higher molecular weight, hydrophobicity, and the presence of phenolic and aromatic compounds in the dissolved/colloidal matter (Barret *et al.*, 2010). Specifically, the molecular weight of the compound and the existence of chlorine atoms enhance the sorption potential due to the presence of an intramolecular hydrophobic environment (Barret *et al.*, 2010).

Environmental conditions also play a crucial role in fungal-mediated removal processes. pH, temperature, and oxygen availability directly impact fungal growth, metabolism, and enzymatic activities. Fungi typically exhibit specific pH and temperature optima for optimal growth and enzyme production (Ashe *et al.*, 2016; Naghdi *et al.*, 2018; Singh *et al.*, 2021). Deviations from these optimal conditions can hinder fungal activity and reduce the efficiency of contaminant degradation. In their study, Esterhuizen *et al.* (2021) examined the factors influencing the ability of the fungus *Mucor hiemalis* and *P. chrysosporium* to degrade acetaminophen (APAP), a commonly used pharmaceutical compound that poses a significant environmental concern due to its widespread use and subsequent presence in various water bodies. Additionally, they noticed that whereas *Phanerochaete chrysoporium* demonstrated superior APAP remediation without pH adjustment, *Mucor hiemalis's* remediation was improved by 12% with pH modification. In fact, the acidic environment favored its metabolic activity, as it produces lignin peroxidase enzymes optimized at pH levels between 4 and 4.5, potentially contributing to APAP degradation (Esterhuizen *et al.*, 2021). A recent study conducted by Kang *et al.*, (2021) aimed to explore the capacity of *Bjerkandera* spp. TBB-03 in synthesizing indigenous fungal enzymes and its effectiveness in breaking down APAP. The findings of this research shed light on the potential of *Bjerkandera* spp. TBB-03 as a promising candidate for bioremediation applications targeting pharmaceuticals and personal care products. Ligninolytic processes were found to generate radical intermediates, facilitating radical polymerization and simplifying the precipitation-based removal of Acetaminophen (APAP). These oxidative coupling processes enabled the rapid polymerization-based elimination of APAP, irrespective of temperature conditions. The optimal temperature for the removal of bisphenol A (BPA) was found to be 40°C, as higher temperatures increased the reaction rates of the catalytic cycle. The removal behavior of three PPCPs at different pH values was assessed, revealing complete removal of APAP within 2 h at pH 5–7, with the best performance observed at neutral pH and reduced efficacy at higher pH. In conclusion, TBB-03 laccase effectively regulated the oxidation of various PPCPs, but its performance and activity were influenced by external factors such as pH and temperature (Kang *et al.*, 2021). Thus, maintaining favorable environmental conditions is essential for maximizing fungal activity and remediation efficiency.

Nutrient availability, particularly from carbon and nitrogen sources, is another critical factor influencing fungal-mediated removal. Carbon sources provide the energy required for fungal growth and enzymatic activities, while nitrogen sources are essential for protein synthesis and enzyme production

(Nitsche *et al.*, 2013; Peng *et al.*, 2021). In their study, Badia-Fabregat *et al.* (2015) investigated the degradation of pharmaceutically active compounds by *T. versicolor* using 1.5 L air-pulsed fluidized bioreactors. They discovered that the introduction of external nutrients significantly improved the degradation capabilities of *T. versicolor*. Impressive degradation rates ranging from 80% to 100% were achieved for challenging-to-biodegrade compounds, including diclofenac, ciprofloxacin, ofloxacin, sulfamethoxazole, furosemide, atenolol, valsartan, and gemfibrozil. In contrast, Esterhuizen *et al.* (2021) found that nitrogen limitation stimulated APAP removal by *M. hiemalis*, as the fungus allocated energy to extracellular enzyme production. Therefore, understanding the nutrient requirements of the fungal species being employed and optimizing the nutrient composition of the growth medium can enhance fungal growth and remediation potential.

In addition to the target contaminants, environmental conditions, and nutrient availability, the presence of co-contaminants can influence the remediation process (Mukherjee *et al.*, 2022). Co-contaminants may include heavy metals, persistent organic pollutants, or other chemical compounds commonly found in sludge and wastewater. These co-contaminants can interact with fungal degradation processes, either synergistically enhancing or inhibiting removal efficiency (Rahman, 2020; Zhou *et al.*, 2015). Understanding these interactions is crucial for predicting the overall remediation outcomes and developing strategies to mitigate any potential negative effects.

In conclusion, fungal-mediated processes offer great promise for the removal of ECs from sludge and wastewater. However, optimizing the efficacy of these processes requires careful consideration of various factors. By understanding the composition and characteristics of the contaminants, optimizing environmental conditions, ensuring proper nutrient availability, and accounting for the presence of co-contaminants, researchers and practitioners can enhance fungal-mediated remediation strategies. These efforts pave the way for sustainable and effective wastewater treatment approaches, contributing to the preservation of water resources and environment.

8.5 APPLICATIONS OF FUNGAL-MEDIATED TECHNOLOGY FOR EC REMOVAL

The environmental impact of pharmaceuticals and personal care products is linked to adverse effects on ecological systems as a consequence of the use of particular inexpensive components. Even at low concentrations, the presence of pharmaceutical-derived ECs in water can pose a long-term risk to human health and aquatic ecosystems (Tran *et al.*, 2019).

Antibiotics like ciprofloxacin, erythromycin, roxithromycin, and ofloxacin have been discovered in quantities as high as 6.7 g/L (Verlicchi *et al.*, 2012), and other drugs detected in effluent waters include antihypertensives, beta-blockers, diuretics, and lipid regulators, indicating that wastewater treatment plants (WWTPs) are deficient in the degradation of these compounds (Gogoi *et al.*, 2018; Jjemba, 2006; Petrović *et al.*, 2003).

Sulfonamides represent the widespread antibiotics used to treat or prevent infectious diseases in humans, and higher dosages are being used to treat or prevent infectious diseases in livestock and cattle husbandry (Boxall *et al.*, n.d.). Their acetylated metabolites have been detected in different environmental samples, such as influent and effluent wastewater samples, rivers, and sediments, and they have reached levels up to 1.10 ng/g in the WWTP sludge (Cui *et al.* 2020; García-Galán *et al.* 2011). A significant portion of these wastes is subsequently released into sewage (Cui *et al.*, 2020). García-Galán *et al.* (2011) used a fungal treatment with *T. versicolor* during the degradation of sulfamethazine. By the end of the process, the degradation of preexisting sulfonamides was highly efficient (about 100%), since none were discovered in the treated sludge. In fact, *T. versicolor* has been demonstrated to break down the medicines naxopren and carbamazepin (Rodríguez-Rodríguez *et al.*, 2010) and ibuprofen (Marco-Urrea *et al.*, 2010) obtained from wastewater. In addition, *T. versicolor* has lately received attention for its ability to degrade a variety of pharmaceuticals and personal care products, including nonsteroidal antiinflammatory drugs, lipid regulators, and antiepileptic pharmaceuticals (Marco-Urrea *et al.*, 2010). In the current circumstances, the degradation of several spiking pharmaceuticals and personal care products (naproxen and carbamazepine) by *T. versicolor* in sludge cultures revealed interesting results for potential applications (Rodríguez-Rodríguez *et al.*, 2010). The degradation capacity of *T. versicolor* was additionally demonstrated in sewage sludge sterilized systems, where 100% removal was accomplished for sulfamethazine, sulfapyridine, and sulfathiazole. This demonstrates the potential application of the fungus for bioremediation purposes.

In 2015, a project was conducted by Díaz-Cruz *et al.* (2015) to investigate the applications of environmentally friendly technology based on fungal-mediated treatment for the degradation of ingredients in personal care products, which are frequently detected at relevant concentrations in the aquatic environment. The reported removal efficacy varied greatly depending on the experimental setup, organic substance, and kind of fungal. The methods and factors governing fungi break down, particularly white-rot fungus and their specialized lignin-modifying enzymes, were thoroughly reviewed and explained. The WRF have been used in water and soil bioremediation processes (Rodríguez-Rodríguez *et al.*, 2013), and they are composed of an eco-physiological group of fungi capable of degrading lignin (Hale & Eaton, 1985; Rodríguez-Rodríguez *et al.*, 2013). Due to the low specificity of this enzymatic machinery, other targets, including a large number of contaminating compounds such as pharmaceutical compounds and antibiotics, can be degraded (Asgher *et al.*, 2008; Cruz-Morató *et al.*, 2013; Marco-Urrea *et al.*, 2010).

Walters *et al.* (2010) received funding from the National Institute of Environmental Health Sciences (NIEHS) to report the incidence and loss of various pharmaceutical compounds from biosolid–soil mixtures subjected to ambient outdoor settings for three years. Some compounds, such as diphenhydramine, fluoxetine, thiabendazole, and triclosan, showed no detectable loss during the monitoring period, while others, such as azithromycin, carbamazepine, ciprofloxacin, doxycycline, tetracycline, 4-epitetracycline, gemfibrozil, norfloxacin, and triclosan, had half-life estimates ranging from 182 to 3466 days. These findings emphasize the potential use of *T. versicolor* to lessen the impact of biosolids once discharged into the environment, perhaps lowering pharmaceutical compound concentrations in considerably shorter treatment periods.

It has been reported that microbially induced calcite has a high removal efficiency of many divalent metal cations and radionuclides such as Pb, Cd, Co, and Sr. It is a promising *in situ* remediation technology for environmental heavy metal contamination because it is extremely efficient, profitable, and ecologically acceptable (Fujita *et al.*, 2004; Lauchnor *et al.*, 2013). In fact, Li *et al.* (2013) used different isolates to assess their capability for removal of heavy metals including cobalt. It was found that the isolates could successfully remove the contaminations ranging from 88% to 99% in a short period of time (24 h). The results show that *Sporosarcina sp.* and *Terrabacter tumescens* had the highest removal for cobalt.

Actually, microbially induced calcite has been wildly investigated by employing bacteria for the mineralization of heavy metals such as chromium and lead. However, the process of metal remediation from solutions and soil using fungi is still not entirely defined. In 2017, Qian *et al.* published one of the few papers defining fungus-induced calcite precipitation in heavy metal cleanup. The fungal strain utilized in this study (*P. chrysogenum*) was isolated from cement sludge and subsequently used to biomineralize chromate and lead from an aqueous solution (Qian *et al.*, 2017). As an outcome, an increase in the carbonate-bound fraction of metals in soil was reported when this fungal strain was used for metal remediation in soil. In polluted soil, the proportion of exchangeable chromium decreased from 41.60% to 1.95%, whereas the percentage of exchangeable plomb decreased from 41.27% to 2.19%.

The Engineering Research Council of Canada (NSERC) financed a study related to the reaction of plants to heavy metal stress and pointed the way to solutions that could make this technology more viable (Gamalero *et al.*, 2009). As an alternative to traditional physical and chemical methods of environmental cleanup, scientists have created phytoremediation systems, which use plants to remove or render harmless a variety of contaminants. Plant growth-promoting bacteria and arbuscular-mycorrhizal fungi can both be used to help with phytoremediation and plant development in metal-contaminated soils (Gamalero *et al.*, 2009). As well as that, pesticides containing halogenated chemicals, such as chlorophenols, were widely utilized in the last century. They are soluble in organic solvents and only marginally soluble in water (Tomasini & León-Santiesteban, 2019).

Some fungal strains from various taxonomic backgrounds can eliminate, degrade, and even mineralize chlorophenols in liquid and solid matrices. A study linked 2-CP para-oxidation to the extracellular laccase activity produced by the white-rot fungus *Trametes versicolor*, which converted 2-CP into 2-chloro-1,4-benzoquinone (Grey *et al.*, 1998). Laccase has been shown to not only oxidize 2-CP but also polymerize monochlorophenols. Purified laccase from *Rhizoctonia praticola* was able to synthesize various dimeric, trimeric, and tetrameric compounds from 2-CP and 4-CP via oxidative coupling mechanisms (Sjoblad and Bollag, 1977). Laccase's capacity to polymerize monochlorophenols provides the possibility that it may be employed in the production of new substances with potential uses across a range of sectors. Another study showed the efficiency of the degradation activity of *P. chrysosporium* against 2-chlorophenol by allowing the fungus to grow on and into porous support particles suspended in an agitated fermenter (Armenante *et al.*, 1992).

There have been very few investigations onto the deterioration of ultraviolet (UV) filters in both liquid and solid media. There needs to be more data on the biodegradation of UV filters by fungi. Recently, an attempt was made to fill this gap by analyzing the potential of the white-rot fungus *T. versicolor* to destroy specific UV filters (Badia-Fabregat *et al.*, 2012). Rodríguez-Rodríguez *et al.* (2013) showed high percentages of degradation of several UV filters and some of their metabolites in a solid-phase treatment of sewage sludge by *T. versicolor*. The near-complete removal of those compounds was attributed to fungal biotransformation because the treatments were performed in sterile conditions. The use of the whole fungus, *T. versicolor*, to destroy specific UV filters showed that the degradation mechanisms are only beneficial if they do not result in the formation of new compounds with increased toxicity or bioaccumulation capacity (Badia-Fabregat *et al.*, 2012; Gago-Ferrero *et al.*, 2012). To have a complete picture of the process, it is required to identify and characterize the derivatives made during the transformation processes, as well as assess the potential toxicity of both the source chemicals and the degradation products formed.

REFERENCES

M. K. Abd Elnabi, Elkaliny N. E., Elyazied M. M., Azab S. H., Elkhalifa S. A., Elmasry S., Mouhamed M. S., Shalamesh E. M., Alhorieny N. A., Abd Elaty A. E., Elgendy I. M., Etman A. E., Saad K. E., Tsigkou K., Ali S. S., Kornaros M. and Mahmoud Y. A.-G. (2023). Toxicity of heavy metals and recent advances in their removal: a review. *Toxics*, **11**(7), 580, https://doi.org/10.3390/toxics11070580

Agrawal N., Verma P. and Shahi S. K. (2018). Degradation of polycyclic aromatic hydrocarbons (phenanthrene and pyrene) by the ligninolytic fungi Ganoderma lucidum isolated from the hardwood stump. *Bioresources and Bioprocessing*, **5**(1), 1–9, https://doi.org/10.1186/s40643-018-0197-5

Akerman-Sanchez G. and Rojas-Jimenez K. (2021). Fungi for the bioremediation of pharmaceutical-derived pollutants: a bioengineering approach to water treatment. *Environmental Advances*, **4**, pp. 100071, https://doi.org/10.1016/j.envadv.2021.100071

Alamo A. C., Pariente M. I., Sanchez-Bayo A., Puyol D., Rodriguez R., Orales V. M. *et al.* (2021). Assessment of trametes versicolor, isochrysis galbana, and purple phototrophic bacteria for the removal of pharmaceutical compounds in hospital wastewater. *Advanced Environment Engineering Res.* **02**, 1, https://doi.org/10.21926/aeer.2104027

Alharbi S. K., Nghiem L. D., van de Merwe J. P., Leusch F. D. L., Asif M. B., Hai F. I. and Price W. E. (2019). Degradation of diclofenac, trimethoprim, carbamazepine, and sulfamethoxazole by laccase from *Trametes versicolor*: transformation products and toxicity of treated effluent. *Biocatalysis and Biotransformation*, **37**(6), 399–408, https://doi.org/10.1080/10242422.2019.1580268

Anacletus F., Nwauche K. and Monago C. C. I. (2017). Mineral and heavy metal composition of crude oil polluted soil amended with non-ionic surfactant (triton X-100) and white rot fungus (*Pleurotus ostratus*). *Journal of Environmental & Analytical Toxicology*, **07**(03), 1–3, https://doi.org/10.4172/2161-0525.1000449

Angeles-de Paz G., Ledezma-Villanueva A., Robledo-Mahón T., Pozo C., Calvo C., Aranda E. and Purswani J. (2023). Assembled mixed co-cultures for emerging pollutant removal using native microorganisms from sewage sludge. *Chemosphere*, **313**, 137472, https://doi.org/10.1016/j.chemosphere.2022.137472

Armenante P. M., Lewandowski G. and Haq I. U. (1992). Mineralization of 2-chlorophenol by *P. chrysosporium* using different reactor designs. *Hazardous Waste & Hazardous Materials*, **9**(3), 213–229, https://doi.org/10.1089/hwm.1992.9.213

Arun K. B., Madhavan A., Tarafdar A., Sirohi R., Anoopkumar A. N., Kuriakose L. L., Awasthi M. K., Binod P., Varjani S. and Sindhu R. (2023). Filamentous fungi for pharmaceutical compounds degradation in the environment: a sustainable approach. *Environmental Technology and Innovation*, **31**, 103182, https://doi.org/10.1016/j.eti.2023.103182

Asgher M., Bhatti H. N., Ashraf M. and Legge R. L. (2008). Recent developments in biodegradation of industrial pollutants by white rot fungi and their enzyme system. *Biodegradation*, **19**(6), 771–783, https://doi.org/10.1007/s10532-008-9185-3

Ashe B., Nguyen L. N., Hai F. I., Lee D. J., van de Merwe J. P., Leusch F. D. L., Price W. E. and Nghiem L. D. (2016). Impacts of redox-mediator type on trace organic contaminants degradation by laccase: degradation efficiency, laccase stability and effluent toxicity. *International Biodeterioration and Biodegradation*, **113**, 169–176, https://doi.org/10.1016/j.ibiod.2016.04.027

Aydin S. (2016). Enhanced biodegradation of antibiotic combinations via the sequential treatment of the sludge resulting from pharmaceutical wastewater treatment using white-rot fungi *Trametes versicolor* and *Bjerkandera adusta*. *Applied Microbiology and Biotechnology*, **100**(14), 6491–6499, https://doi.org/10.1007/s00253-016-7473-0

Baborová P., Möder M., Baldrian P., Cajthamlová K. and Cajthaml T. (2006). Purification of a new manganese peroxidase of the white-rot fungus *Irpex lacteus*, and degradation of polycyclic aromatic hydrocarbons by the enzyme. *Research in Microbiology*, **157**(3), 248–253, https://doi.org/10.1016/j.resmic.2005.09.001

Badia-Fabregat M., Rodríguez-Rodríguez C. E., Gago-Ferrero P., Olivares A., Piña B., Díaz-Cruz M. S., Vicent T., Barceló D. and Caminal G. (2012). Degradation of UV filters in sewage sludge and 4-MBC in liquid medium by the ligninolytic fungus *Trametes versicolor*. *Journal of Environmental Management*, **104**, 114–120, https://doi.org/10.1016/j.jenvman.2012.03.039

Badia-Fabregat M., Lucas D., Gros M., Rodríguez-Mozaz S., Barceló D., Caminal G. and Vicent T. (2015). Identification of some factors affecting pharmaceutical active compounds (PhACs) removal in real wastewater. Case study of fungal treatment of reverse osmosis concentrate. *Journal of Hazardous Materials*, **283**, 663–671, https://doi.org/10.1016/j.jhazmat.2014.10.007

Badia-Fabregat M., Lucas D., Pereira M. A., Alves M., Pennanen T., Fritze H., Rodríguez-Mozaz S., Barceló D., Vicent T. and Caminal G. (2016). Continuous fungal treatment of non-sterile veterinary hospital effluent: pharmaceuticals removal and microbial community assessment. *Appl. Microbiol. Biotechnol.* **100**, 2401–2415. http://dx.doi.org/10.1007/s00253-015-7105-0

Bala S., Garg D., Thirumalesh B. V., Sharma M., Sridhar K., Inbaraj B. S. and Tripathi M. (2022). Recent strategies for bioremediation of emerging pollutants: a review for a green and sustainable environment. *Toxics*, **10**(8), 484, https://doi.org/10.3390/toxics10080484

Baldrian P., Gabriel J., Nerud F. and Zadrăzil F. 2000. Influence of cadmium and mercury on activities of ligninolytic enzymes and degradation of polycyclic aromatic hydrocarbons by Pleurotus ostreatus in soil. *Appl. Environ. Microbiol.* **66**, 2471–2478, https://doi.org/10.1128/aem.66.6.2471-2478.2000

Barret M., Carrère H., Latrille E., Wisniewski C. and Patureau D. (2010). Micropollutant and sludge characterization for modeling sorption equilibria. *Environmental Science & Technology*, **44**(3), 1100–1106, https://doi.org/10.1021/es902575d

Batista-García R. A., Kumar V. V., Ariste A., Tovar-Herrera O. E., Savary O., Peidro-Guzmán H., González-Abradelo D., Jackson S. A., Dobson A. D. W., del Sánchez-Carbente M. R., Folch-Mallol J. L., Leduc R. and Cabana H. (2017). Simple screening protocol for identification of potential mycoremediation tools for the elimination of polycyclic aromatic hydrocarbons and phenols from hyperalkalophile industrial effluents. *Journal of Environmental Management*, **198**, 1–11, https://doi.org/10.1016/j.jenvman.2017.05.010

Becker D., Varela Della Giustina S., Rodriguez-Mozaz S., Schoevaart R., Barceló D., de Cazes M., Belleville M.-P, Sanchez-Marcano J., de Gunzburg J., Couillerot O., Völker J., Oehlmann J and Wagner M. (2016). Removal of antibiotics in wastewater by enzymatic treatment with fungal laccase – Degradation of compounds does not always eliminate toxicity. *Bioresource Technology*, **219**, 500–509, https://doi.org/10.1016/j.biortech.2016.08.004

Beltrán-Flores E., Sarrà M., and Blánquez P. (2021). Pesticide bioremediation by Trametes versicolor: Application in a fixed-bed reactor, sorption contribution and bioregeneration. *Science of The Total Environment*, **794**, 148386, https://doi.org/10.1016/j.scitotenv.2021.148386

Benghazi L., Record E., Suárez A., Gomez-Vidal J. A., Martínez J. and de la Rubia T. (2014). Production of the Phanerochaete flavido-alba laccase in *Aspergillus niger* for synthetic dyes decolorization and biotransformation. *World Journal of Microbiology and Biotechnology*, **30**(1), 201–211, https://doi.org/10.1007/s11274-013-1440-z

Bhambri A., Karn S. K. and Singh R. K. (2021). In-situ remediation of nitrogen and phosphorus of beverage industry by potential strains *Bacillus* sp. (BK1) and Aspergillus sp. (BK2). *Scientific Reports*, **11**(1), 12243, https://doi.org/10.1038/s41598 021 91539 y

Bhandari S., Poudel D. K., Marahatha R., Dawadi S., Khadayat K., Phuyal S., Shrestha S., Gaire S., Basnet K., Khadka U. and Parajuli N. (2021). Microbial enzymes used in bioremediation. *Journal of Chemistry*, **2021**, 1–17, https://doi.org/10.1155/2021/8849512

Boxall A. B. A., Fogg L. A., Blackwell P. A., Kay P., Pemberton E. J. and Croxford A. (2004). Veterinary medicines in the environment. In: Reviews of Environmental Contamination and Toxicology, Springer, New York, (Vol. **180**), Springer, New York, pp. 1–91, https://doi.org/10.1007/0-387-21729-0_1

Buchicchio A., Bianco G., Sofo A., Masi S. and Caniani D. (2016). Biodegradation of carbamazepine and clarithromycin by Trichoderma harzianum and Pleurotus ostreatus investigated by liquid chromatography – high-resolution tandem mass spectrometry (FTICR MS-IRMPD). *Science of the Total Environment*, ***557–558***, 733–739, https://doi.org/10.1016/j.scitotenv.2016.03.119

Cai Y., Yu H., Ren L., Ou Y., Jiang S., Chai Y., Chen A., Yan B., Zhang J., and Yan Z. (2023). Treatment of amoxicillin-containing wastewater by Trichoderma strains selected from activated sludge. *Science of The Total Environment*, ***867***, 161565, https://doi.org/10.1016/j.scitotenv.2023.161565

Crini G., Lichtfouse E., Wilson L., Morin-Crini N. and Wilson L. D. (2018). Adsorption-oriented processes using conventional and non-conventional adsorbents for wastewater treatment adsorption-oriented processes using conventional and non-conventional adsorbents for wastewater treatment. Green adsorbents for pollutant removal adsorption-oriented processes using conventional and non-conventional adsorbents for wastewater treatment. *Environmental Chemistry for aSustainable World*, **18**, 978, https://doi.org/10.1007/978-3-319-92111-2_2ï

Cruz-Morató C., Rodríguez-Rodríguez C. E., Marco-Urrea E., Sarrà M., Caminal G., Vicent T., Jelić A., García-Galán M. J., Pérez S., Díaz-Cruz M. S., Petrović M. and Barceló D. (2013). Biodegradation of pharmaceuticals by fungi and metabolites identification. In: Handbook of Environmental Chemistry (Vol. 24). Springer-Verlag, Berlin Heidelberg, pp. 165–213, https://doi.org/10.1007/698_2012_158

Cui P., Bai Y., Li X., Peng Z., Chen D., Wu Z., Zhang P., Tan Z., Huang K., Chen Z., Liao H. and Zhou S. (2020). Enhanced removal of antibiotic resistance genes and mobile genetic elements during sewage sludge composting covered with a semi-permeable membrane. *Journal of Hazardous materials*, **396**, p.122738.

Daccò C., Girometta C., Asemoloye M. D., Carpani G., Picco A. M. and Tosi S. (2020). Key fungal degradation patterns, enzymes and their applications for the removal of aliphatic hydrocarbons in polluted soils: A review. *International Biodeterioration & Biodegradation*, **147**, p.104866.

da Coelho-Moreira J. S., Brugnari T., Sá-Nakanishi A. B., Castoldi R., de Souza C. G. M., Bracht A. and Peralta R. M. (2018). Evaluation of diuron tolerance and biotransformation by the white-rot fungus *Ganoderma lucidum*. *Fungal Biology*, **122**(6), 471–478, https://doi.org/10.1016/j.funbio.2017.10.008

Dalecka B., Juhna T. and Rajarao G. K. (2020). Constructive use of filamentous fungi to remove pharmaceutical substances from wastewater. *Journal of Water Process Engineering*, **33**, 100992, https://doi.org/10.1016/j.jwpe.2019.100992

del Álamo A. C., Pariente M. I., Vasiliadou I., Padrino B., Puyol D., Molina R. and Martínez F. (2018). Removal of pharmaceutical compounds from urban wastewater by an advanced bio-oxidation process based on fungi *Trametes versicolor* immobilized in a continuous RBC system. *Environmental Science and Pollution Research*, **25**(35), 34884–34892, https://doi.org/10.1007/s11356-017-1053-4

Deshmukh R., Khardenavis A. A. and Purohit H. J. (2016). Diverse metabolic capacities of fungi for bioremediation. *Indian Journal of Microbiology*, **56**(3), 247–264, https://doi.org/10.1007/s12088-016-0584-6

de Souza P. M., de Assis Bittencourt M. L., Caprara C. C., de Freitas M., de Almeida R. P. C., Silveira D., Fonseca Y. M., Filho E. X. F., Pessoa Junior A. and Magalhães P. O. (2015). A biotechnology perspective of fungal proteases. *Brazilian Journal of Microbiology*, **46**(2), pp. 337–346, https://doi.org/10.1590/S1517-838246220140359

Dhiman N., Jasrotia T., Sharma P., Negi S., Chaudhary S., Kumar R., Mahnashi M. H., Umar A. and Kumar R. (2020). Immobilization interaction between xenobiotic and Bjerkandera adusta for the biodegradation of atrazine. *Chemosphere*, 127060, https://doi.org/10.1016/j.chemosphere.2020.127060

Díaz-Cruz M. S., Gago-Ferrero P., Badia-Fabregat M., Caminal G., Vicent T. and Barceló D. (2015). Fungal-mediated biodegradation of ingredients in personal care products. *Handbook of Environmental Chemistry*, **36**, 295–318, https://doi.org/10.1007/698_2014_329

Drevinskas T., Mickiene R., Maruška A., Stankevičius M., Tiso N., Mikašauskaite J., Ragažinskiene O., Levišauskas D., Bartkuviene V., Snieškiene V., Stankevičiene A., Polcaro C., Galli E., Donati E., Tekorius T., Kornyšova O. and Kaškoniene V. (2016). Downscaling the *in vitro* test of fungal bioremediation of polycyclic aromatic hydrocarbons: Methodological approach. *Analytical and Bioanalytical Chemistry* **408**, 1043–1053, https://doi.org/10.1007/s00216-015-9191-3

Espinosa-Ortiz E. J., Rene E. R. and Gerlach R. (2022). Potential use of fungal-bacterial co-cultures for the removal of organic pollutants. *Critical Reviews in Biotechnology*, **42**(3), pp. 361–383, https://doi.org/10.1080/07388551.2021.1940831

Esterhuizen M., Behnam Sani S., Wang L., Kim Y. J. and Pflugmacher S. (2021). Mycoremediation of acetaminophen: culture parameter optimization to improve efficacy. *Chemosphere*, **263**, 128117, https://doi.org/10.1016/j.chemosphere.2020.128117

Feng N. X., Yu J., Xiang L., Yu L. Y., Zhao H. M., Mo C. H., Li Y. W., Cai Q. Y., Wong M. H. and Li Q. X. (2019). Co-metabolic degradation of the antibiotic ciprofloxacin by the enriched bacterial consortium XG and its bacterial community composition. *Science of the Total Environment*, **665**, 41–51, https://doi.org/10.1016/j.scitotenv.2019.01.322

Fujita Y., Redden G. D., Ingram J. C., Cortez M. M., Ferris F. G. and Smith R. W. (2004). Strontium incorporation into calcite generated by bacterial ureolysis. *Geochimica et Cosmochimica Acta*. **68**, e3261–e3270.

Gago-Ferrero P., Badia-Fabregat M., Olivares A., Piña B., Blánquez P., Vicent T., Caminal G., Díaz-Cruz M. S. and Barceló D. (2012). Evaluation of fungal- and photo-degradation as potential treatments for the removal of sunscreens BP3 and BP1. *Science of the Total Environment*, **427–428**, 355–363, https://doi.org/10.1016/j.scitotenv.2012.03.089

Gamalero E., Lingua G., Berta G. and Glick B. R. (2009). Beneficial role of plant growth promoting bacteria and arbuscular mycorrhizal fungi on plant responses to heavy metal stress. *Canadian Journal of Microbiology*, **55**(5), 501–514, https://doi.org/10.1139/W09-010

García-Galán M. J., Rodríguez-Rodríguez C. E., Vicent T., Caminal G., Díaz-Cruz M. S. and Barceló D. (2011). Biodegradation of sulfamethazine by *Trametes versicolor*: removal from sewage sludge and identification of intermediate products by UPLC-QqTOF-MS. *Science of the Total Environment*, **409**(24), 5505–5512, https://doi.org/10.1016/j.scitotenv.2011.08.022

Gogoi A., Mazumder P., Tyagi V. K., Tushara Chaminda G. G., An A. K. and Kumar M. (2018). Occurrence and fate of emerging contaminants in water environment: a review. *Groundwater for Sustainable Development*, **6**, 169–180, https://doi.org/10.1016/j.gsd.2017.12.009

Grey R., Höfer C. and Schlosser D. (1998). Degradation of 2-chlorophenol and formation of 2-chloro-1,4-benzoquinone by mycelia and cell-free crude culture liquids of *Trametes versicolor* in relation to extracellular laccase activity 1). *Journal of Basic Microbiology*, **38**(5–6), 371–82.

Gros M., Cruz-Morato C., Marco-Urrea E., Longrée P., Singer H., Sarrà M., Hollender J., Vicent T., Rodriguez-Mozaz S. and Barceló D. (2014). Biodegradation of the X-ray contrast agent iopromide and the fluoroquinolone antibiotic ofloxacin by the white rot fungus Trametes versicolor in hospital wastewaters and identification of degradation products. *Water Research*, **60**, 228–241, https://doi.org/10.1016/j.watres.2014.04.042

Guo X. L., Zhu Z. W. and Li H. L. (2014). Biodegradation of sulfamethoxazole by *Phanerochaete chrysosporium*. *Journal of Molecular Liquids*, **198**, 169–172, https://doi.org/10.1016/j.molliq.2014.06.017

Gupta S. K., Rachna Singh B., Mungray A. K., Bharti R., Nema A. K., Pant K. K. and Mulla S. I. (2022). Bioelectrochemical technologies for removal of xenobiotics from wastewater. *Sustainable Energy Technologies and Assessments*, **49**, 101652, https://doi.org/10.1016/j.seta.2021.101652

Hadibarata T., Tachibana S. and Itoh K. (2009). Biodegradation of chrysene, an aromatic hydrocarbon by *Polyporus* sp. S133 in liquid medium. *Journal of Hazardous Materials*, **164**(2–3), 911–917, https://doi.org/10.1016/j.jhazmat.2008.08.081

Hale M. D. and Eaton R. A. (1985). Oscillatory growth of fungal hyphae in wood cell walls. *Transactions of the British Mycological Society*, **84**(2), 277–288, https://doi.org/10.1016/S0007-1536(85)80079-6

Hu K., Peris A., Torán J., Eljarrat E., Sarrà M., Blánquez P. and Caminal G. (2020). Exploring the degradation capability of Trametes versicolor on selected hydrophobic pesticides through setting sights simultaneously on culture broth and biological matrix. *Chemosphere*, **250**, 126293, https://doi.org/10.1016/j.chemosphere.2020.126293

Huang C., Zeng G., Huang D., Lai C., Xu P., Zhang C., Cheng M., Wan J., Hu L. and Zhang Y. (2017). Effect of Phanerochaete chrysosporium inoculation on bacterial community and metal stabilization in lead-contaminated agricultural waste composting. *Bioresource Technology*, **243**, 294–303, https://doi.org/10.1016/j.biortech.2017.06.124

Jafari M., Danesh Y. R., Goltapeh E. M. and Varma A. (2013). Bioremediation and Genetically Modified Organisms, Springer, New York, pp. 433–451, https://doi.org/10.1007/978-3-642-33811-3_19

Jain A., Yadav S., Nigam V. K. and Sharma S. R. (2017). Fungal-Mediated Solid Waste Management: a Review, Springer, New York, pp. 153–170, https://doi.org/10.1007/978-3-319-68957-9_9

Jjemba P. K. (2006). Excretion and ecotoxicity of pharmaceutical and personal care products in the environment. *Ecotoxicology and Environmental Safety*, **63**(1), 113–130, https://doi.org/10.1016/j.ecoenv.2004.11.011

Kalyani D., Tiwari M. K., Li J., Kim S. C., Kalia V. C., Kang Y. C. and Lee J. K. (2015). A highly efficient recombinant laccase from the yeast Yarrowia lipolytica and its application in the hydrolysis of biomass. *PLoS ONE*, ***10***(3), e0120156, https://doi.org/10.1371/journal.pone.0120156

Kang B. R., Kim S. Y., Kang M. and Lee T. K. (2021). Removal of pharmaceuticals and personal care products using native fungal enzymes extracted during the ligninolytic process. *Environmental Research*, **195**, 110878, https://doi.org/10.1016/j.envres.2021.110878

Kothawale S. S., Kumar L. and Singh S. P. (2023). Role of organisms and their enzymes in the biodegradation of microplastics and nanoplastics: a review. *Environmental Research*, **232**, 116281, https://doi.org/10.1016/j.envres.2023.116281

Kour D., Kaur T., Devi R., Yadav A., Singh M., Joshi D., Singh J., Suyal D. C., Kumar A., Rajput V. D., Yadav A. N., Singh K., Singh J., Sayyed R. Z., Arora N. K. and Saxena A. K. (2021). Beneficial microbiomes for bioremediation of diverse contaminated environments for environmental sustainability: present status and future challenges. *Environmental Science and Pollution Research*, **28**(20), 24917–24939, https://doi.org/10.1007/s11356-021-13252-7

Kum H., Lee S., Ryu S. and Choi H. T. (2011). Degradation of endocrine disrupting chemicals by genetic transformants with two lignin degrading enzymes in *Phlebia tremellosa*. *Journal of Microbiology*, **49**(5), 824–827, https://doi.org/10.1007/s12275-011-1230-y

Kumar A., Yadav A. N., Mondal R., Kour D., Subrahmanyam G., Shabnam A. A., Khan S. A., Yadav K. K., Sharma G. K., Cabral-Pinto M., Fagodiya R. K., Gupta D. K., Hota S. and Malyan S. K. (2021). Myco-remediation: a mechanistic understanding of contaminants alleviation from natural environment and future prospect. *Chemosphere*, **284**, 131325, https://doi.org/10.1016/j.chemosphere.2021.131325

Lauchnor E. G., Schultz L. N., Bugni S., Mitchell A. C., Cunningham A. B. and Gerlach R. (2013). Bacterially induced calcium carbonate precipitation and strontium coprecipitation in a porous media flow system. *Environment Science Technology*, **47**, e1557–e1564.

Leitão A. L. (2009). Potential of penicillium species in the bioremediation field. *International Journal of Environmental Research and Public Health*, **6**(4), 1393–1417, https://doi.org/10.3390/ijerph6041393

Levin L., Papinutti L. and Forchiassin F. (2004). Evaluation of Argentinean white rot fungi for their ability to produce lignin-modifying enzymes and decolorize industrial dyes. *Bioresource Technology*, **94**(2), 169–176, https://doi.org/10.1016/j.biortech.2003.12.002

Li M., Cheng X. and Guo, H. (2013). Heavy metal removal by biomineralization of urease producing bacteria isolated from soil. *International Biodeterioration & Biodegradation*, **76**, 81–85.

Li B., Chen Y., Zhang Z., Qin G., Chen T. and Tian S. (2020). Molecular basis and regulation of pathogenicity and patulin biosynthesis in Penicillium expansum. *Comprehensive Reviews in Food Science and Food Safety*, **19**(6), 3416–3438.

Li X., Xu Q. M., Cheng J. S. and Yuan Y. J. (2016). Improving the bioremoval of sulfamethoxazole and alleviating cytotoxicity of its biotransformation by laccase producing system under coculture of Pycnoporus sanguineus and Alcaligenes faecalis. *Bioresource Technology*, **220**, 333–340, https://doi.org/10.1016/j.biortech.2016.08.088

Ma K. and Ruan Z. (2015). Production of a lignocellulolytic enzyme system for simultaneous bio-delignification and saccharification of corn stover employing co-culture of fungi. *Bioresource Technology*, **175**, 586–593.

Malik G., Arora R., Chaturvedi R. and Paul M. S. (2022). Implementation of genetic engineering and novel omics approaches to enhance bioremediation: a focused review. *Bulletin of Environmental Contamination and Toxicology*, **108**(3), 443–450, https://doi.org/10.1007/s00128-021-03218-3

Malik S., Bora J., Nag S., Sinha S., Mondal S., Rustagi S., Hazra R., Kumar H., Rajput V. D., Minkina T., Sadier N. S. and Almutary A. G. (2023). Fungal-based remediation in the treatment of anthropogenic activities and pharmaceutical-pollutant-contaminated wastewater. *Water*, **15**(12), 2262, https://doi.org/10.3390/w15122262

Maqsood Q., Sumrin A., Waseem R., Hussain M., Imtiaz M. and Hussain N. (2023). Bioengineered microbial strains for detoxification of toxic environmental pollutants. *Environmental Research*, **227**, 115665, https://doi.org/10.1016/j.envres.2023.115665

Marco-Urrea E., Pérez-Trujillo M., Blánquez P., Vicent T. and Caminal G. (2010). Biodegradation of the analgesic naproxen by *Trametes versicolor* and identification of intermediates using HPLC-DAD-MS and NMR. *Bioresource Technology*, **101**(7), 2159–2166, https://doi.org/10.1016/j.biortech.2009.11.019

Marshall H., Meneely J. P., Quinn B., Zhao Y., Bourke P., Gilmore B. F., Zhang G. and Elliott C. T. (2020). Novel decontamination approaches and their potential application for post-harvest aflatoxin control. *Trends in Food Science and Technology*, **106**, 489–496, https://doi.org/10.1016/j.tifs.2020.11.001

Mir-Tutusaus J. A., Caminal G. and Sarrà M. (2018). Influence of process variables in a continuous treatment of non-sterile hospital wastewater by *Trametes versicolor* and novel method for inoculum production. *Journal of Environmental Management*, **212**, 415–423, https://doi.org/10.1016/j.jenvman.2018.02.018

Mohapatra S., Menon N. G., Padhye L. P., Tatiparti S. S. V. and Mukherji S. (2021). Natural Attenuation of Pharmaceuticals in the Aquatic Environment and Role of Phototransformation, pp. 65–94, Springer, New York City, https://doi.org/10.1007/978-981-15-4599-3_3

Mukherjee A. G., Wanjari U. R., Eladl M. A., El-Sherbiny M., Elsherbini D. M. A., Sukumar A., Kannampuzha S., Ravichandran M., Renu K., Vellingiri B., Kandasamy S. and Gopalakrishnan A. V. (2022). Mixed contaminants: occurrence, interactions, toxicity, detection, and remediation. *Molecules*, **27**(8), 2577, https://doi.org/10.3390/molecules27082577

Naghdi M., Taheran M., Brar S. K., Kermanshahi-pour A., Verma M. and Surampalli R. Y. (2018). Removal of pharmaceutical compounds in water and wastewater using fungal oxidoreductase enzymes. *Environmental Pollution*, **234**, 190–213, https://doi.org/10.1016/j.envpol.2017.11.060

Nguyen L. N., Hai F. I., Yang S., Kang J., Leusch F. D. L., Roddick F., Price W. E. and Nghiem L. D. (2013). Culture of bacteria and white-rot fungi.

Nitsche B. M., Burggraaf-Van Welzen A. M., Lamers G., Meyer V. and Ram A. F. J. (2013). Autophagy promotes survival in aging submerged cultures of the filamentous fungus *Aspergillus niger*. *Applied Microbiology and Biotechnology*, **97**(18), 8205–8218, https://doi.org/10.1007/s00253-013-4971-1

Palli L., Castellet-Rovira F., Pérez-Trujillo M., Caniani D., Sarrà-Adroguer M. and Gori R. (2017). Preliminary evaluation of *Pleurotus ostreatus* for the removal of selected pharmaceuticals from hospital wastewater. *Biotechnology Progress*, **33**(6), 1529–1537, https://doi.org/10.1002/btpr.2520

Pathak N., Tran V. H., Merenda A., Johir M. A. H., Phuntsho S. and Shon H. (2020). Removal of organic micro-pollutants by conventional membrane bioreactors and high-retention membrane bioreactors. *Applied Sciences (Switzerland)*, **10**(8), 2969, https://doi.org/10.3390/APP10082969

Peng Y. Y., He S. and Wu F. (2021). Biochemical processes mediated by iron-based materials in water treatment: enhancing nitrogen and phosphorus removal in low C/N ratio wastewater. *Science of the Total Environment*, **775**, 145137, https://doi.org/10.1016/j.scitotenv.2021.145137

Petrović M., Gonzalez S. and Barceló D. (2003). Analysis and removal of emerging contaminants in wastewater and drinking water. *TrAC – Trends in Analytical Chemistry*, **22**(10), 685–696, https://doi.org/10.1016/S0165-9936(03)01105-1

Pezzella C., Macellaro G., Sannia G., Raganati F., Olivieri G., Marzocchella A., Schlosser D. and Piscitelli A. (2017). Exploitation of Trametes versicolor for bioremediation of endocrine disrupting chemicals in bioreactors. *PLoS One 12*, e0178758, http://dx.doi.org/10.1371/journal.pone.0178758

Pironti C., Ricciardi M., Proto A., Bianco P. M., Montano L. and Motta O. (2021). Endocrine-disrupting compounds: An overview on their occurrence in the aquatic environment and human exposure. *Water*, **13**(10), 1347.

Pozdnyakova N., Dubrovskaya E., Chernyshova M., Makarov O., Golubev S., Balandina S. and Turkovskaya O. (2018). The degradation of three-ringed polycyclic aromatic hydrocarbons by wood-inhabiting fungus *Pleurotus ostreatus* and soil-inhabiting fungus *Agaricus bisporus*. *Fungal Biology*, **122**(5), 363–372, https://doi.org/10.1016/j.funbio.2018.02.007

Priyadarshini E., Priyadarshini S. S., Cousins B. G. and Pradhan N. (2021). Metal-fungus interaction: review on cellular processes underlying heavy metal detoxification and synthesis of metal nanoparticles. *Chemosphere*, **274**, 129976, https://doi.org/10.1016/j.chemosphere.2021.129976

Purnomo A. S., Mori T., Kamei I., Nishii T. and Kondo R. (2010). Application of mushroom waste medium from *Pleurotus ostreatus* for bioremediation of DDT-contaminated soil. *International Biodeterioration and Biodegradation*, **64**(5), 397–402, https://doi.org/10.1016/j.ibiod.2010.04.007

Purnomo A. S., Sariwati A. and Kamei I. (2020). Synergistic interaction of a consortium of the brown-rot fungus Fomitopsis pinicola and the bacterium Ralstonia pickettii for DDT biodegradation. *Heliyon*, **6**(6), e04027, https://doi.org/10.1016/j.heliyon.2020.e04027

Qian X., Fang C., Huang M. and Achal V. (2017). Characterization of fungal-mediated carbonate precipitation in the biomineralization of chromate and lead from an aqueous solution and soil. *Journal of Cleaner Production*, **164**, 198–208, https://doi.org/10.1016/j.jclepro.2017.06.195

Rafeeq H., Afsheen N., Rafique S., Arshad A., Intisar M., Hussain A., Bilal M. and Iqbal H. M. N. (2023). Genetically engineered microorganisms for environmental remediation. *Chemosphere*, **310**, 136751, https://doi.org/10.1016/j.chemosphere.2022.136751

Rahman Z. (2020). An overview on heavy metal resistant microorganisms for simultaneous treatment of multiple chemical pollutants at co-contaminated sites, and their multipurpose application. *Journal of Hazardous Materials*, **396**), 122682, https://doi.org/10.1016/j.jhazmat.2020.122682

Ranieri D., Colao M. C., Ruzzi M., Romagnoli G. and Bianchi M. M. (2009). Optimization of recombinant fungal laccase production with strains of the yeast *Kluyveromyces lactis* from the pyruvate decarboxylase promoter. *FEMS Yeast Research*, **9**(6), 892–902, https://doi.org/10.1111/j.1567-1364.2009.00532.x

Roccatano D. (2015). Structure, dynamics, and function of the monooxygenase P450 BM-3: insights from computer simulations studies. *Journal of Physics: Condensed Matter*, **27**(27), p. 273102.

Rodríguez-Rodríguez C. E., Marco-Urrea E. and Caminal G. (2010). Degradation of naproxen and carbamazepine in spiked sludge by slurry and solid-phase *Trametes versicolor* systems. *Bioresource Technology*, **101**(7), 2259–2266, https://doi.org/10.1016/j.biortech.2009.11.089

Rodríguez-Rodríguez C. E., Barón E., Gago-Ferrero P., Jelić A., Llorca M., Farré M., Díaz-Cruz M. S., Eljarrat E., Petrović M., Caminal G., Barceló D. and Vicent T. (2012a). Removal of pharmaceuticals, polybrominated flame retardants and UV-filters from sludge by the fungus *Trametes versicolor* in bioslurry reactor. *Journal of Hazardous Materials*, **233–234**, 235–243, https://doi.org/10.1016/j.jhazmat.2012.07.024

Rodríguez-Rodríguez C. E., Jelić A., Pereira M. A., Sousa D. Z., Petrović M., Alves M. M., Barceló D., Caminal G. and Vicent T. (2012b). Bioaugmentation of sewage sludge with *Trametes versicolor* in solid-phase biopiles produces degradation of pharmaceuticals and affects microbial communities. *Environmental Science and Technology*, **46**(21), 12012–12020, https://doi.org/10.1021/es301788n

Rodríguez-Rodríguez C. E., Caminal G., Vicent T., Díaz-Cruz M. S., Eljarrat E., Farré M., de Alda M. J. L., Petrović M. and Barceló D. (2013). Fungal-mediated degradation of emerging pollutants in sewage sludge. In: Handbook of Environmental Chemistry (Vol. 24). Springer-Verlag, Berlin Heidelberg, pp. 137–164, https://doi.org/10.1007/698_2012_159

Rubin B. E., Diamond S., Cress B. F., Crits-Christoph A., Lou Y. C., Borges A. L., Shivram H., He C., Xu M., Zhou Z., Smith S. J., Rovinsky R., Smock D. C. J., Tang K., Owens T. K., Krishnappa N., Sachdeva R., Barrangou R., Deutschbauer A. M. and … Doudna J. A. (2022). Species- and site-specific genome editing in complex bacterial communities. *Nature Microbiology*, **7**(1), 34–47, https://doi.org/10.1038/s41564-021-01014-7

Singh A. K., Bilal M., Iqbal H. M. N. and Raj A. (2021). Lignin peroxidase in focus for catalytic elimination of contaminants – a critical review on recent progress and perspectives. *International Journal of Biological Macromolecules*, **177**, 58–82, https://doi.org/10.1016/j.ijbiomac.2021.02.032

Sjoblad R. D. and Bollag J. M. (1977). Oxidative coupling of aromatic pesticide intermediates by a fungal phenol oxidase. *Applied and Environmental Microbiology*, **33**(4), 906–910.

Soares P. R. S., Birolli W. G., Ferreira I. M. and Porto A. L. M. (2021). Biodegradation pathway of the organophosphate pesticides chlorpyrifos, methyl parathion and profenofos by the marine-derived fungus *Aspergillus sydowii* CBMAI 935 and its potential for methylation reactions of phenolic compounds. *Marine Pollution Bulletin*, **166**, 112185, https://doi.org/10.1016/j.marpolbul.2021.112185

Somu P., Narayanasamy S., Gomez L. A., Rajendran S., Lee Y. R. and Balakrishnan D. (2022). Immobilization of enzymes for bioremediation: a future remedial and mitigating strategy. *Environmental Research*, **212**, 113411, https://doi.org/10.1016/j.envres.2022.113411

Stenholm Å, Hedeland M., Arvidsson T. and Pettersson C. E. (2019). Removal of diclofenac from a non-sterile aqueous system using *Trametes versicolor* with an emphasis on adsorption and biodegradation mechanisms. *Environmental Technology (United Kingdom)*, **40**(19), 2460–2472, https://doi.org/10.1080/09593330.2018.1444098

Suenaga H., Watanabe T., Sato M., Ngadiman and Furukawa K. (2002). Alteration of regiospecificity in biphenyl dioxygenase by active-site engineering. *Journal of Bacteriology*, **184**(13), 3682–3688, https://doi.org/10.1128/JB.184.13.3682-3688.2002

Tomasini A. amd León-Santiesteban H. H. (2019). The role of the filamentous fungi in bioremediation. In: Fungal Bioremediation: Fundamentals and Applications, 3–21, Tomasini, A. and H. H. León-Santiesteban (eds), CRC Press, Taylor & Francis. https://doi.org/10.1201/9781315205984

Torán J., Blánquez P. and Caminal G. (2017). Comparison between several reactors with *Trametes versicolor* immobilized on lignocellulosic support for the continuous treatments of hospital wastewater. *Bioresource Technology*, **243**, 966–974, https://doi.org/10.1016/j.biortech.2017.07.055

Tormo-Budowski R., Cambronero-Heinrichs J. C., Durán J. E., Masís-Mora M., Ramírez-Morales D., Quirós-Fournier J. P. and Rodríguez-Rodríguez C. E. (2021). Removal of pharmaceuticals and ecotoxicological changes in wastewater using *Trametes versicolor*: a comparison of fungal stirred tank and trickle-bed bioreactors. *Chemical Engineering Journal*, **410**, 128210, https://doi.org/10.1016/j.cej.2020.128210

Tran N. H., Urase T. and Kusakabe O. (2010). Biodegradation characteristics of pharmaceutical substances by whole fungal culture *Trametes versicolor* and its laccase. *Journal of Water and Environment Technology*, **8**(2), 125–140.

Tran N. H., Reinhard M., Khan E., Chen H., Nguyen V. T., Li Y., Goh S. G., Nguyen Q. B., Saeidi N. and Gin K. Y. H. (2019). Emerging contaminants in wastewater, stormwater runoff, and surface water: application as chemical markers for diffuse sources. *Science of the Total Environment*, **676**, 252–267, https://doi.org/10.1016/j.scitotenv.2019.04.160

Turovskiy I. S. and Mathai P. K. (2006). Wastewater sludge processing. John Wiley & Sons, New Jersey.

Vaksmaa A., Guerrero-Cruz S., Ghosh P., Zeghal E., Hernando-Morales V. and Niemann H. (2023). Role of fungi in bioremediation of emerging pollutants. *Frontiers in Marine Science*, **10**, 1–21, https://doi.org/10.3389/fmars.2023.1070905

Verlicchi P., Al Aukidy M. and Zambello E. (2012). Occurrence of pharmaceutical compounds in urban wastewater: removal, mass load and environmental risk after a secondary treatment-A review. *Science of the Total Environment*, **429**, 123–155, https://doi.org/10.1016/j.scitotenv.2012.04.028

Vicent T., Caminal G., Eljarrat, E. and Barceló D. (2013). The Handbook of Environmental Chemistry: Emerging Organic Contaminants in Sludges: Analysis, Fate and Biological Treatment (The Handbook of Environmental Chemistry; Vol. 24). Springer-Verlag. https://doi.org/10.1007/698_2012_208

Walters E., McClellan K. and Halden R. U. (2010). Occurrence and loss over three years of 72 pharmaceuticals and personal care products from biosolids-soil mixtures in outdoor mesocosms. *Water Research*, **44**(20), 6011–6020, https://doi.org/10.1016/j.watres.2010.07.051

Wan J., Zeng G., Huang D., Huang C., Lai C., Li N., Wei Z., Xu P., He X., Lai M. and He Y. (2015). The oxidative stress of *Phanerochaete chrysosporium* against lead toxicity. *Applied Biochemistry and Biotechnology*, **175**(4), 1981–1991, https://doi.org/10.1007/s12010-014-1397-x

Wani K. A., Mamta Shuab R. and Lone R. A. (2017). Earthworms and associated microbiome: natural boosters for agro-ecosystems. In: Probiotics in Agroecosystem, V. Kumar, M. Kumar and S. Sharma (eds), Springer, Singapore, ISBN: 9789811040580. https://doi.org/10.1007/978-981-10-4059-7_25

Zhou Z., Chen Y., Liu X., Zhang K. and Xu H. (2015). Interaction of copper and 2,4,5-trichlorophenol on bioremediation potential and biochemical properties in co-contaminated soil incubated with *Clitocybe maxima*. *RSC Advances*, **5**(53), 42768–42776, https://doi.org/10.1039/C5RA04861C

doi: 10.2166/9781789063752_0159

Chapter 9

Tracing the pathways: the journey of emerging contaminants from wastewater into the environment

Purusottam Tripathy[1], *Charu Juneja*[1,2], *Abhishek Sharma*[1,2], *Om Prakash*[1] *and Sukdeb Pal*[1,2*]

[1]*Wastewater Technology Division, CSIR-National Environmental Engineering Research Institute, Nagpur 440020, India*
[2]*Academy of Scientific and Innovative Research (AcSIR), Ghaziabad 201002, India*
**Corresponding author: s_pal@neeri.res.in*

ABSTRACT

Emerging contaminants (ECs) are pervasive in the environment and have gained more prominence over the past few years. ECs such as pharmaceuticals, personal care products, endocrine disrupting compounds, and their active congeners, whose occurrences at trace levels in untreated/treated wastewater are detrimental to human health and the natural biota. The search for these chemicals' origins has frequently led to wastewater treatment facilities as a portal of entry for pollutants into the environment. Emerging pollutants enter wastewater via a variety of routes, including hazardous spills, farm runoff, excretion via urine and feces, and consumer product disposal and usage. Another source could be items like shampoo, toothpaste, soap, and disinfection washes which contain biologically active components that, when used, release these pollutants into the sewage system and are subsequently transferred to a wastewater treatment plant. Since these contaminants were introduced or detected relatively recently, there is an existential knowledge gap about their fate, impacts, and behaviors, as well as treatment strategies for their effective removal. The word "emerging pollutants," which is frequently used in environmental debates, refers to new health risks that emerge as a result of microlevel exposure to existing pollutants. Emerging contaminants have a variety of physicochemical characteristics that influence their exposure, fate, persistence, and toxicity in the environment. In the present chapter, a brief overview of emerging pollutants, their main categories, occurrences, points of discharge, and toxicity in natural and engineered systems will be provided, followed by an illustration of physicochemical properties and detected micropollutant concentrations that have raised concerns in recent years. This chapter also focuses on existing research that provides credible and quantitative information on ECs in various water sources, as well as the removal efficacy of different treatment techniques for different emerging contaminants.

Keywords: emerging contaminants, micropollutant, PPCPs, fate and transport, environment

9.1 BACKGROUND

The extensive prevalence of water and environmental contamination poses a threat to the well-being and sustainability of the interconnected ecosystems that support all life. The growing concern regarding pollution has prompted individuals worldwide to dedicate their efforts to investigating various types of pollutants, including biological entities like microbes, metallic substances such as nutrients, heavy metals, and trace organic micropollutants (Chen & Zhou, 2014; De la Cruz *et al.*, 2012). Over the past few decades, research on the characteristics of wastewater has brought to notice the environmental presence of numerous newly identified anthropogenic compounds. These trace compounds, predominantly organic, are commonly referred to as "emerging pollutants". An emerging contaminant refers to a contaminant that has a new origin, an alternative pathway to human exposure, or novel treatment techniques. These contaminants are classified based on their perceived, potential, or actual risks to the environment and humans. They can be instigated from various sources such as industrial activities, municipal (domestic) wastewater, agricultural practices, hospital waste, or laboratory effluents. In aquatic environments, the concentration of emerging contaminants (ECs) typically shows variation ranging from parts per billion (ppb) to parts per trillion (ppt). The detrimental effects they have on both aquatic and terrestrial organisms, as well as human health, have become a matter of concern among experts and the general public. The notable classes of emerging contaminants encompass pharmaceuticals and personal care products, pesticides, surfactants and their congeners, nanomaterials, plasticizers, and flame retardants. Certain endocrine-disrupting chemicals (EDCs) have been recognized as a subset of these emerging contaminants. They possess the capacity to interfere with the regular functioning of the endocrine system, which is found in a wide array of organisms, including amphibians, snails, fish, birds, crustaceans, humans, and other species. However, emerging contaminants are not limited to the aforementioned categories and can also include nanomaterials (NMs), metabolites of contaminants, engineered genes, illegal drugs, and so on. The organic trace pollutants, which are not currently regulated are categorized as emerging micropollutants and have been discovered or identified more recently due to advancements in analytical technologies. Nanomaterials, for instance, have an impact on bacterial biomass during wastewater treatment, resulting in reduced biological activity and subsequently decreasing the efficiency of emerging contaminant removal (Wang *et al.*, 2012).

ECs have been detected in various water sources, including groundwater, drinking water, surface water, and WWTP effluent discharge. These contaminants can be found in all these water reservoirs and pose potential risks to both the environment and human health (Samaras *et al.*, 2013; Yang *et al.*, 2014). Municipal wastewater is considered a significant contributor to the release of emerging contaminants into the environment, along with other sources such as point and non-point sources, industries, stormwater runoff, domestic wastewater, and WTPs. Additionally, there is growing attention toward the management of sludge due to the presence of high levels of emerging contaminants in it. The existing design of wastewater treatment plants (WWTPs) has limitations in effectively removing emerging contaminants and their metabolites, resulting in their release into rivers or streams. Thus, there is a need for further research and improvements in wastewater treatment technologies to address the removal of emerging contaminants. Significant progress has been made in the field of wastewater technologies concerning nutrient removal. However, the focus on the performance of these technologies in removing emerging contaminants is still an ongoing area of study (Molinos-Senante *et al.*, 2012). Limited data are available regarding the ecotoxicological impacts of ECs on surface water bodies as well as their removal efficacy in water treatment processes. In wastewater, pharmaceutical compounds can belong to various classes, including veterinary/human antibiotics, prescribed/non-prescribed drugs, and certain steroids. PCPs encompass chemicals present in consumer goods, such as galaxolide and tonalide. EDCs have the potential to disrupt endocrine systems by exhibiting estrogenic or androgenic activities even at lower concentrations. The environmental presence of these emerging contaminants leads to disruptions in both physiological and reproductive metabolism, increasing the risk of cancer, developing antibiotic-resistant bacteria, and enhancing toxicity from the mixture of chemicals. It is important to note that these pollutants are generally not monitored extensively and are not currently regulated in potable water sources (Noguera-Oviedo & Aga, 2016). The current lack of knowledge primarily pertains to the long-term impacts of emerging contaminants, which have been relatively unexplored thus far. Despite the significant discharge of human medicines and pharmaceuticals into the environment, there is a notable absence of comprehensive controls for ecological risk assessment. It is crucial to implement effective wastewater treatment processes before their release into the environment. Therefore, further research is necessary to understand the occurrence of ECs at lower concentrations (~ng/L level) in wastewater, their fate and transformation during wastewater treatment, and their potential implications for drinking water production (Gogoi *et al.*, 2018).

9.2 EMERGING (MICRO)POLLUTANTS IN THE ENVIRONMENT

ECs encompass a range of inorganic and organic micro-pollutants including pharmaceuticals, PCPs, polycyclic aromatic hydrocarbons (PAHs), pesticides, surfactants, synthetic organic dyes, perfluorinated compounds, heavy metal ions, plasticizers, flame retardants, and more. These compounds are generated as a result of human activities in various sectors such as domestic, healthcare units, agriculture, and industry (Goel, 2006). These contaminants widely spread throughout the environment and pose critical risks to both humans and wildlife due to their physicochemical properties. They are challenging to detect and have diverse activities and sources of production. Even at low concentrations, their presence can cause disruption in the endocrine system, chronic toxicity, and contribute to the development of pathogen resistance. The potential adverse effects of these contaminants underline the need for thorough monitoring and mitigation measures (Houtman, 2010).

A wide range of chemical and microbial substances, previously not regarded as contaminants, are currently being discovered in different environmental contexts, including areas where they were never intentionally introduced. This occurrence can be primarily attributed to their ability to persist during long-distance transportation. The origins and routes through which these newly identified contaminants are introduced into the environment are becoming more closely associated with the waste and wastewater produced by agricultural, industrial, and municipal practices. Despite arising from similar industrial, commercial, and domestic activities as conventional contaminants, these pollutants possess distinct characteristics that necessitate changes in the conventional approach to pollution control and prevention. Chemical micropollutants often originate from the degradation of organic compounds, leading to the accumulation of persistent metabolites (Kolpin *et al.*, 2002). Additionally, the disposal of such products into the natural environment contributes to their presence. Moreover, shifts in agricultural practices toward intensive farming and the application of sludge or manure on agricultural fields can result in the leaching of pollutants into groundwater and surface water, thereby posing health concerns (Gavrilescu *et al.*, 2015).

9.2.1 Pharmaceuticals

Pharmaceuticals encompass a diverse range of compounds that can be categorized as acidic, basic, or neutral. They are commonly used to treat various conditions, including pain, inflammation, and other medical ailments. The prevalence of pharmaceuticals in the environment varies, with some compounds like naproxen being prescription drugs, while others like caffeine are found in everyday consumables such as coffee, tea, and chocolate. Pharmaceuticals play a crucial role in human healthcare and veterinary medicine, serving purposes such as nutrition, therapy, diagnostic aids, and preventive medicine. However, the extensive use of pharmaceutical products, including prescribed and non-prescribed drugs, antibiotics, and hormones has led to their widespread detection in the aquatic environment, including surface water and groundwater. These pharmaceuticals are significant emerging organic contaminants that are found in trace amounts in water sources worldwide (Chinnaiyan *et al.*, 2018). These pharmaceuticals have adverse effects on human health, as well as on fish farming, livestock, and poultry. While researchers have studied over 3000 chemicals used in therapeutic products, only a small proportion of these have been examined in the field at ng/L doses. This knowledge gap raises concerns about the potential negative effects of these pharmaceuticals on human health and wildlife. In livestock farming, organic fertilizers such as urine and manure are used to enhance productivity. However, these substances indirectly impact the environment and can reach living organisms through food consumption. Frequently mentioned pharmaceuticals found in wastewater encompass a range of substances such as antacids, analgesics, antibiotics, clofibric acid, tranquilizers, β-blockers, stimulants, lipid-lowering drugs, steroids, nitroglycerin, antidepressants, antipyretics, propranolol, salicylic acid, and anti-inflammatory drugs (Richardson & Kimura, 2017).

Synthetic or natural hormones are indeed significant ecological contaminants due to their androgenic and estrogenic effects on the ecosystem. Both organic and inorganic hormones can have adverse impacts on the environment. Some examples of these hormones include 17β-estradiol, 17α-estradiol, norethindrone, estrone, equilenin, equiline, estriol, and mestranol. These hormones can enter the atmosphere through agricultural practices and are not fully eliminated during wastewater treatment processes. Consequently, they can persist in the aquatic environment, posing risks to aquatic life and potentially affecting human health as well. The estrogenic and androgenic properties of these hormones can disrupt the endocrine systems of ecology, leading to adverse effects on reproduction, development, and overall ecosystem health. Therefore, the presence of hormones in the environment is a matter of concern and requires attention in terms of pollution control and mitigation efforts.

9.2.2 Antidepressants

Antidepressants belong to a class of pharmaceuticals that impact neurotransmitters, which are chemicals used by nerves in the brain to communicate. Neurotransmitters such as serotonin, dopamine, and norepinephrine play vital roles in this communication process. The mechanism of action for antidepressants is thought to involve the inhibition of neurotransmitter release or modulation of their activity. The specific antidepressant compounds are O-dimethyl venlafaxine, venlafaxine, citalopram, and dimethyl citalopram. Venlafaxine is categorized as a serotonin-norepinephrine reuptake inhibitor (SNRI) and is frequently prescribed to address conditions such as depression, depression accompanied by anxiety symptoms, panic disorder, social anxiety disorder, and generalized anxiety disorder in adults. O-dimethyl venlafaxine, a major active metabolite of venlafaxine, also functions as an SNRI. Desvenlafaxine, a synthetic form of O-dimethyl venlafaxine, has been approved by Health Canada since 2009 for the treatment of depression. On the other hand, Citalopram is a selective serotonin reuptake inhibitor (SSRI) used to manage depression. It is also prescribed for treating conditions such as premenstrual dysphoric disorder, post-traumatic stress disorder, anxiety disorder, panic disorder, and obsessive–compulsive disorder. Dimethyl citalopram, an active metabolite of citalopram, also functions as an SSRI. These antidepressant medications act on the neurotransmitter systems in the brain to alleviate symptoms associated with various mental health conditions (Wilson & Ashraf, 2018).

9.2.3 Personal care products (PCPs)

PCPs encompass a broad range of household chemicals that are commonly used for health, beauty, cleaning, or odor control purposes. These chemicals can be found in various personal care products such as hair and skincare products, soaps, sunscreens, cosmetics, fragrances, and lotions. PCPs also extend to a class of chemicals present in a wide range of consumer goods, including toothpaste and kitchen utensils. PCPs are typically utilized for their antimicrobial and antifungal properties, serving to maintain hygiene and prevent microbial growth. These products are used in substantial quantities worldwide, contributing to the increasing release of these pollutants into the environment. As PCPs are extensively used in everyday life, their presence in the environment is becoming more significant. The continuous use and disposal of personal care products contribute to the contamination of water bodies, soil, and air. It is important to consider the environmental impact of PCPs and explore ways to mitigate their release and potential adverse effects on humans and ecosystems (Kim *et al.*, 2016). Indeed, many substances found in personal care products are bioactive and have the potential to accumulate in the environment and living organisms. These substances can pose risks to both the environment and human health. The bioactive nature of PCP substances means that they can interact with biological systems. This can lead to various ecological effects when these substances are released into the environment. For example, certain PCP compounds may disrupt the endocrine system of organisms, interfere with reproductive functions, or cause toxicity in aquatic organisms. Additionally, the bioaccumulative nature of these substances means that they can build up in the tissues of organisms over time. This bioaccumulation can occur through various pathways, such as ingestion or absorption. Over time, the accumulated levels of PCP substances can reach concentrations that are harmful to both the ecosystem and humans. The potential harm to the environment and humans emphasizes the need for responsible use, disposal, and regulation of PCPs. It is important to consider the potential long-term effects and to explore alternatives that are less harmful to the environment and human health (Claudia & Magrini, 2017). The potential emerging pollutants in personal care products (PCPs) include antiseptics, perfume pollutants such as ultraviolet (UV) filters, galaxolide, preservatives like diethyl phthalate, pest repellants and disinfectant pollutants like Triclosan (TCS) and triclocarban (TCC). TCS and TCC are commonly found in wastewater samples at higher concentrations. These substances are antibacterial and antifungal agents that are widely used in consumer goods like toothpaste, soaps, body washes, and disinfectants. These products typically contain between 0.1% of TCS and 2% of TCC by weight (Gatidou *et al.*, 2007). TCS and TCC exert their effects on bacteria by interacting with the enoyl-acyl carrier protein reductase enzyme (ENR), which is present in their cell membrane. This interaction inhibits the synthesis of fatty acids. Parabens, including methyl, ethyl, butyl, propyl, benzyl, isobutyl, and isopropyl hydroxybenzoates are antimicrobial preservatives commonly used in cosmetics, pharmaceuticals, and certain food products (North, 2004).

9.2.4 Polycyclic aromatic hydrocarbons (PAHs)

PAHs are abundant compounds that occur naturally in fossil fuels (Cao *et al.*, 2017). The incomplete combustion of wood, coal, gas, and oil is also one of the primary sources of PAHs. PAHs are recognized for their ability to undergo bioconcentration, leading to their rapid entry into the food chain (Yang *et al.*, 2022). The United States Environmental Protection Agency (US EPA) has classified 16 specific PAHs as significant contaminants of concern (Li *et al.*, 2019). PAHs are persistent compounds due to their lipophilic (fat-loving) and hydrophobic (water-repellent) characteristics. These properties enable them to remain in the environment for extended periods. PAHs have low volatility, making them resistant to burning, and non-biodegradable. As the molecular mass of PAHs increases, their solubility in an aqueous solution decreases logarithmically. PAHs with five or more rings, due to their lower solubility and low volatility, are commonly found in a granular form. They tend to be attached to contaminated soil, air, or sediment particulates. On the other hand, PAHs with fewer rings are more easily soluble in water, making them readily available for biological uptake and degradation. In general, PAHs with higher numbers of rings are more persistent in the environment compared to those with lower rings. The increased molecular complexity of higher ring PAHs contributes to their reduced solubility and resistance to degradation, leading to their longer persistence in environmental systems

9.2.5 Phthalate esters (PAEs)

PAEs are frequently employed as additives to enhance the flexibility of specific polyvinyl chloride (PVC) resins. They are also utilized in various other resins like cellulose, polyurethanes, and vinyl acetate. The low volatility, fluidity, and stability of phthalate esters make them well-suited as plasticizers (Jorgensen, 2008). Phthalate esters are derived from phthalic anhydride and are blended with plastics to enhance properties such as plasticity, resilience, and transparency (Thomas & Brogat, 2022). These derivatives find applications in various end-user products, including agricultural adjuvants, resin houses, toys, cosmetics, soaps and laundry detergents, and more. PAEs are characterized by their poor water solubility, which plays a crucial role in their biodegradability, aquatic toxicity, and distribution in the environment. Despite their low solubility, PAEs can be readily absorbed by organic residues and solid surfaces in environmental systems. This slow and continuous accumulation and release of

PAEs can have implications for the ecological conditions of water systems. The presence of PAEs in wastewater treatment facilities and sludge-amended soils is also affected by this phenomenon. Given their widespread use, the accumulation of PAEs in various compartments of ecosystems has become a concern. In agricultural soils, the accumulation of PAEs can lead to contamination of the food chain, including vegetables, and result in indirect or direct human exposure (Vickers, 2017).

9.2.6 Pesticides

Pesticides encompass a collection of organic pollutants categorized according to their distinctive physiochemical properties, which include fungicidal agents, herbicidal compounds, bacteriostatic substances, and insecticidal compounds. These substances are widely employed in the agroindustry to manage and regulate harmful insects, weeds, and microorganisms, among other factors. Nowadays, pesticides are often detected in groundwater which leads to toxicity, and health effects on living beings due to their high octanol–water ($K_{ow} > 3$) values. Generally, dichloro di-phenyl tri-chloroethane (DDT) and hexachlorocyclohexane are the most consuming pesticides rather than those of phorate, chlorpyriphos, Atrazine, methyl parathione, and Bentazone (Poonia *et al.*, 2021).

9.2.7 Endocrine active compounds

EACs are a diverse group containing both natural and synthetic compounds that have the ability to disrupt hormone systems in animals' bodies. The identification and control of detrimental effects caused by estrogenic pesticides and drugs (known as xenoestrogens) in living beings commenced during the mid-1900s. In the late 1980s and early 1990s, substantial evidence emerged regarding the endocrine-disrupting effects on reproductive systems caused by human-derived xenoestrogens present in wastewater effluents, even at nano-scale concentrations measured in parts per billion and parts per trillion (Watkinson *et al.*, 2007). Various studies have been done on the commotion of sex determination and rations, transformation in procreant behavior, and contraceptive-like actions in both sexes. Comparable disruptions have also been reported in amphibians, reptiles, birds, and mammals through various pathways of exposure. Research on accidental exposures and correlative studies examining the impact of estrogenic compounds found in meals, synthetic polymers (plastics), PPCPs, and other sources have been conducted on cultivated cells, gnawing, and humans. It has been observed that developing organisms, including the human fetus, exhibit greater sensitivity to exogenous estrogenic chemicals compared to adults (Yu *et al.*, 2009). Research conducted on gnawers and humans has shown that a single exposure to estrogenic compounds during growth not only affects the exposed generation but also induces enduring alterations that can be inherited by subsequent generations, without requiring additional exposures.

9.2.8 Surfactants and food additives

Surfactants, which come under synthetic organic compounds, are used globally in the production of domestic and commercial goods such as emulsifiers, cleaning agents, dyes, pesticides, and PPCPs (Mandaric *et al.*, 2016). Surfactants are classified into three categories based on their chemical properties: cationic, anionic, and dipolar surfactants. High global demand has led to significant production of surfactants such as lignin, benzene sulfonates, and alcohol ethoxylates. Synthetic sweeteners like saccharin and sucralose, which find extensive usage in food products, PPCPs, enter wastewater primarily through human excretion. Metabolized sweeteners not only pose environmental pollution but also exhibit long-term persistence, contributing to their persistency impact (Mahmood *et al.*, 2022).

9.2.9 Musks

Musks are a group of compounds renowned for their aromatic qualities. They are widely utilized across various cosmetics and detergents. Musks are classified into three classes: aromatic nitro, polycyclic, and macrocyclic musks. Polycyclic musks are used in cleansing products such as shampoos, hair care, and detergents. Polycyclic musks are applied topically to human skin, their subsequent release into the environment occurs without undergoing any metabolic alterations. Cosmetics, being extensively utilized among these products, can pose a potential risk to human beings, wildlife, and the environment, even when present in low quantities (Wilson & Ashraf, 2018).

9.3 EC IN AN AQUEOUS ENVIRONMENT

9.3.1 Classification and sources of EC

Approximately 70% of the numerous ECs identified in samples belong to the category of pharmacologically active chemicals and PPCPs, while the remaining 30% comprise industrial and agrochemical substances (Das *et al.*, 2017). Globally, more than 200 pharmacologically active chemicals have been detected in river streams, with the highest documented abundance being 6 ppm for ciprofloxacin (Hughes *et al.*, 2013). Moreover, tamoxifen was identified between 25 and 38 ng/L in river streams (Ferrando-Climent *et al.*, 2014). Jones and group detected multiple antibiotics, hormones, antidepressants, lipid regulators, analgesic compounds, and chemotherapy drugs in the range between 0.04 and 6.3 μg/L. In contrast, commonly used chemicals such as sunscreen agents and preservatives are frequently detected at concentration levels exceeding 1000 ng/L (Petrie *et al.*, 2015). Kasprzyk-Hordern and group detected 4-benzophenone (sunscreen agent) within the spectrum of 3597–5790 ng/L (Kasprzyk-Hordern *et al.*, 2009). However, several ECs have been found to have EC50 values below 1 mg/L. Based on their EC50 values, these contaminants are classified as harmful to aquatic organisms when the EC50 ranges from 10 to 100 mg/L, toxic when it ranges from 1 to 10 mg/L, and highly toxic when it is below 1 mg/L.

ECs have various pathways to enter aquatic and subsurface environments, which could be categorized into five main sources: domestic, industrial, hospital, and agrochemical wastes. Domestic wastes are found to be one of the prominent sources of PPCPs in the environment. The metabolized drugs inside the human body could come to water and wastewater systems via human urines and feces. Additionally, human activities contribute to the introduction of PPCPs into the environment, including substances like cleansing agents, sunscreen, toothpaste, and various others. However, the production units including PPCPs, biocides, and agrochemicals are significant sources of ECs in the environment. Additionally, clinical effluents contain drugs, ARGs/genes, and pharmaceutical byproducts are also major contributors to ECs. Livestock rearing and cultivation activities are additional significant contributor to ECs, especially in terms of hormones and biocides employed for enhanced productivity (Barbosa *et al.*, 2016). Additionally, landfill leaching, irrigation activities, and aquaculture discharge significantly contributed ECs in surroundings. Table 9.1 summarizes the functions, occurrence, and adverse effects of emerging contaminants present in the environment.

9.3.2 Occurrence of EC in different water matrix

Significant disparities in the levels of ECs across various aquatic systems have been documented, primarily attributable to factors such as dilution, persistency, treatment effectiveness, and other variables. It was observed that ECs are omnipotent in various water systems such as wastewater, sewage, sludge, ground, surface water, and potable water. During morning period a significant increase in antibiotic concentration was observed, which was attributed to the accumulation of ingested drugs in urine during sleep, indicating an intra-day variation (Coutu *et al.*, 2013). Similarly, an elevated bioactive byproduct of cocaine was identified on weekends in European countries. However, seasonal variation of ECs in wastewater was compared, and observed that a substantial concentration of sunscreen agents and pholcodine was found in summer and cold, respectively (Petrie *et al.*, 2015). The majority of ECs in treatment systems were found to occur within the concentration range of 0.1–10 μg/L. However, the concentration of ibuprofen and caffeine occurred in between 3.73–603 and 50 μg/L, respectively (Luo *et al.*, 2014; Zhou *et al.*, 2010).

Table 9.1 Occurrence and effects of emerging contaminants in the environment.

ECs	Function	Sources	Concentration of EC in WWTP Effluent (ng/L)	Removal Efficiency (%)	Adverse Effects
Antibiotics (sulfonamides, clarithromycin, roxithromycin, tetracycline, penicillin)	Antimicrobial substances are agents that prevent infection by either destroying or impeding the growth of bacteria	Domestic wastewater, industry effluent, pharmaceutical, hospital effluent, effluent from aqua culture, and livestock farms	42–276	<0–99	Induce antibiotic resistance in microbial strains, modify the structure of microbial communities, and lead to decreased populations of algae, bacteria, nematodes, and other organisms
Fire retardants (polybrominated diphenyl ethers (PBDEs))	Utilized in various applications such as paints, plastics, televisions, and building materials to increase the resistance and make them less prone to catch fire easily.	Domestic wastewater and industrial effluent	1–150	86–96	Impact the brain and nervous system, disrupt hormone activity, and influence reproduction and fertility
Endocrine-disrupting chemicals (EDCs) (Bisphenol phthalates)	Group of chemicals used as plasticizers, plastics, industrial lubricants/solvents, and so on.	Drinking water, surface water, sediments, secondary sludge and soil	331	32–100	Interfere with the endocrine system, exhibit estrogenic effects in rats, result in feminizing side effects in men, and can contribute to birth defects and developmental delays
Nonsteroidal anti-inflammatory drugs (ibuprofen, diclofenac)	Reduce inflammation, pain, and fever		394–647	<0–98	Associated with an increased risk of gastrointestinal ulcers, kidney diseases, and gill alterations in rainbow trout.
Lipid regulators (gemfibrozil clorfibric acid)	Regulation of levels of cholesterol and triglycerides in the blood		5.3–137.7	27.7–71.8, 0–100	Inhibit bioluminescence and impede the growth of microalgae.
Beta-blockers (metoprolol, atenolol)	Management of abnormal heart rhythms involves the inhibition of the hormones adrenaline and noradrenaline		166–843	<0–96, <0–58.7	They affect the reproduction and growth of fishes, inhibit receptor discharge in the gills, and have an impact on breeding cycles and activity rhythms in trout.

(Continued)

Anticonvulsants (carbamazepine)	Treat epileptic seizures and mood disorders		482	0–83	Induce oxidative stress in rainbow trout and have an impact on the central nervous system
Hormones (testosterone estrone)	Coordination and control of various hormones and signaling pathways to ensure proper metabolic function, stable internal conditions, and appropriate progression of sexual maturation.	Domestic wastewater, Hospital effluent, Sewage treatment plants	2–15	0–100	Affects fertility and reproduction, feminization of males, masculinization of females, and reduced fertility in fish.
Polyaromatic hydrocarbons (PAH) (pyrene anthracene)	Used in the manufacture of pesticides, plastics, dyes, and so on.	Agricultural runoffs, sewage treatment plants, surface water, sediments and soils	14–4700	63–69	Cardiovascular diseases, carcinogenic effects, and poor fetal development
Per-fluorinated alkylated substances (PFAs) (per-fluoro octanoic acid)	Used in paints, emulsion polishes, polymerization and coatings		15–<1500	~95	liver damage, Thyroid disease, kidney cancer, developmental effects on an unborn child, and reduced response to vaccines
Nano materials (nanocomposites, nanoparticles)	Used in a variety of, products, manufacturing processes, and healthcare.	Industrial effluent	–	–	Affects respiratory systems, poses risks to wildlife, and contributes to environmental toxicity.
Viable but nonculturable microbes (*Vibrio cholerae, Yersinia pestis*)	Significant implications in pathogenesis and bioremediation	Aquaculture effluent, Agricultural runoff, and surface water	–	–	Affect the gastrointestinal systems and immune systems, and they are toxic to the environment and wildlife

9.3.2.1 Surface water

The concentration of ECs in surface water is found to be lower than WWTP effluents due to the dilution of ECs. Rainfall exerts a dual influence on the concentration of ECs in surface water. On the other hand, rainfall is also subjected to higher ECs via chemical leaching from paints, building and pavement materials, and sewage runoff due to flash floods. African surface water consisted of multiple pesticides which varied from 0.06 ng/L to 9 μg/L, respectively (K'oreje *et al.*, 2020).

9.3.2.2 Groundwater

The presence of ECs in groundwater is found to be lower as compared to surface water. There are various point and non-point sources such as landfill leachate, infiltration from farming areas, aquifer recharge using non-potable water, and percolation from sewage systems are the main sources of groundwater contamination (Luo *et al.*, 2014; Stepien *et al.*, 2013). The octanol–water partition coefficient (K_{ow}) plays a crucial role in determining the extent of groundwater pollution. This is because the soil serves as the core dissemination pathway for ECs to reach the groundwater. Usually, the log K_{ow} value of less than 2.5 indicates high hydrophilic mobility of ECs. If the log K_{ow} falls between 2.5 and 4, it signifies medium mobility. On the other hand, a log K_{ow} value greater than four suggests low mobility or high retention of ECs within the soil matrix. Triclosan having a high log K_{ow} value (>4), conserved within the soil, while Trimethoprim, with a lower log K_{ow} value (<1.5) leached to groundwater (Dougherty *et al.*, 2010; Petrie *et al.*, 2015). Corresponding to surface water, various PPCPs such as caffeine, sulfamethoxazole, and carbamazepine were also detected (<100 ng/L) in groundwater (Luo *et al.*, 2014). In South African countries, substantial amounts of paracetamol (18 μg/L) and nevirapine (1.6 μg/L) were detected in groundwater. Similarly, steroid hormones are also detected in groundwater in US land sites (Karnjanapiboonwong *et al.*, 2011).

9.3.2.3 Drinking water

Due to the fact that the majority of ECs are not detectable in drinking water, there is a scarcity of publications available on this topic. Various treatment systems employed in the purification of potable water are believed to play a crucial role in maintaining the ECs under non-detectable limits. Alike to groundwater, the majority of ECs detected in drinking water are generally present below 100 ng/L. However, the concentration of non-phenol (100 ng/L) and carbamazepine (600 ng/L) was detected lower than PNEC values of 330 and 25 000 ng/L, respectively (Kleywegt *et al.*, 2011). Most of the ECs in drinking water are detected at permissible limits, it is important to note that potential detrimental effects from synergistic interactions and transformed by-products of these contaminants cannot be overlooked. Therefore, regular monitoring and examination of the potability of drinking water is necessary to ensure the well-being of consumers.

9.3.2.4 Wastewaters

The release of treated wastewater from WWTPs stands as the main contributor to the presence of ECs in the environment. Unfortunately, most treatment methods are not engineered to eliminate ECs, leading to their release into the surface water system. Around 70 pharmaceutical compounds, including NSAIDs, β-blockers, antidepressants, and carbamazepine, have been extensively studied. These compounds are widely prescribed, with annual usage exceeding 1000 kg, and can be found in influent wastewater from various sources. The removal of these pharmaceuticals during wastewater treatment exhibits a wide range, with some experiencing low removal rates of less than 50%, while others achieve high removal rates exceeding 80%. This variation can be attributed to the diverse physicochemical properties of the compounds and their varying susceptibility to biological degradation processes. Due to partial removal during wastewater treatment, pharmaceutical compounds are frequently detected in receiving surface waters at concentrations ranging from nanograms per liter (ng/L) to milligrams per liter (mg/L). Fifteen illegal drugs and legal stimulants were identified in wastewater out of which tramadol concentration was found to be highest, that is, 7731 ng/L (Kasprzyk-Hordern *et al.*, 2008).

However, the widespread social unacceptability of illicit drugs within the community greatly reduces the likelihood of their presence in the water treatment system. The hallucinogen like $C_{11}H_{15}NO_2$ and the stimulant $C_{17}H_{21}NO_4$ were identified in riverine systems at concentrations of 25 and 17 ng/L, respectively. On the other hand, the high usage of licit chemicals in the community frequently contributes to ECs in water systems. Most studies conducted on EC analysis primarily focus on the aqueous phase, which involves analyzing pre-filtered samples. However, there has been a notable lack of measurements made on the particulate phase, such as sludge or suspended particulate matter.

9.3.2.5 Other matrix

In contradistinction to the dissolved phase, emerging contaminants in particulate phases such as biosolids (sludge) have rarely been documented. This is likely attributed to the complex composition of sludge and the absence of highly sophisticated analytical methods capable of detecting ECs within the sludge matrix. The utilization of sludge as a fertilizer is a prevalent practice in the agriculture field and currently, there are no regulations in place to oversee the application of sludge on agricultural land concerning emerging contaminants. As a result, ECs in the sludge matrix have the potential to contaminate the groundwater and soil. Hence, obtaining information regarding the presence and distribution of these contaminants in such matrices is crucial for formulating an effective strategy for sludge management. The concentrations of the designated contaminants in the sludge varied between 0 (below the limit of detection) and approximately 5000 ng/g of the dry weight of the sludge. The concentration of emerging contaminants is highly influenced by their pattern of usage, and physicochemical properties such as ionization state and hydrophobicity (Tran *et al.*, 2018). To cite an instance, hydrophobic emerging contaminants such as miconazole and bisphenol A have been found in higher concentrations in the sludge matrix attributed to their affinity for the particulate phase. On the other hand, ciprofloxacin (fluoroquinolones) is absorbed into sludge through the coulombic interaction, as they exist as zwitterions under environmental pH conditions ranging from 6 to 8. Additionally, the higher concentrations of ibuprofen in sludge can be attributed to the hydrolysis of its conjugates. The anaerobic digestion of sludge has been proven to be an effective process reducing the concentration of ibuprofen as well as other extensively studied emerging contaminants such as carbamazepine, ketoprofen, ethinylestradiol, and so on (Martín *et al.*, 2012).

9.3.3 Pathways of ECs

Once released into the environment, ECs undergo various natural modification processes, such as photolysis, dispersion, transformation, and adsorption onto suspended particles. These processes contribute to the reduction of EC concentrations in surface water bodies (Pal *et al.*, 2010). However, traditional WWTPs lack effective treatment strategies to address the entry of ECs into the system. Consequently, partially degraded, or unchanged ECs find their way into aquatic systems via the discharge of effluent and sludge from WWTPs s after treatment (Luo *et al.*, 2014; Pesqueira *et al.*, 2020). Moreover, the utilization of treated wastewater for agricultural practices substantially introduces ECs into the soil, leading to further uptake of these contaminants by the crops (Paz *et al.*, 2016). Because soil, groundwater, and other environmental matrices are interrelated, the occurrence of emerging pollutants in any of the domains can have an accumulative and detrimental effect on the entire environment. As a result, WWTPs are regarded as one of the main anthropogenic sources that emit emerging contaminants into the surroundings, in addition to industrial discharges, hospital effluents agricultural runoffs, aquaculture effluents, and landfill leachates. Without proper identification and implementation of appropriate treatment technologies, the management of emerging contaminants is compromised, thereby posing a risk to the sustainable utilization of treated wastewater.

The environmental fate of emerging contaminants can be evaluated using the source-path-receptor model, which also facilitates the risk assessment of target receptors. Figure 9.1 illustrates various pathways of emerging contaminants originating from different sources. Emerging contaminants predominantly derive from industrial effluents, agricultural runoff, effluent from hospitals,

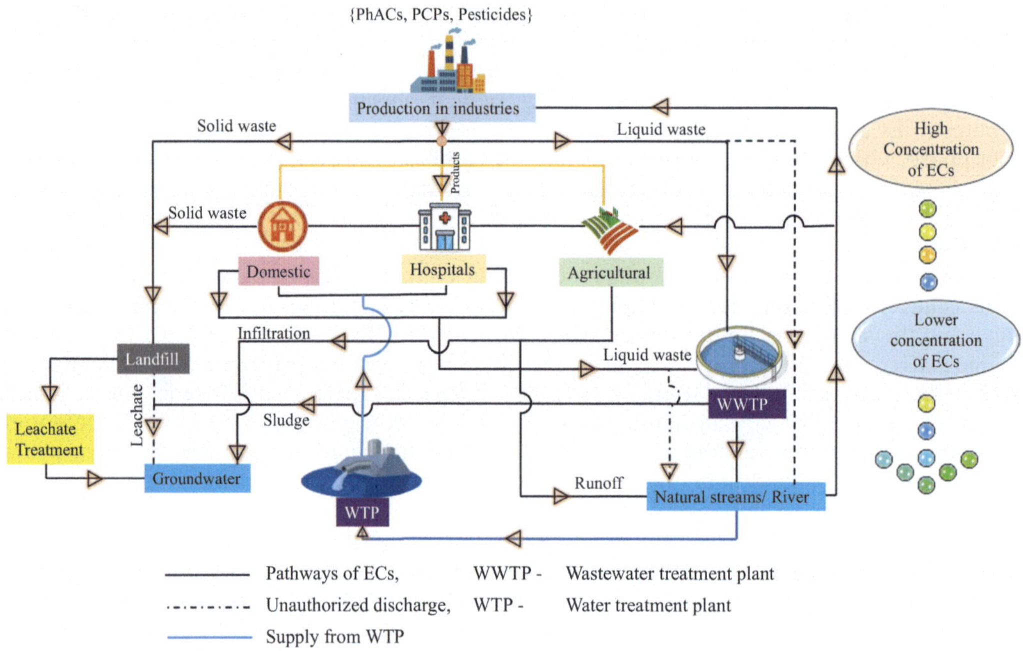

Figure 9.1 Schematic representation of emerging pollutants' origin and pathways with their varied concentration from the source to disposal locations.

wastewater from laboratories, leachate from landfill disposal sites, and to a lesser extent, domestic wastewater. Domestic sewage comprises partially metabolized or unmetabolized pharmaceutical and personal care products (PhACs) that are excreted by humans. Furthermore, the occurrence of emerging contaminants in water bodies is influenced by the population's consumption patterns. Conventional WWTPs also release emerging contaminants in the effluent as they are inadequate for effectively treating such pollutants. Septic tanks and landfill sites are marked by the presence of highly concentrated leachate, which is the major contributor to emerging contaminants in groundwater. This is particularly notable in areas where the aquifers have a high percolation rate and the groundwater table is less deep (Ramakrishnan *et al.*, 2015). Additionally, the employment of fertilizers and other chemicals in agricultural practices is another major factor leading to the contamination of groundwater and surface water. It is important to note that most emerging contaminants are highly distributed in the environment and are resistant to biodegradation or hydrolysis under normal environmental conditions. Ebele's team conducted a study that revealed elevated concentrations of pharmaceuticals and personal care products (PhACs) in various organisms such as goldfish (*Carassius auratus*), snails, and mosquitofish (*Gambusia holbrooki*) (Ebele *et al.*, 2017). These findings suggest the potential biomagnification of ECs within aquatic ecosystems after being released into the environment.

9.4 GLOBAL OCCURRENCE OF SOME IMPORTANT ECS

ECs typically consist of elements such as nitrogen, hydrogen, carbon, fluorine, sulfur, and fluorine. The presence, behavior, and movement of ECs in the environment are influenced by various factors, including water characteristics, sources of contamination, climatic conditions, available treatment methods, physicochemical properties of the compounds, and, socioeconomic factors (Vickers, 2017). Among all ECs, pharmaceuticals and personal care products (PhACs) are the most commonly detected contaminants, often found in higher concentrations. This is primarily attributed to their polar nature and substantial use by consumers, which enables them to persist in the environment and exhibit greater mobility. It was reported that in 2020, over 50% of the world population will be dependent on compulsory medicine consumption daily (Aitken & Kleinrock, 2015). According to Statista, the maximum sales of pharmaceuticals were observed in North America followed by Europe. Among different analgesics, ibuprofen was reported with a higher concentration in North America (75.8 μg/L) as compared to Asia (26.45 μg/L), Australia (10.3 μg/L), and Europe (33.76 μg/L) in the influent of WWTPs. Naproxen was found in the range of 0.08–25 μg/L, in countries like USA, India, China, Northern America, and different European countries (Mandaric et al., 2017; Singh *et al.*, 2014; Vickers, 2017). Whereas, the anti-inflammatory drug diclofenac has been detected in municipal wastewater ranging from 0.11 to 25.68 μg/L with a mean value of 2 μg/L. High ketoprofen (16.2 μg/L) concentration was found in WWTP influents in India. In 2020, it was reported that more than 50% of the global population would rely on regular medication consumption (Aitken & Kleinrock, 2015). As stated in Statista data, the highest pharmaceutical sales, followed by those in Europe, were recorded in North America. When different analgesics were investigated in WWTPs, ibuprofen exhibited higher concentrations in North America (75.8 μg/L) compared to Asia (26.45 μg/L), Australia (10.3 μg/L), and Europe (33.76 μg/L). Naproxen, another commonly used medication, has been found in the range of 0.08–25 μg/L in nations such as India, Northern America, USA, and China, Northern America, and various European countries. Additionally, the anti-inflammatory drug diclofenac was detected in municipal wastewater with concentrations ranging from 0.11 to 25.68 μg/L, with an average value of 2 μg/L. Notably, India exhibited a high concentration of ketoprofen (16.2 μg/L) in WWTP influents.

9.5 FATE OF ECS IN ENVIRONMENTAL WATERS

9.5.1 Human metabolites

Parent compounds are frequently eliminated from the human body alongside various accompanying metabolites. To illustrate, ibuprofen is discharged in its unaltered drug form (1%), along with multiple metabolites: (+)–2–4′-(2–Hydroxy-2-methylpropyl)-phenyl propionic acid (25%), (+)-2–40-(2–carboxypropyl)-phenyl propionic acid (37%), and conjugated ibuprofen (14%) (Kasprzyk Hordern *et al.*, 2008). It has been investigated during preliminary studies involving raw sewage samples and activated sludge samples for steroid estrogens (17α – EE2 3-glucuronide, estriol 16α-glucuronide, and estrone 3-glucuronide) (Gomes *et al.*, 2009), the presence of carbamazepine was majorly detected (Vieno *et al.*, 2007). Moreover, it is imperative to analyze metabolites due to their potential occurrence at significantly higher concentrations than the respective parent chemical, and their potential pharmacological activity. A primary metabolite of carbamazepine, namely carbamazepine epoxide, was detected in the influent of wastewater at concentrations ranging from 880 to 4026 ng/L, on the other hand, the parent compound was found at levels below 1.5 113 ng/L (Huerta Fontela *et al.*, 2010) and these substances can persist even after the secondary wastewater treatment. Given their discharge into the environmental matrix and the potential for further bio-transformation into the parent environmental compound, it is imperative to determine these metabolites to accurately assess the associated ecological risk.

9.5.2 Microbial transformation

Numerous emerging pollutants experience microbial-mediated transformations during secondary treatment as well as in the natural environment. Biodegradation is commonly considered as the primary pathway for the elimination of certain emerging contaminants from water systems (wastewater and surface waters). Nevertheless, it can lead to the generation of various transformation products. Unfortunately, these products have not been extensively investigated due to the limitations of screening approaches employed for known compounds, such as low-resolution mass spectrometry utilizing triple quadrupoles technology, which could not effectively identify these products. Additionally, the scarcity of available standards for these transformation products is attributed to the limited knowledge regarding the bio-transformation pathways associated with them. In the laboratory-scale studies of activated sludge, Helbling et al. successfully detected previously unexplored biological transformation products for several pharmaceuticals, namely valsartan, bezafibrate, diazepam, oseltamivir, and levetiracetam through the utilization of high-resolution mass spectrometry employing linear ion trap-orbitrap technology (Helbling *et al.*, 2010). Furthermore, degradation products have been detected at naturally occurring concentrations in the effluents of activated sludge. In a study, a non-targeted screening approach using quadrupole time of flight (QTOF) mass spectrometry was employed to successfully detect transformation products of acetaminophen (P-aminophenol) and azithromycin (Gomes *et al.*, 2009). The identification of these transformation products is of utmost importance as they exhibit higher toxicity compared to the parent compound, as exemplified by P-aminophenol. Hence, the removal of the parent environmental compound does not guarantee the elimination of its associated toxicity. Despite various parent compounds are found in wastewater, it is plausible to anticipate the presence of various transformation products in the effluent as well as the receiving water systems.

9.5.3 Physicochemical processes

Physical–chemical processes can also aid in the removal of EC from surface and wastewater. During wastewater treatment, removal from the aqueous phase will take place on biomass, or when present in the river environment, by discharge to sediments. This is most likely accurate for a small number of ECs, though. For example, if a state of equilibrium is achieved between biomass or sediment and the aquatic environment for a specific environmental contaminant (EC), there will be no net exchange between the two phases and no removal from the aqueous phase. Consequently, sorption is not effective in eliminating them. Studies have demonstrated this for certain ECs like steroid estrogens in activated sludge processes (Petrie *et al.*, 2014). However, antibiotics such as ofloxacin and ciprofloxacin, due to their strong affinity for solid organic matter, are removed through sorption during wastewater treatment (Petrie *et al.*, 2014). Hence, understanding the influence of physicochemical parameters on the sorption of these ECs is crucial. Additionally, the impact of dissolved organic matter on the fate of ECs in the environment must be considered, as the binding to dissolved organic materials can aid in retaining ECs in the aqueous phase of environmental matrices. Furthermore, the development of organic matter complexes containing EC may prevent the EC from being recognized during analysis. ECs are subject to photolysis when they are present in an aqueous environment. In river water, photolysis has been demonstrated to effectively break down several ECs, including naproxen, ketoprofen, E2, propranolol, EE2, ibuprofen, and gemfibrozil. Ketoprofen's half-life was 4 min, while gemfibrozil and ibuprofens were 15 h. This spectrum of susceptibility to photolysis breakdown is explained by variations in their bond formation. For example, the presence of two aromatic rings in the carbonyl moiety of ketoprofen results in the formation of a highly reactive triple state, making it susceptible to breakdown through photolysis. As a result, photolysis can be instrumental in removing numerous (ECs) from surface waters. However, it is important to note that the degradation of the parent compound through photolysis does not guarantee complete mineralization, and various transformation products can be formed, similar to those produced through biological degradation. Therefore, the absence of the parent molecule may not result in a decrease in toxicity. It is possible to hypothesize that the presence of particles and dissolved organic matter in environmental waters will slow down the rate of EC deterioration by reducing the brightness of the sun. Depending on the particular EC, humic acid (a tiny molecular weight charged molecule) either slowed down or accelerated the rate of decomposition (West & Rowland, 2012). Indirect photolysis may be the cause of increased deterioration when humic acid or nitrates are present. Wastewater contains OH radicals and organic materials (triplet excited state), which makes some ECs more amenable to indirect photolysis (Ryan *et al.*, 2011). Further research is needed to determine the effects of other environmental elements on EC photolysis in particular environmental settings.

9.6 ENVIRONMENTAL MONITORING OF ECS

9.6.1 Sampling mode and strategy

Sampling plays a crucial role in the monitoring of ECs in wastewater and the environment, as it enables the collection of representative data. To account for hydraulic retention time (HRT) and assess the performance of EC removal in treatment processes, relevant grab samples can be utilized. This approach is particularly suitable for systems like trickling filters, which operate at HRTs of approximately 2 hours (Petrie *et al.*, 2014). However, for systems such as activated sludge, which commonly have longer HRTs of 6 h, this method may not be feasible. Additionally, daily grab samples, although commonly collected, do not provide a comprehensive understanding of the treatment process performance. To ensure stability and obtain a composite sample representative of a system over a longer period, a sampling strategy employing flow or volume proportional sampling is necessary. This involves collecting samples over a period of 24 h, considering the flow rate or volume proportionality (Hillebrand *et al.*, 2013). Deploying sampling equipment, such as samplers and flow measuring devices, at strategic locations within WWTPs or rivers presents logistical challenges, but it is essential to obtain representative data during environmental monitoring. Passive samplers could be a potential alternative, although further research is needed to assess their suitability for absorbing more polar substances like ECs (Mills *et al.*, 2014). Ideally, real-time sensors should be employed in situ to enhance monitoring accuracy. The frequency of repeat sampling efforts throughout the year should also be considered. To capture the dynamics of seasonal variations, it is recommended to conduct a minimum of two sampling events per year, representing summer and winter conditions. This approach ensures that the monitoring reflects the changes that occur throughout different seasons. This will make it possible to determine the seasonal trends of EC usage and the effect of temperature on the operation of WWTPs. The sample plan must take into account achieving total mass balances for the concerned WWTP or section of the river system. Analyzing waste/recycled sludge and river sediment is crucial for figuring out how ECs behave in these kinds of environments. All sampling positions' particle phase analyses are part of this. It is true that obtaining this for final effluents over the course of a thorough sampling program will be challenging. However, considering the absence of a prior study done here, it is valuable to determine final effluent particle phase concentrations at least once during the sampling session.

9.6.2 Analysis methods

It is advised to use analytical techniques that can identify ECs down to the enantiomeric level. However, because chiral stationary phases are so specialized and the mechanism of separation is poorly understood, establishing multi-residue separations with them is challenging. Additionally, because of their maximum back pressure, which is typically less than 2000 psi, chiral columns can only be used in high-performance liquid chromatography mode. Consequently, the ability to process samples and achieve turnover rates is limited by the commonly observed 60-minute run times (Bagnall *et al.*, 2012, 2013; López-Serna *et al.*, 2013). To overcome this constraint, it would be advantageous to utilize stationary phases consisting of smaller particle sizes, specifically below 2 mm. This would allow for comparable performance in terms of run time and column efficiency to ultraperformance liquid chromatography (UPLC). Prior to their development, it is advised to ascertain the enantiomeric fraction for as many substances as feasible using relatively quick achiral UPLC procedures backed by chiral separations. At relatively quick analytical periods (~10 min), targeted UPLC approaches can simultaneously determine up to 100 ECs in distinct environmental matrices (Gracia-Lor *et al.*, 2011; Gros *et al.*, 2012; López-Serna *et al.*, 2011). These multi-residue methods, which are used to analyze ECs, should ideally be dynamic so they may carry out targeted (quantitative) determinations while also doing the non-targeted (qualitative) screening. The use of high-resolution mass spectrometers that enable retrospective analysis and can perform both targeted and non-targeted screening, such as QTOF or Orbitrap technology, is advantageous. With the aid of such technologies, compounds that were later identified as being of interest but were not initially included in the targeted screening can be quickly added. A successful chromatographic separation is essential for non-targeted screening. To separate a wide variety of target ECs exhibiting extremes in physiochemical properties (molecular weight, hydrophobicity, etc.), the chromatography process needs to be optimized. It will be possible to find unidentified substances with significant concentrations by combining screening in both negative and positive ionization modalities. However, due to the unknown behavior of ECs is a big question, non-targeted screening has a number of drawbacks. They might not thus be retrieved at the time of sample preparation or during the ionization at the time of analysis. It is also necessary to support chemical analysis with cutting-edge bioanalytical methods (like metabolomics). At the molecular level, a metabolomics approach provides valuable insights into the functioning and health of organisms. Unlike traditional toxicological assays that focus on a limited number of indicator species and endpoints such as growth, death, and reproduction, metabolomics allows for the detection of detailed information that would otherwise be overlooked. To comprehensively understand the impact of ECs and their concentrations on ecological systems, it is essential to conduct lengthy, multigenerational studies across various trophic levels. These studies should aim to mimic environmental conditions and replicate EC concentrations observed in the environment. This approach facilitates a better understanding of how the observed concentrations of ECs in the environment affect the ecology.

9.7 POLICY AND LEGISLATION (INDIA)

Figure 9.2 shows the consumption pattern of different emerging contaminants in India. The National Environment Policy (NEP) of 2006 represents the most recent manifestation of the government's

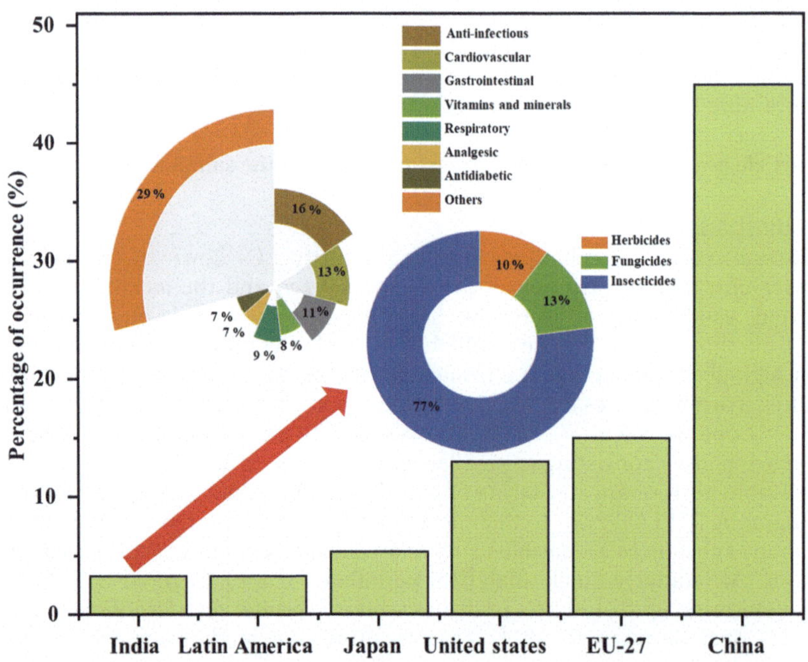

Figure 9.2 Consumption status of pesticides and antibiotics in India (*Source*: Puri *et al.* 2023).

commitment to enhancing environmental conditions and fostering national economic progress. This approach emphasizes the integration of environmental considerations into all developmental processes, safeguarding crucial environmental values, and identifying necessary legal and policy adjustments (Gani & Kazmi, 2016; Richards *et al.*, 2021). Alongside the NEP, the Environment Protection Act (EPA) was introduced as a comprehensive national environmental law. The EPA outlines measures for environmental protection in the domains of air, water, and land. It also establishes a framework for central government coordination with the state authorities established under prior legislation, such as the Water Act and Air Act (Compliance, 2006).

Currently, Indian sewage treatment facilities discharge their effluents into rivers, traditional treatment methods are insufficient in eliminating ECs, and adequate waste management practices are lacking. There is no existing legal legislation specifically addressing ECs in India. The regulations concerning drinking water in India are only partially covered by the Indian Standards (IS 10500) (Mathew & Kanmani, 2020).

9.8 CONCLUSIONS AND FUTURE OUTLOOK

Environmental laws are likely to be expanded to include a variety of ECs with municipal origins. However, there is still a lack of thorough knowledge regarding what happens to them during wastewater treatment and in the environment. The reported elimination of ECs by WWTP contains uncertainty because of the shortcomings of the current sampling techniques. Therefore, it is necessary to reevaluate, using appropriate sample techniques, the removal efficacy of various WWTP process types under varying operational situations. This will make it easier to determine the steps needed for EC improvement. The use of novel treatment techniques will rise in response to the growing trend of enhancing sustainability and decreasing energy consumption for wastewater treatment. An example of a promising treatment approach that can inadvertently generate energy is the use of algal ponds for secondary effluent polishing. However, very few studies have kept track of how well they perform in terms of removing EC. To assess the fate and removal of ECs during treatment, taking into account their probable integration into the standard WWTPs flow sheet, additional research on these process types is required. Now, environmental monitoring must adopt a comprehensive strategy. This entails figuring out what happens to ECs and how they affect the environment over their whole life cycle, which includes the terrestrial environment. In-depth case studies of supplemented soils in real-world settings are required to examine leaching, runoff, the effect on the quality of nearby surface waters, soil deterioration, the toxicity to terrestrial creatures, and potential uptake by plants and entry into the human food chain. Monitoring other contaminated environmental compartments, such as river sediments, can be done using a similar strategy. Finally, the revision and creation of a more accurate environmental risk assessment will be made possible by the combined use of chemical and biological studies to better analyze the environmental impact of ECs.

ACKNOWLEDGMENTS

Director, CSIR-NEERI is thankfully acknowledged for giving the opportunity to pursue work in CSIR-NEERI, Nagpur, India. AS and CJ acknowledge the University Grant Commission, New Delhi, India for providing Senior and Junior Research Fellowship. The article is checked for plagiarism using the iThenticate software and recorded in the Knowledge Resource Center, CSIR-NEERI, Nagpur for anti-plagiarism.

REFERENCES

Aitken M. and Kleinrock M. (2015). Global Medicines Use in 2020: Outlook and Implications. Parsippany, IMS Institute for Healthcare Informatics, New Jersey, USA.

Bagnall J., Evans S., Wort M., Lubben A. and Kasprzyk-Hordern B. (2012). Using chiral liquid chromatography quadrupole time-of-flight mass spectrometry for the analysis of pharmaceuticals and illicit drugs in surface and wastewater at the enantiomeric level. *Journal of Chromatography A*, **1249**, 115–129, https://doi.org/10.1016/j.chroma.2012.06.012

Bagnall J., Malia L., Lubben A. and Kasprzyk-Hordern B. (2013). Stereoselective biodegradation of amphetamine and methamphetamine in river microcosms. *Water Research*, **47**, 5708–5718, https://doi.org/10.1016/j.watres.2013.06.057

Barbosa M. O., Moreira N. F., Ribeiro A. R., Pereira M. F. and Silva A. M. (2016). Occurrence and removal of organic micropollutants: an overview of the watch list of EU Decision 2015/495. *Water Research*, **94**, 257–279, https://doi.org/10.1016/j.watres.2016.02.047

Cao H., Chao S., Qiao L., Jiang Y., Zeng X. and Fan X. (2017). Urbanization-related changes in soil PAHs and potential health risks of emission sources in a township in Southern Jiangsu, China. *Science of the Total Environment*, **575**, 692–700, https://doi.org/10.1016/j.scitotenv.2016.09.106

Chen K. and Zhou J. (2014). Occurrence and behavior of antibiotics in water and sediments from the Huangpu River, Shanghai, China. *Chemosphere*, **95**, 604–612, https://doi.org/10.1016/j.chemosphere.2013.09.119

Chinnaiyan P., Thampi S. G., Kumar M. and Mini K. (2018). Pharmaceutical products as emerging contaminant in water: relevance for developing nations and identification of critical compounds for Indian environment. *Environmental Monitoring and Assessment*, **190**, 1–13, https://doi.org/10.1007/s10661-018-6672-9

Claudia J. and Magrini G. (2017). Cosmetic ingredients as emerging pollutants of environmental and health concern. A mini-review. *Cosmetics*, **4**, 11–29, https://doi.org/10.3390/cosmetics4020011

Compliance E. (2006). Enforcement in India: rapid assessment. A report prepared by OECD Programme jointly with the Secretariat of the Asian Environmental Compliance and Enforcement Network (AECEN).

Coutu S., Wyrsch V., Wynn H. K., Rossi L. and Barry D. A. (2013). Temporal dynamics of antibiotics in wastewater treatment plant influent. *Science of the Total Environment*, **458**, 20–26, https://doi.org/10.1016/j.scitotenv.2013.04.017

Das S., Ray N. M., Wan J., Khan A., Chakraborty T. and Ray M. B. (2017). Micropollutants in wastewater: fate and removal processes. *Physico-Chemical Wastewater Treatment and Resource Recovery*, **3**, 75–117.

De la Cruz N., Giménez J., Esplugas S., Grandjean D., De Alencastro L. and Pulgarin C. (2012). Degradation of 32 emergent contaminants by UV and neutral photo-Fenton in domestic wastewater effluent previously treated by activated sludge. *Water Research*, **46**, 1947–1957, https://doi.org/10.1016/j.watres.2012.01.014

Dougherty J. A., Swarzenski P. W., Dinicola R. S. and Reinhard M. (2010). Occurrence of herbicides and pharmaceutical and personal care products in surface water and groundwater around Liberty Bay, Puget Sound, Washington. *Journal of Environmental Quality*, **39**, 1173–1180, https://doi.org/10.2134/jeq2009.0189

Ebele A., Abou-Elwafa Abdallah M. and Harrad S. (2017) Pharmaceuticals and personal care products (PPCPs) in the freshwater aquatic environment. *Emerging Contaminants*, **3**(1), 1–16, https://doi.org/10.1016/j.emcon.2016.12.004

Ferrando-Climent L., Rodriguez-Mozaz S. and Barceló D. (2014). Incidence of anticancer drugs in an aquatic urban system: from hospital effluents through urban wastewater to natural environment. *Environmental Pollution*, **193**, 216–223, https://doi.org/10.1016/j.envpol.2014.07.002

Gani K. M. and Kazmi A. A. (2016). Phthalate contamination in aquatic environment: a critical review of the process factors that influence their removal in conventional and advanced wastewater treatment. *Critical Reviews in Environmental Science and Technology*, **46**, 1402–1439, https://doi.org/10.1080/10643389.2016.1245552

Gatidou G., Thomaidis N. S., Stasinakis A. S. and Lekkas T. D. (2007). Simultaneous determination of the endocrine disrupting compounds nonylphenol, nonylphenol ethoxylates, triclosan and bisphenol A in wastewater and sewage sludge by gas chromatography–mass spectrometry. *Journal of Chromatography A*, **1138**, 32–41, https://doi.org/10.1016/j.chroma.2006.10.037

Gavrilescu M., Demnerová K., Aamand J., Agathos S. and Fava F. (2015). Emerging pollutants in the environment: present and future challenges in biomonitoring, ecological risks and bioremediation. *New Biotechnology*, **32**, 147–156, https://doi.org/10.1016/j.nbt.2014.01.001

Goel P. (2006). Water Pollution: Causes, Effects and Control. New Age International, New Delhi, India.

Gogoi A., Mazumder P., Tyagi V. K., Chaminda G. T., An A. K. and Kumar M. (2018). Occurrence and fate of emerging contaminants in water environment: a review. *Groundwater for Sustainable Development*, **6**, 169–180, https://doi.org/10.1016/j.gsd.2017.12.009

Gomes R. L., Scrimshaw M. D. and Lester J. N. (2009). Fate of conjugated natural and synthetic steroid estrogens in crude sewage and activated sludge batch studies. *Environmental Science & Technology*, **43**, 3612–3618, https://doi.org/10.1021/es801952h

Gracia-Lor E., Sancho J. V. and Hernández F. (2011). Multi-class determination of around 50 pharmaceuticals, including 26 antibiotics, in environmental and wastewater samples by ultra-high performance liquid chromatography–tandem mass spectrometry. *Journal of Chromatography A*, **1218**, 2264–2275, https://doi.org/10.1016/j.chroma.2011.02.026

Gros M., Rodríguez-Mozaz S. and Barceló D. (2012). Fast and comprehensive multi-residue analysis of a broad range of human and veterinary pharmaceuticals and some of their metabolites in surface and treated waters by ultra-high-performance liquid chromatography coupled to quadrupole-linear ion trap tandem mass spectrometry. *Journal of Chromatography A*, **1248**, 104–121, https://doi.org/10.1016/j.chroma.2012.05.084

Helbling D. E., Hollender J., Kohler H. E., Singer H. and Fenner K. (2010). High-throughput identification of microbial transformation products of organic micropollutants. *Environmental Science & Technology*, **44**, 6621–6627, https://doi.org/10.1021/es100970m

Hillebrand O., Musallam S., Scherer L., Nödler K. and Licha T. (2013). The challenge of sample-stabilisation in the era of multi-residue analytical methods: a practical guideline for the stabilisation of 46 organic micropollutants in aqueous samples. *Science of the Total Environment*, **454**, 289–298, https://doi.org/10.1016/j.scitotenv.2013.03.028

Houtman C. J. (2010). Emerging contaminants in surface waters and their relevance for the production of drinking water in Europe. *Journal of Integrative Environmental Sciences*, **7**, 271–295, https://doi.org/10.1080/1943815X.2010.511648

Huerta-Fontela M., Galceran M. T. and Ventura F. (2010). Fast liquid chromatography–quadrupole-linear ion trap mass spectrometry for the analysis of pharmaceuticals and hormones in water resources. *Journal of Chromatography A*, **1217**, 4212–4222, https://doi.org/10.1016/j.chroma.2009.11.007

Hughes S. R., Kay P. and Brown L. E. (2013). Global synthesis and critical evaluation of pharmaceutical data sets collected from river systems. *Environmental Science & Technology*, **47**, 661–677, https://doi.org/10.1021/es3030148

Jorgensen S. E. (2008). Encyclopedia of ecology, edited by Sven Erik Jorgensen, Brian D. Fath.

Karnjanapiboonwong A., Suski J. G., Shah A. A., Cai Q., Morse A. N. and Anderson T. A. (2011). Occurrence of PPCPs at a wastewater treatment plant and in soil and groundwater at a land application site. *Water, Air, & Soil Pollution*, **216**, 257–273, https://doi.org/10.1007/s11270-010-0532-8

Kasprzyk-Hordern B., Dinsdale R. M. and Guwy A. J. (2008). The occurrence of pharmaceuticals, personal care products, endocrine disruptors and illicit drugs in surface water in South Wales, UK. *Water Research*, **42**, 3498–3518, https://doi.org/10.1016/j.watres.2008.04.026

Kasprzyk-Hordern B., Dinsdale R. M. and Guwy A. J. (2009). The removal of pharmaceuticals, personal care products, endocrine disruptors and illicit drugs during wastewater treatment and its impact on the quality of receiving waters. *Water Research*, **43**, 363–380, https://doi.org/10.1016/j.watres.2008.10.047

Kim E., Jung C., Han J., Her N., Park C. M., Jang M., Son A. and Yoon Y. (2016). Sorptive removal of selected emerging contaminants using biochar in aqueous solution. *Journal of Industrial and Engineering Chemistry*, **36**, 364–371, https://doi.org/10.1016/j.jiec.2016.03.004

Kleywegt S., Pileggi V., Yang P., Hao C., Zhao X., Rocks C., Thach S., Cheung P. and Whitehead B. (2011). Pharmaceuticals, hormones and bisphenol A in untreated source and finished drinking water in Ontario, Canada–occurrence and treatment efficiency. *Science of the Total Environment*, **409**, 1481–1488, https://doi.org/10.1016/j.scitotenv.2011.01.010

Kolpin D. W., Furlong E. T., Meyer M. T., Thurman E. M., Zaugg S. D., Barber L. B. and Buxton H. T. (2002). Pharmaceuticals, hormones, and other organic wastewater contaminants in US streams, 1999–2000: a national reconnaissance. *Environmental Science & Technology*, **36**, 1202–1211, https://doi.org/10.1021/es011055j

K'oreje K. O., Okoth M., Van Langenhove H. and Demeestere K. (2020). Occurrence and treatment of contaminants of emerging concern in the African aquatic environment: literature review and a look ahead. *Journal of Environmental Management*, **254**, 109752, https://doi.org/10.1016/j.jenvman.2019.109752

Li X., Qu C., Bian Y., Gu C., Jiang X. and Song Y. (2019). New insights into the responses of soil microorganisms to polycyclic aromatic hydrocarbon stress by combining enzyme activity and sequencing analysis with metabolomics. *Environmental Pollution*, **255**, 113312, https://doi.org/10.1016/j.envpol.2019.113312

López-Serna R., Petrović M. and Barceló D. (2011). Development of a fast instrumental method for the analysis of pharmaceuticals in environmental and wastewaters based on ultra high performance liquid chromatography (UHPLC)–tandem mass spectrometry (MS/MS). *Chemosphere*, **85**, 1390–1399, https://doi.org/10.1016/j.chemosphere.2011.07.071

López-Serna R., Kasprzyk-Hordern B., Petrović M. and Barceló D. (2013). Multi-residue enantiomeric analysis of pharmaceuticals and their active metabolites in the Guadalquivir River basin (South Spain) by chiral liquid chromatography coupled with tandem mass spectrometry. *Analytical and Bioanalytical Chemistry*, **405**, 5859–5873, https://doi.org/10.1007/s00216-013-6900-7

Luo Y., Guo W., Ngo H. H., Nghiem L. D., Hai F. I., Zhang J., Liang S. and Wang X. C. (2014). A review on the occurrence of micropollutants in the aquatic environment and their fate and removal during wastewater treatment. *Science of the Total Environment*, **473**, 619–641, https://doi.org/10.1016/j.scitotenv.2013.12.065

Mahmood T., Momin S., Ali R., Naeem A. and Khan A. (2022). Technologies for removal of emerging contaminants from wastewater. *Wastewater Treatment*, **159**, 1–20, http://dx.doi.org/10.5772/intechopen.104466

Mandaric L., Celic M., Marcé R. and Petrovic M. (2016). Introduction on emerging contaminants in rivers and their environmental risk. Emerging Contaminants in River Ecosystems: Occurrence and Effects Under Multiple Stress Conditions, **590–591**, 3–25.

Mandaric L., Diamantini E., Stella E., Cano-Paoli K., Valle-Sistac J., Molins-Delgado D., Bellin A., Chiogna G., Majone B. and Diaz-Cruz M. S. (2017). Contamination sources and distribution patterns of pharmaceuticals and personal care products in Alpine rivers strongly affected by tourism. *Science of the Total Environment*, **590**, 484–494, https://doi.org/10.1016/j.scitotenv.2017.02.185

Martín J., Camacho-Muñoz M. D., Santos J. L., Aparicio I. and Alonso E. (2012). Distribution and temporal evolution of pharmaceutically active compounds alongside sewage sludge treatment. Risk assessment of sludge application onto soils. *Journal of Environmental Management*, **102**, 18–25, https://doi.org/10.1016/j.jenvman.2012.02.020

Mathew R. A. and Kanmani S. (2020). A review on emerging contaminants in Indian waters and their treatment technologies. *Nature Environment and Pollution Technology*, **19**, 549–562, https://doi.org/10.46488/NEPT.2020.v19i02.010

Mills G. A., Gravell A., Vrana B., Harman C., Budzinski H., Mazzella N. and Ocelka T. (2014). Measurement of environmental pollutants using passive sampling devices – an updated commentary on the current state of the art. *Environmental Science: Processes & Impacts*, **16**, 369–373, https://doi.org/10.1039/C3EM00585B

Molinos-Senante M., Garrido-Baserba M., Reif R., Hernández-Sancho F. and Poch M. (2012). Assessment of wastewater treatment plant design for small communities: environmental and economic aspects. *Science of the Total Environment*, **427**, 11–18, https://doi.org/10.1016/j.scitotenv.2012.04.023

Noguera-Oviedo K. and Aga D. S. (2016). Lessons learned from more than two decades of research on emerging contaminants in the environment. *Journal of Hazardous Materials*, **316**, 242–251, https://doi.org/10.1016/j.jhazmat.2016.04.058

North K. D. (2004). Tracking polybrominated diphenyl ether releases in a wastewater treatment plant effluent, Palo Alto, California. *Environmental Science & Technology*, **38**, 4484–4488, https://doi.org/10.1021/es049627y

Pal A., Gin K. Y.-H., Lin A. Y.-C. and Reinhard M. (2010). Impacts of emerging organic contaminants on freshwater resources: review of recent occurrences, sources, fate and effects. *Science of the Total Environment*, **408**, 6062–6069, https://doi.org/10.1016/j.scitotenv.2010.09.026

Paz A., Tadmor G., Malchi T., Blotevogel J., Borch T., Polubesova T. and Chefetz B. (2016). Fate of carbamazepine, its metabolites, and lamotrigine in soils irrigated with reclaimed wastewater: sorption, leaching and plant uptake. *Chemosphere*, **160**, 22–29, https://doi.org/10.1016/j.chemosphere.2016.06.048

Pesqueira J. F., Pereira M. F. R. and Silva A. M. (2020). Environmental impact assessment of advanced urban wastewater treatment technologies for the removal of priority substances and contaminants of emerging concern: a review. *Journal of Cleaner Production*, **261**, 121078, https://doi.org/10.1016/j.jclepro.2020.121078

Petrie B., McAdam E. J., Lester J. N. and Cartmell E. (2014). Obtaining process mass balances of pharmaceuticals and triclosan to determine their fate during wastewater treatment. *Science of the Total Environment*, **497**, 553–560, https://doi.org/10.1016/j.scitotenv.2014.08.003

Petrie B., Barden R. and Kasprzyk-Hordern B. (2015). A review on emerging contaminants in wastewaters and the environment: current knowledge, understudied areas and recommendations for future monitoring. *Water Research*, **72**, 3–27, https://doi.org/10.1016/j.watres.2014.08.053

Poonia T., Singh N. and Garg M. (2021). Contamination of arsenic, chromium and fluoride in the Indian groundwater: a review, meta-analysis and cancer risk assessment. *International Journal of Environmental Science and Technology*, **18**, 2891–2902, https://doi.org/10.1007/s13762-020-03043-x

Puri M., Gandhi K. and Kumar M. S. (2023). Emerging environmental contaminants: a global perspective on policies and regulations. *Journal of Environmental Management*, **332**, 117344, https://doi.org/10.1016/j.jenvman.2023.117344

Ramakrishnan A., Blaney L., Kao J., Tyagi R. D., Zhang T. C. and Surampalli R. Y. (2015). Emerging contaminants in landfill leachate and their sustainable management. *Environmental Earth Sciences*, **73**, 1357–1368, https://doi.org/10.1007/s12665-014-3489-x

Richards L. A., Kumari R., White D., Parashar N., Kumar A., Ghosh A., Kumar S., Chakravorty B., Lu C. and Civil W. (2021). Emerging organic contaminants in groundwater under a rapidly developing city (Patna) in northern India dominated by high concentrations of lifestyle chemicals. *Environmental Pollution*, **268**, 115765, https://doi.org/10.1016/j.envpol.2020.115765

Richardson S. D. and Kimura S. Y. (2017). Emerging environmental contaminants: challenges facing our next generation and potential engineering solutions. *Environmental Technology & Innovation*, **8**, 40–56, https://doi.org/10.1016/j.eti.2017.04.002

Ryan C. C., Tan D. T. and Arnold W. A. (2011). Direct and indirect photolysis of sulfamethoxazole and trimethoprim in wastewater treatment plant effluent. *Water Research*, **45**, 1280–1286, https://doi.org/10.1016/j.watres.2010.10.005

Samaras V. G., Stasinakis A. S., Mamais D., Thomaidis N. S. and Lekkas T. D. (2013). Fate of selected pharmaceuticals and synthetic endocrine disrupting compounds during wastewater treatment and sludge anaerobic digestion. *Journal of Hazardous Materials*, **244**, 259–267, https://doi.org/10.1016/j.jhazmat.2012.11.039

Singh K. P., Rai P., Singh A. K., Verma P. and Gupta S. (2014). Occurrence of pharmaceuticals in urban wastewater of north Indian cities and risk assessment. *Environmental Monitoring and Assessment*, **186**, 6663–6682, https://doi.org/10.1007/s10661-014-3881-8

Stepien D., Regnery J., Merz C. and Püttmann W. (2013). Behavior of organophosphates and hydrophilic ethers during bank filtration and their potential application as organic tracers. A field study from the Oderbruch, Germany. *Science of the Total Environment*, **458**, 150–159, https://doi.org/10.1016/j.scitotenv.2013.04.020

Thomas O. and Brogat M. (2022). Organic constituents. In: UV-Visible Spectrophotometry of Waters and Soils, O. Thomas and C. Burgess (eds), Elsevier, Amsterdam, Netherlands, pp. 95–160.

Tran N. H., Reinhard M. and Gin K. Y.-H. (2018). Occurrence and fate of emerging contaminants in municipal wastewater treatment plants from different geographical regions – a review. *Water Research*, **133**, 182–207, https://doi.org/10.1016/j.watres.2017.12.029

Vickers N. J. (2017). Animal communication: when I'm calling you, will you answer too? *Current Biology*, **27**, R713–R715, https://doi.org/10.1016/j.cub.2017.05.064

Vieno N., Tuhkanen T. and Kronberg L. (2007). Elimination of pharmaceuticals in sewage treatment plants in Finland. *Water Research*, **41**, 1001–1012, https://doi.org/10.1016/j.watres.2006.12.017

Wang Y., Westerhoff P. and Hristovski K. D. (2012). Fate and biological effects of silver, titanium dioxide, and C60 (fullerene) nanomaterials during simulated wastewater treatment processes. *Journal of Hazardous Materials*, **201**, 16–22, https://doi.org/10.1016/j.jhazmat.2011.10.086

Watkinson A., Murby E. and Costanzo S. (2007). Removal of antibiotics in conventional and advanced wastewater treatment: implications for environmental discharge and wastewater recycling. *Water Research*, **41**, 4164–4176, https://doi.org/10.1016/j.watres.2007.04.005

West C. E. and Rowland S. J. (2012). Aqueous phototransformation of diazepam and related human metabolites under simulated sunlight. *Environmental Science & Technology*, **46**, 4749–4756, https://doi.org/10.1021/es203529z

Wilson M. and Ashraf M. A. (2018). Study of fate and transport of emergent contaminants at waste water treatment plant. *Environmental Contaminants Reviews*, **1**, 01–12, https://doi.org/10.26480/ecr.01.2018.01.12

Yang G. C., Yen C.-H. and Wang C.-L. (2014). Monitoring and removal of residual phthalate esters and pharmaceuticals in the drinking water of Kaohsiung City, Taiwan. *Journal of Hazardous Materials*, **277**, 53–61, https://doi.org/10.1016/j.jhazmat.2014.03.005

Yang X., Hu Z., Liu Y., Xie X., Huang L., Zhang R. and Dong B. (2022). Effect of pyrene-induced changes in root activity on growth of Chinese cabbage (*Brassica campestris* L.), and the health risks caused by pyrene in Chinese cabbage at different growth stages. *Chemical and Biological Technologies in Agriculture*, **9**, 7, https://doi.org/10.1186/s40538-021-00280-1

Yu J., Hu J., Tanaka S. and Fujii S. (2009). Perfluorooctane sulfonate (PFOS) and perfluorooctanoic acid (PFOA) in sewage treatment plants. *Water Research*, **43**, 2399–2408, https://doi.org/10.1016/j.watres.2009.03.009

Zhou H., Wu C., Huang X., Gao M., Wen X., Tsuno H. and Tanaka H. (2010). Occurrence of selected pharmaceuticals and caffeine in sewage treatment plants and receiving rivers in Beijing, China. *Water Environment Research*, **82**, 2239–2248, https://doi.org/10.2175/106143010X12681059116653

doi: 10.2166/9781789063752_0181

Chapter 10

Fate and behaviour of pharmaceutical and personal care products in wastewater

*Akanksha Bakshi, Megha Latwal, Sonali, Nitika Sharma, Anamika Sharma, Jatinder Kaur Katnoria and Avinash Kaur Nagpal**

Department of Botanical and Environmental Sciences, Guru Nanak Dev University, Amritsar, Punjab 143005, India
**Corresponding author: avinash.botenv@gndu.ac.in*

ABSTRACT

Nowadays, pharmaceuticals and personal care products (PPCPs) are being used by almost every section of society and are classified as 'emerging pollutants' due to their adverse effects on human and environmental health. They consist of different chemical compounds, hormones, and human and veterinarian prescription drugs. These are discharged into wastewater and find their way to aquatic ecosystems and even drinking water. A rising environmental problem is the presence of pharmaceuticals, hormones, and personal care products in sources of drinking and surface water. These substances have been identified in samples of surface water, groundwater and even drinking water in quantities ranging from parts-per-trillion (ng/L) to parts-per-billion (μg/L). Traditional sewage treatment plants and industrial wastewater treatment plants (WWTPs) can get rid of common pollutants like pathogens, nutrients, and organic matter but they fail to remove PPCPs which causes them to be released into the aquatic environment. Individual PPCPs can be successfully removed using a variety of treatment methods, such as membrane filtering, granular activated carbon, and advanced oxidation procedures. The use of a membrane bioreactor might also be an attractive means for dealing with pharmaceuticals. Due to their negative impacts on the ecosystem, information on their fate and interaction is essential for their management. Therefore, the present study focuses on the sources, types, effects, monitoring and suitable removal techniques for different PPCPs.

10.1 INTRODUCTION

Pharmaceutical and personal care products, abbreviated as PPCPs, are a group of numerous compounds or chemicals that include veterinary and human medication, nutraceuticals, bioactive food supplements, and common household goods such as soaps, shampoos, toothpaste, cosmetic products and so on. Despite their numerous advantages, these compounds are expected to pose potentially hazardous side effects. These chemical compounds are typically released either directly or indirectly into wastewater systems and have a negative impact on the environment (Kumar *et al.*, 2023). The occurrence of PPCPs in wastewater systems has recently attracted significant global attention and has become a topic of growing concern as they affect both human health and ecosystems. Unlike conventional pollutants, PPCPs can have specific biological effects, and their presence in water bodies can lead to unintended consequences. Some PPCP compounds remain unnoticed in the environment and unregulated. United States Environmental Protection Agency has listed some of these unmonitored PPCPs as priority pollutants (Anand *et al.*, 2022). Among the various PPCP sources, wastewater treatment plants (WWTPs) are primarily responsible for the presence of these chemicals in the ecosystem (Silori & Tauseef, 2022). Traditional wastewater treatment facilities like membrane filtration, advanced oxidation and adsorption are primarily employed to get rid of common pollutants like nutrients, organic matter and pathogens. However, they often fail to remove PPCPs, which causes them to be released into aquatic system (Kumar *et al.*, 2022). This results in the contamination of the majority of aquatic ecosystems. Once ingested, these PPCPs have the potential to disrupt the endocrine system and increase the risk of developing antimicrobial resistance, which results in the loss of the effectiveness of the majority of common antibiotics (Kumar *et al.*, 2022). Hence, there is an urgency to regulate the PPCP concentrations in different water bodies like wastewater and drinkable water. Prior to being disposed of into the environment, these substances must first undergo proper treatment. When PPCPs enter wastewater systems, complicated processes determine their fate and behaviour including their transformation, environmental effects, persistence and movement. Some countries have already started the treatment of wastewater containing PPCPs in their sewage treatment plants (STPs) and wastewater treatment plants (WWTPs) via sorption or biodegradation (Guerrero-Gualan *et al.*, 2023). They can also undergo chemical changes via conjugate cleavage or pass through the system without any change. The concentration of PPCPs following treatment could go down, up, or stay same as untreated sewage (Agnihotri & Thathola, 2019).

For the purpose of removing PPCPs from water bodies, a number of different processes have already been covered by many reviewers, including adsorption, ozonation, UV oxidation, membrane filtration, biological processes, and Fenton oxidation (John *et al.*, 2022; Kumar *et al.*, 2022; Zhang *et al.*, 2022). Other than these, constructed wetlands (CWs) are also used to remove these chemicals from the water bodies. CWs are considered as most sustainable and environmentally friendly method used to remove different PPCPs. In CWs, macrophytes and associated microbial assemblages effectively degrade/transform various PPCPs (Kumar *et al.*, 2023). Treatment plants are not completely capable of eliminating or neutralizing these PPCPs and regulatory effluent standards do not include any PPCP concentration parameters, hence their release into the environment remains unchecked and unmonitored (Bavumiragira & Yin, 2022). In addition, some PPCPs have the potential to transform into metabolites or other secondary products with potentially different properties and increased environmental risks. PPCP removal effectiveness of various processes depends on the reaction mechanism and PPCPs' chemical structures. Due to the wide range of chemical and physical properties of PPCPs, none of these processes can completely eliminate them. However, their combined effects may make it easier to get rid of these PPCPs from the water system (Loganathan *et al.*, 2023).

Understanding the fate and behaviour of PPCPs in wastewater is critical for developing effective mitigation strategies. To do this, it is necessary to examine their occurrence and concentration in wastewater influents, effluents, and sludge as well as their behaviour throughout various wastewater treatment processes. Additionally, identification of transformation products and their potential toxicity is crucial for assessing the overall environmental risk associated with PPCPs (Liu *et al.*, 2020). This chapter represents a significant advancement over prior literature as it incorporates the latest research, advanced methodologies as well as provides a comprehensive overview of the fate and behaviour of PPCPs in wastewater, highlighting the challenges and potential solutions for their removal and mitigation. By improving our understanding of the behaviour and potential risks of these emerging contaminants, we can work towards developing more efficient treatment technologies and implementing appropriate regulatory measures to safeguard water resources and protect both human and ecological health. Figure 10.1 shows various sources of PPCPs and their pathway in environment.

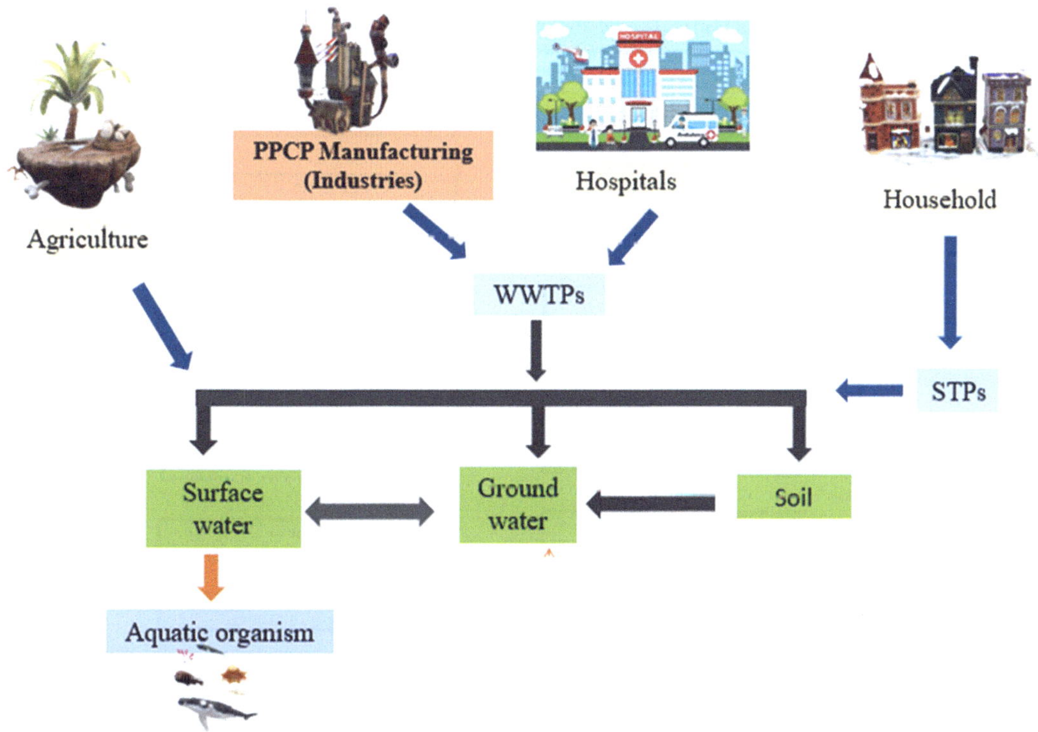

Figure 10.1 Schematic diagram showing various PPCP sources and their pathways in the environment. *STPs – sewage treatment plants; WWTPs – wastewater treatment plants.

10.2 MAJOR CATEGORIES OF PPCPS

In 2005, the European Union (EU) initiated the 'Norman Project' to monitor the emerging pollutants being reported around the world and found more than 1036 contaminants and their products which were classified into 30 categories. PPCPs were classified as a major category of emerging pollutants (Ricky & Shanthakumar, 2022). As a consequence of increasing utilization of PPCPs, more of these products are being released into wastewater through wash-off, urine, and feces as parent compounds, conjugates or their derivatives (Bavumiragira & Yin, 2022). Major categories of PPCPs are presented in Figure 10.2. In the following two sub-sections, categories of pharmaceutical products and personal care products are being discussed separately.

10.2.1 Categories of pharmaceutical products

Pharmaceutical products consist of active ingredients (chemically manufactured or natural substances) present in prescription and non-prescription drugs, veterinary medications, and illegal drugs. These include antibiotics, stimulants, hormones and steroids, non-steroidal anti-inflammatory drugs (NSAIDs), antihypertensives, lipid regulators and antidepressants.

Antibiotics are utilized for curing infections caused by bacteria in humans and animals and are passed out from body either in metabolized or unmetabolized forms (Sodhi *et al.*, 2021). Even after excretion, they do not get fully degraded in the environment and persist in water bodies. Some of the antibiotics that have been reported in wastewaters worldwide include erythromycin from Croatia

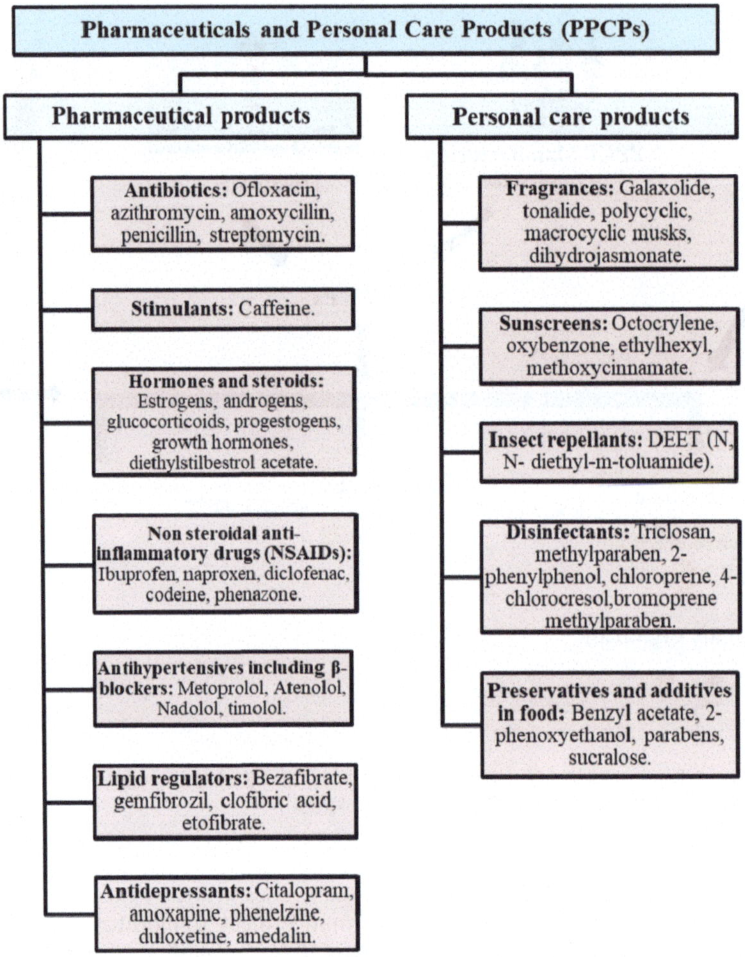

Figure 10.2 Major categories of PPCPs.

(Senta *et al.*, 2019); ciprofloxacin and ofloxacin from Spain (Bijlsma *et al.*, 2021); and tetracycline from Beijing (Zhang *et al.*, 2018).

Caffeine, 1,3,7-trimethylxanthine, is the oldest known stimulant which is a natural alkaloid belonging to the family of xanthines. Its structure is similar to adenosine and impedes its activity by acting on its receptors (Cerveny *et al.*, 2022). Adenosine has a crucial role in sleep–wake cycle, locomotive and psychological activities. As a pharmaceutical, caffeine is used to prevent sleepiness, treat pain, circulatory system failure and respiratory issues. It is the most commonly detected contaminant in wastewater because of its high consumption and presence in chocolates and beverages such as tea, coffee, sodas, and so on. Its presence in water is considered as an indicator of water pollution caused due to human activities (Júnior *et al.*, 2019).

Hormones are used to maintain physical growth and sexual wellness in organisms (Wang & Wang, 2016). Utilization of steroids and hormones in livestock agriculture has emerged as a major problem since they contaminate the water resources. Steroids and hormones, which are frequently found in WWTPs, include androgens, estrogens, progestogens, and growth hormones (Guerrero-Gualan *et al.*,

2023). Estrogens are the most commonly reported hormones in sewage, wastewater, and surface water (Damkjaer *et al.*, 2018; Yazdan *et al.*, 2022).

Non-steroidal anti-inflammatory drugs (NSAIDs) are among the commonly used drugs to alleviate pain and inflammation in patients suffering from arthritis, menstrual cramps, postoperative conditions, and so on. Among different types of NSAIDs, diclofenac, naproxen, ketoprofen, and ibuprofen are the most common drugs detected in aquatic systems (Mussa *et al.*, 2022). Due to their toxicity and persistence, European Union (EU) regards NSAIDs as high priority pharmaceuticals (Rastogi *et al.*, 2021). NSAIDs have been reported from different WWTPs like WWTPs of Durban, South Africa (Madikizela & Chimuka, 2017); and WWTPs of Mangalore, India (Thalla & Vannarath, 2020). Besides parent compounds, metabolites of NSAIDs have also been reported in different waste and surface waters (Březinova *et al.*, 2018).

Two main lipid regulators used to control the levels of triglycerides and cholesterol in blood are fibrates and statins. Fibrates function by lowering the levels of fatty acids and triglycerides whereas statins decrease cholesterol level. Fibrates are the derivatives of fibric acid. Clofibric acid is a persistent compound that is detected in water years after its emission (Rosal *et al.*, 2010). It is also the most commonly reported drug among fibrates in water. Commonly used statins worldwide include atorvastatin, lovastatin, pitavastatin, pravastatin, simvastatin and fluvastatin (Blonç *et al.*, 2023). Compounds such as bezafibrate, clofibrate, gemfibrozil, atorvastatin and simvastatin are the most prevalent pharmaceuticals in water systems (Ulvi *et al.*, 2022). Ramírez-Morales *et al.* (2020) reported presence of gemfibrozil in wastewater samples from 11 WWTPs in Costa Rica. Lipid regulators have also been reported from wastewaters of Mahdia, Tunisia (Afsa *et al.*, 2020); Sri Lanka (Goswami *et al.*, 2022); and Jiangsu Province, China (Liu *et al.*, 2023).

Antihypertensives drugs used for the treatment of high blood pressure which is the common cause of cardiovascular diseases (Subedi *et al.*, 2017). These include angiotensin converting enzyme inhibitors, diuretics, β-blockers (blockers of β-adrenergic receptors), angiotensin II receptor antagonists and calcium channel blockers. Increase in their concentration in water systems stems from increase in the number of patients with cardiovascular diseases. Reuse of treated wastewater in agricultural fields often leads to addition of these pollutants to soil affecting soil microbes and organisms. Presence of antihypertensives has been reported from six STPs in Bavaria (Bayer *et al.*, 2014); two WWTPs of Albany, New York (Subedi and Kannan, 2015); three WWTPs from Foshan and Guangzhou, China (Huang *et al.*, 2018); and WWTPs of two major cities of Colombia (Botero-Coy *et al.*, 2018).

Antidepressants regulate mood by targeting specific neurotransmitters. On the basis of their mode of action, they can be classified as tricyclic antidepressants (TCA) for example amoxapine; selective serotonin reuptake inhibitors (SSRIs) for example citalopram; selective noradrenaline reuptake inhibitors (NARI) for example amedalin; serotonin-norepinephrine reuptake inhibitors (SNRIs) for example duloxetine; and monoamine oxidase inhibitors (MAOIs) for example isocarboxazid (Chen *et al.*, 2022). Occurrence of antidepressants has been reported from two WWTPs of Albany, New York (Subedi & Kannan, 2015); two municipal WWTPs and Rhine River in Germany (Schlüsener *et al.*, 2015); and municipal WWTPs of Tehran, Iran (Golbaz *et al.*, 2023). High concentrations of venlafaxine have been reported from two WWTPs in Greece (Ofrydopoulou *et al.*, 2022). In China also, venlafaxine and its metabolite were found in WWTPs at Shanghai (Ma *et al.*, 2018).

10.2.2 Categories of personal care products (PCPs)

PCPs are a large group of compounds including both inert as well as active ingredients used for personal care and hygiene (Figure 10.2). PCPs mainly include dyes, make-up products, oral hygiene products, preservatives, disinfectants, fragrances, sunlight UV filters, food additives, supplements, and insect repellents (Ricky & Shanthakumar 2022). PCPs, now considered as emerging contaminants, have been detected in trace concentrations (ranging from ng/L to mg/L) in industrial wastewater as well as sewage and sludges (Anand *et al.*, 2022), and different ground and surface waters (Cooney *et al.*, 2023; Nozaki *et al.*, 2023). They are thought to escape STPs and WWTPs (in their original

Table 10.1 Active constituents of common PCPs.

Fragrances	Sunscreens	Insect Repellents	Disinfectants	Preservatives and Additives
Nitro musks Musk ambrette Musk alpha Musk ketone Musk moskene Musk tibetene **Polycyclic musks** Celestolide Fixolide Galaxolide Traseolide Versalide **Alicyclic musks** Cyclomusk Helvetolide Romandolide **Macrocyclic musks** Ambrettolide Cyclopentadecanolide Ethylene brassilate Globalide Velvione	Benzophenone-3 (BP-3) 4-methyl-benzylidene camphor (4- MBC) Octocrylene (OC) 2-ethyl-hexyl-4- tri methoxycinnamate (EHMC) Oxybenzone	N,N-diethyl-m-toluamide (DEET)	Benzyl dimethyl dodecyl ammonium chloride Dioctyl dimethyl ammonium chloride Ethanol Methylparaben 2-phenyl phenol Isopropanol Triclosan	Benzyl acetate. Bronidox Iodopropynyl butyl carbamate (IPBC) Propylparaben Sucralose 2,4,4-trichloro-2 -hydroxy diphenyl ether
Agnihotri and Thathola (2019)	Agnihotri and Thathola (2019); Ricky and Shanthakumar (2022)	Merel and Snyder (2016)	Al-Baldawi *et al.* (2021); Ricky and Shanthakumar (2022)	Al-Baldawi *et al.* (2021)

or biologically altered forms) and make their way to both surface and groundwater systems causing environmental contamination (Anand *et al.*, 2022). Active constituents of common personal care products are given in Table 10.1.

Fragrances (compounds with sweet odour) are important constituents of perfumes, deodorants, shampoos, conditioners, and many cleaning products used in everyday life. Different kinds of musk fragrances (aromatic compounds) are used in the perfume industry. Natural musk is obtained from animals, particularly from male musk deer or from a cat with a musk civet. Nitro musks (musk xylene and musk ketone) and polycyclic musks (Galaxolide HHCB and tonalide ATHN) come under synthetic musks. HHCB is the most widely used synthetic musk recognized as an emerging contaminant and affects human health (Li *et al.*, 2021). Other musks include macrocyclic musks and alicyclic musks. Tasselli and Guzzella (2020) detected galaxolide and its metabolite galaxolidone in the sludge of a WWTP in Northern Italy. Tasselli *et al.* (2021) investigated the presence of PCPs (HHCB, AHTN, celestolide, etc.) in municipal and industrial sludge and wastewater in Milan, Northern Italy. Galaxolide, tonalide, musk xylene, and musk ketone were found in abundance from sewage sludge samples collected in 55 WWTPs in the Czech Republic (Košnář *et al.*, 2021).

Sunscreen products contain various chemicals which act as protectants against UV radiation. Commonly known as UV filters, they are effective against harmful UV rays and are divided into two categories viz. organic filters and inorganic filters. Organic filters include benzophenone 3 (BP-3), 2-ethylhexyl 4-methoxycinnamate (octinoxate, EHMC), oxybenzone, avobenzone, and inorganic ones including zinc oxide (ZnO) and titanium dioxide (TiO2) (Tsui *et al.*, 2014). (BP-3), 4-methyl benzylidene camphor (4-MBC), and EHMC are the most common UV filters found in daily activity products such as cosmetics (Tran *et al.*, 2022). In an earlier study, BP3 was detected from WWTP influents from southern California (USA) with a concentration of 10 μg/l (Mao *et al.*, 2019). Benzophenones (BPs) were detected in water samples from Huangpu River, China (Wang *et al.*, 2021). Tran *et al.* (2022) reported BP-3 and BP-4 to be the most common UV filters found in wastewater. Hsieh *et al.* (2023) detected triclosan and BP-3 from three WWTPs in southern Taiwan.

Insect repellents are the most commonly detected contaminants in water samples. N N-diethyl-m-toluamide (DEET), an insect repellent and was developed by the US Army in 1946 (Kitchen *et al.*, 2009). DEET and its products are potential toxicants, which is a matter of concern with regard to human health and the environment (Xu *et al.*, 2022). Merel and Snyder (2016) reviewed the literature on the occurrence and fate of DEET in water across the globe. The study concluded that DEET had been detected in almost all types of aquatic ecosystems including ground and surface waters as well as wastewaters from different parts of the globe including America, Asia, Africa, Europe and Oceania. Güzel (2021) detected DEET in the Seyhan River (Turkey) and its concentration was relatively higher in autumn than in summer.

Disinfectants are used for surface cleaning, instrument disinfection and so on. Alcohol, aldehydes and chlorine-containing compounds are the main active ingredients. Methylparaben, 2-phenyl phenol, triclosan and alcohols (isopropanol and ethanol) are the most commonly used chemicals as disinfectants. The concentrations of these compounds in urban runoff and groundwater have increased significantly in recent years (Rodriguez-Narvaez *et al.*, 2017). One of the most popular biocides is triclosan (TRC), which is detected in aquatic environments at high rates.

Preservatives are used in PCPs mainly cosmetics, food, and beverages as they protect these products against microbial growth. Table 10.1 gives the active constituents of preservatives. Triclosan and triclocarban are prime active ingredients widely used in personal hygiene products (Musee, 2018). Due to the insufficient removal from WWTPs, cosmetic preservatives have been widely detected in aquatic environments and sewage sludge. Parabens are the most extensively used preservatives in PCPs due to their high stability (Penrose & Cobb, 2023).

10.3 OCCURRENCE OF PPCPS IN WATER ECOSYSTEM

PPCPs though enter surface waters at low doses, but long-term exposure has the potential to harm aquatic organisms. Several studies carried out around the globe have shown PPCP contamination of surface water. Liu *et al.* (2020) reported the presence of 50 PPCPs in the Chinese aquatic system in concentrations between ng/L and g/L. Bisognin *et al.* (2021) demonstrated presence of 13 PPCPs including paracetamol, caffeine, metronidazole and sulphamethoxazole in the effluent and sludge samples from STPs in southern Brazil. Nozaki *et al.* (2023) found 43 PPCPs including triclosan, triclocarban, chlorpheniramine, diphenhydramine, and chlorpheniramine in freshwater samples from different rivers, lakes and/or ponds from three Asian countries viz. India, Indonesia and Vietnam. Balakrishna *et al.* (2017) reviewed studies from India that showed the presence of PPCPs like atenolol, triclosan, carbamazepine, trimethoprim, ibuprofen, acetaminophen, caffeine and triclocarban in different rivers. Sharma *et al.* (2019) reported occurrence of 15 PPCPs in the Ganga River and groundwater of several sites along the river. Out of 15 PPCPs, caffeine was found to be most prevalent. Singh and Suthar (2021) showed the occurrence of different PPCPs like triclosan, caffeine, acetaminophen and tetracycline in Ganga River at Rishikesh and Haridwar cities. Yu *et al.* (2023) detected the presence of artificial sweeteners in Yellow River of China. Okoye *et al.* (2022) reported the occurrence of PPCPs in surface waters of Africa.

Groundwater is a critical water source with significant environmental concerns as it provides water for human consumption, ecosystem demands and irrigation. Different concentrations of various PPCPs have been found to contaminate groundwater. In a recent study, different antibiotics like erythromycin, trimethoprim, clindamycin and ofloxacin were detected in the groundwater of the Bialka River valley in Poland (Lenart-Boroń *et al.*, 2022). Cooney *et al.* (2023) also noted high concentrations of antibiotics such as cephalexin, ampicillin, sulphamethoxazole, trimethoprim, tetracyclin, oxytetracycline and erythromycin in groundwater of Riviera Maya, Mexico. Caffeine has been found in many groundwater samples from all over the world and enters into the water system through various household activities and human urine. The presence of caffeine has been shown in various effluents, landfills, septic tanks and wastewater that can contaminate the groundwater by natural recycling process (Cerveny *et al.*, 2022; Do *et al.*, 2022; Júnior *et al.*, 2019; Sui *et al.*, 2015). As compared to other PPCPs, consumption of caffeine is much higher in day-to-day life and is expected to be present in higher concentrations in groundwater. However, Sui *et al.* (2015) observed that the concentration of caffeine in groundwater was not higher as compared to other PPCPs. It is expected that caffeine might either be removed during wastewater treatment or it can undergo degradation rapidly. Do *et al.* (2022) noted that 80% of groundwater of Hawpe village in Sri Lanka's Galle district was contaminated with caffeine (7.9 ng/L).

Lipid regulators like bezafibrate, clofibric acid and gemfibrozil have been detected in groundwater, however, their concentration was negligible as compared to other PPCPs like anti-inflammatories, antibiotics, caffeine and so on (Sui *et al.*, 2015; Wang & Wang, 2016). Clofibric acid and gemfibrozil were detected at lower concentrations in groundwater samples from China, Spain and Singapore (López-Serna *et al.*, 2013; Peng *et al.*, 2014; Tran *et al.*, 2014). Some lipid regulators were also found in the groundwater of Barcelona, Spain which included bezafibrate (López-Serna *et al.*, 2013) and clofibric acid (Jurado *et al.*, 2022). Gemfibrozil has been observed in waterbodies of Marmara, Turkey (Korkmaz *et al.*, 2022) and in the surface water of the Beijiang River and surrounding groundwater (Lei *et al.*, 2023).

Some studies have shown the presence of some PPCPs including X-ray film contrast media, musks, sunscreen agents and beta blockers in groundwater (Subedi *et al.*, 2017). Zemann *et al.* (2015) reported the presence of iodinated X-ray contrast media (ICM) in the groundwater of Wadi Shueib, Jordan. However, groundwater samples from Barcelona, Spain were mainly contaminated with tonalide, octocrylene, ethylhexyl methoxycinnamate, propranolol, metoprolol and musk galaxolide (Jurado *et al.*, 2022; López-Serna *et al.*, 2013). In a groundwater sample from Riviera Maya, Mexico, sunscreens were found (Cooney *et al.*, 2023). Other PPCPS like acetaminophen, climbazole, cotinine, carbamazepine, crotamiton, atenolol and lidocaine were also detected in groundwater samples of Spain (Jurado *et al.*, 2022); Poland (Lenart-Boroń *et al.*, 2022); and Sri Lanka (Do *et al.*, 2022).

PPCPs were also found in the effluents from both sewage and wastewater treatment plants due to their extensive use and incomplete removal during the treatment process. Industrial setups produce huge quantities of wastewater. Solid wastes of these PPCPs are generally discarded into landfills and garbage sites. Likewise, effluents of wastewater treatment plants and biosolids produced from sludges also contain large amount of PPCPs which ultimately contaminate agricultural lands (Anand *et al.*, 2022). Conventional wastewater treatment processes are not designed to specifically remove PPCPs, resulting in incomplete removal of these compounds during treatment. PPCPs can persist in the effluent and are often discharged into receiving waters (Anand *et al.*, 2022; Guerrero-Gualan *et al.*, 2023; Kumar *et al.*, 2022). Different PPCPs, including ranitidine hydrochloride, sulphamethoxazole, ibuprofen, ampicillin sodium, ribavirin, and clozapine, have been found in the effluent of Chinese WWTPs (Liu *et al.*, 2022). Similarly, Kumar *et al.* (2023) found Ibuprofen, 1-hexadecanoyl-sn-glycerol (1HSG), acetaminophen, triclocarban, trimethoprim, 19-docosapentaynoic acid, sulphamethoxazole, arbamazepine and caffeine in wastewater effluent of an academic institution of Gandhinagar, Gujrat. Concentrations of PPCPs can range from low (parts per billion (ppb)) to high (parts per million (ppm)) levels (Ebele *et al.*, 2020). The occurrence and concentrations of PPCPs in effluents from STPs and WWTPs can vary depending on several factors including population density, wastewater characteristics, and local usage patterns.

10.4 SOURCES AND FATE OF PPCPS

10.4.1 Sources of PPCPs in wastewater

Anthropogenic activities are the main sources of PPCPs in the environment (Al-Baldawi *et al.*, 2021). The emission of PPCPs in wastewater starts from the manufacturing of pharmaceutical ingredients and PCPs by industries proceeding with the utilization of these products by hospitals, households, animal husbandry, fish rearing and so on (Falahi *et al.*, 2022). Healthcare institutions are important sources of pharmaceutical discard (Anand *et al.*, 2022). The unmetabolized products of the pharmaceutical drugs used by patients in hospitals, households and so on are excreted and discharged into wastewater/ STPs (Al-Baldawi *et al.*, 2021). Veterinary pharmaceuticals used by animal husbandry and aquaculture practices are also responsible for polluting water and soil. Animal excreta containing unmetabolized drugs is released directly into surface waters or converted into manure and applied in fields where it can contaminate the groundwater (Ebele *et al.*, 2020). Alternatively, personal care products are used more extensively and frequently, hence, producing more quantities of waste. These products are discharged from homes into wastewater treatment facilities whereas, households without the facility directly discharge their wastewater into surface water systems such as rivers and streams (Wang & Wang, 2016). Personal care products are lipophilic by nature and adsorb onto sediments (Okoye *et al.*, 2022). Another prevalent practice is dumping expired PPCPs in solid waste which leads to leaching of these contaminants in soil and groundwater. In urban areas, due to a more organized sewage system, PPCP-contaminated wastewater from different sources accumulated in STPs. Since these systems cannot treat PPCP-laden wastewater, these contaminants get discharged into water bodies.

10.4.2 Fate of PPCPs in wastewater

PPCPs present direct and indirect effects on the environment such as the development of bacterial resistance, disturbance of the endocrine system and bioaccumulation in organisms (Nozaki *et al.*, 2023). They are classified as pseudo-persistent contaminants due to their polarity, optical activity and semi-volatility (Liu *et al.*, 2022). Personal care products are mainly used on hair and skin and are discharged (by excretion and washing) into wastewater (Anand *et al.*, 2022). They may further degrade by coming in contact with sunlight (UV radiations), air (oxidation), water, microbes and so on (Anand *et al.*, 2022). As degradation of PPCPs is dependent on their chemical and physical composition, different products will degrade in different proportions. Compounds such as caffeine and paracetamol are widely used but often detected in less quantities due to more biodegradation as compared to carbamazepine and sulphamethoxazole which are persistent (Sui *et al.*, 2015). Due to the absence of treatment techniques for these compounds, most of the treated wastewaters contain PPCPs in the undegraded form which are released into water bodies. When they are continuously released in water bodies, the non-target species bioaccumulate these contaminants over time and pass them into the food chain (biomagnification). Many PPCPs such as ciprofloxacin and galaxolide have been found to be bioaccumulated in aquatic species. The presence of antibacterial compounds has been observed to develop antibiotic resistance in non-target microorganisms (Ricky & Shanthakumar, 2022).

10.5 HARMFUL EFFECTS OF PPCPS

The continuous release of PPCPs into bodies of water, as well as their exposure, may have chronic consequences on aquatic plants and animals (Silori & Tauseef, 2022). The presence of PPCPs in water systems causes genotoxicity, mutagenesis, and ecotoxicity in plants, animals, and humans.

PPCPs reach plants largely through irrigation with recovered wastewater, application of manure and biosolids for agricultural soil fertilization and deposition from volatilized chemicals. PPCPs are absorbed by plants through their aerial tissues and roots by the process of mass flow or diffusion of dissolved compounds. Deposition of vaporized compounds and aerosols, direct contact with irrigation or amendment materials, and translocation from root tissues are all ways for aerial tissues to absorb (Trapp & Legind, 2011). In an experiment conducted by Pérez *et al.* (2023), *Typha latifolia* was exposed to some PPCPs viz. triclosan, gemfibrozil, carbamazepine and fluoxetine and it was observed that triclosan and gemfibrozil were accumulated in roots while carbamazepine and fluoxetine accumulated in leaf tissues. Zeng *et al.* (2022) reported accumulation of PPCPs including amantadine, carbamazepine chlorpheniramine, chlorosibutramine haemosibutramine, N-monomethyl sibutramine, and sibutramine in radish, buckwheat and okra (sprouting seeds). Falahi *et al.* (2022) reported the accumulation of Ibuprofen and paracetamol in the roots and shoots of *Scirpus grossus*. PPCPs accumulated by plants can be metabolized, detoxified, inactivated, and sequestered by the plant's defence mechanism and thus it appears that the majority of PPCPs do not cause phytotoxicity. However, some recent studies have shown that prolonged exposure of plants to PPCPs leads to the overproduction of ROS that is, oxidative damage (Sun *et al.*, 2018). Osma *et al.* (2018) reported that two PPCPs viz. ß-estradiol and gemfibrozil produced a negative effect on wheat seedlings through oxidative stress. Ravichandran and Philip (2022) noted that the deposition of carbamazepine (anti-epileptic drug) on four wetland plants (*Canna indica, Chrysopogon zizanioides, Colocasia esculenta* and *Phragmites australis*) led to oxidative stress in these plants. Similarly, Elveren and Osma (2022) reported an increase in the activity of catalase, peroxidase and superoxide dismutase in *Triticum aestivum* (L.) on exposure to PPCPs like ciprofloxacin, doxylamine succinate, and ibuprofen.

The presence of PPCPs in aquatic systems has been shown to have deleterious effects on aquatic organisms. In a recent study by Chabchoubi *et al.* (2023), NSAIDS like Ibuprofen, diclofenac, paracetamol and ketoprofen were found to adversely affect the embryo and larval growth of a Zebrafish, *Danio rerio*. In another study, indomethacin, ibuprofen and their mixture were observed to affect the rate of food intake and enzymatic activities in a crustacean, *Daphnia magna* (Michalaki & Grintzalis, 2023). Cory *et al.* (2019) reported toxicity of naproxen and its photo-transformation derivatives (NAP-PT1 and NAP-PT2) in *Anaxyrus terrestris* larvae. Diclofenac and its metabolite 4-hydroxydiclofenac were shown to cause deformations and low protein content in the gills of a mollusc, *Mytilus trossulus* (Świacka *et al.*, 2022). Liu *et al.* (2018) documented the toxicology of antibiotic pollution in many aquatic organisms such as tetrogenic effects of macrolides, tetracyclines, sulphonamides, quinolones and so on on aquatic vertebrates such as *Danio rerio* and *Xenopus tropicalis* and toxicity of quinolone antibiotics in tilapia, *Oreochronis niloticus*. Reckless consumption of antibiotics is also one of the major reasons for the increase in the emergence of antibiotic-resistant genes (ARGs) (Langbehn *et al.*, 2021).

Various studies have shown the ability of hormones present in wastewater to cause endocrine disruption in aquatic species. Bahamonde *et al.* (2015) reported intersex conditions in male fishes (*Etheostoma caeruleum*) exposed to municipal wastewater effluent in Grand River, Canada. Plahuta *et al.* (2017) showed the potential of influent and effluent wastewater containing EDCs (estrogens) to hamper the moulting ability of *Asellus aquaticus*. Leese *et al.* (2021) observed changes in the reproductive behaviour of fathead minnows on exposure to EDC containing wastewater effluent. Estrogen and testosterone levels in a fish species, *Oryzias latipes,* were disturbed on exposure to a lipid-regulating medication-gemfibrozil (Lee *et al.*, 2019). Whereas in another study, the presence of gemfibrozil in water affected antioxidant enzymes activity in liver of *Danio rerio* (Falfushynska *et al.*, 2022). The presence of antihypertensives in water is also shown to have deleterious effects on aquatic animals (Pusceddu *et al.*, 2022). The presence of antidepressants in the sewage water has been shown to affect the behaviour of aquatic species. Antonopoulou *et al.* (2022) reported toxicity of Paroxetine, an antidepressant, on freshwater and marine species, bacteria and human lymphocytes.

Antiepileptic drugs (AEDs) are frequently present in various aquatic ecosystems, making aquatic organisms vulnerable to these medications. Salahinejad *et al.* (2023) reviewed literature on effect of some antiepileptic drugs on teleost fishes. It was concluded that AEDs disrupted parasympathetic neurotransmitters, and serotonergic and glutamatergic systems in fishes. Gebuij's *et al.* (2020) studied the concentration dependent effect of an antiepileptic drug (valproic acid) on the survivability of Zebra fish. It was found that the drug exposure led to decrease in bone and cartilage development, and reduction in the length of the ethmoid plate. Organic UV filters are found to be extensively accumulated in aquatic organisms and are known to impact the survivability, development and reproduction of aquatic organisms. Zhou *et al.* (2019) investigated the effect of organic UV filter (OUVF), ethylhexyl methoxy cinnamate (EHMC), on the growth and reproduction of *Danio rerio* (Zebra fish) and the study revealed that EHMC caused a reduction in the number of hatchlings and increase in mortality rate of *Danio rerio* embryo. In another study done was by Zhou *et al.* (2020), OUVFs 4-methyl benzylidene camphor (4-MBC) and sulphanilamide antibiotic, sulphamethoxazole caused slow growth, body weight reduction, impaired vitality and irregular behaviour of *Danio rerio*.

PPCPs in the environment around the world have been linked to negative impacts on human health. Generally, PPCPs enter the human body by drinking water or the ingestion of vegetables, fruits and crops irrigated with contaminated water. NSAIDs, commonly used for treatment of musculoskeletal pain, have direct side-effects on human health such as cardiovascular problems (heart attacks, hypertension), gastrointestinal issues (ulcers, perforation, bleeding) and kidney-related problems (Banerjee & Maric, 2023; Machado *et al.*, 2021). Overconsumption of antibiotics leads to antibiotic-resistant bacteria and antibiotic-resistant genes, posing a great threat to the human health by increasing the chances of infections and difficulty in treating them (Seethalakshmi *et al.*, 2022). EDCs have been shown to affect the synthesis and action of sex steroid hormones which in turn leads to hormone sensitive cancers and infertility in men and women (Yilmaz *et al.*, 2020). Caffeine besides being one of the best-known stimulants, has been linked to many health issues when taken in higher doses. (Tandiono & Budiyanti 2023) reported that coffee intake raised systolic blood pressure (SBP) and diastolic blood pressure (DBP) in 16 male and female teenagers. In another study, caffeine withdrawal caused severe migraine attacks in majority of subjects (Alstadhaug *et al.*, 2020).

10.6 REMOVAL AND MANAGEMENT OF PPCPS FROM WASTEWATER

Personal care items and pharmaceuticals have been identified as emerging polluters of water resources. The wastewater containing PPCPs must be treated so that it does not impact the quality of aquatic as well as terrestrial life. The standard environmental regulation allowed 50 ng/L of pharmaceutical waste in discharged water (Rosman *et al.*, 2018). The majority of pharmaceuticals are discharged through urine and feces to sewage systems and effluents from industrial areas, hospitals and so on. As a result, STPs/ WWTPs acquire varied amounts of different PPCPs (Bavumiragira & Yin, 2022). Different treatment methods have been explored for the removal of PPCPs which include both conventional and advanced processes (Figure 10.3).

10.6.1 Different methods of management

10.6.1.1 Conventional systems

10.6.1.1.1 Chlorine

The most popular conventional method for sanitizing drinking water is still the treatment with chlorine. Studies have shown that chlorine reacts rapidly with amine-containing pharmaceuticals and gives rise to chlorine-containing products (Pinkston & Sedlak, 2004). For instance, chloramines are produced when fluoxetine and metoprolol react with chlorine (Bedner & MacCrehan, 2006a). Acetaminophen, the active ingredient in paracetamol, interacts with chlorine to produce a number of byproducts, two among them (1, 4-benzoquinone and N-acetyl-p-benzoquinone imine) have been recognized as hazardous chemicals (Bedner & MacCrehan, 2006b). But when chlorine is used with UV it is reported to destroy PPCPs like 17a-ethinylestradiol, benzotriazole, carbamazepine, chloramphenicol, diclofenac, iopamidole, metoprolol, and sulphamethoxazole (Pai & Wang, 2022).

10.6.1.1.2 Wastewater treatment plants (WWTPs)

WWTPs have a basic system of biophysical treatments and an additional system made up of an active sludge-based biological reactor. As reviewed by Rivera-Utrilla *et al.* (2013), most of the PPCPs in urban

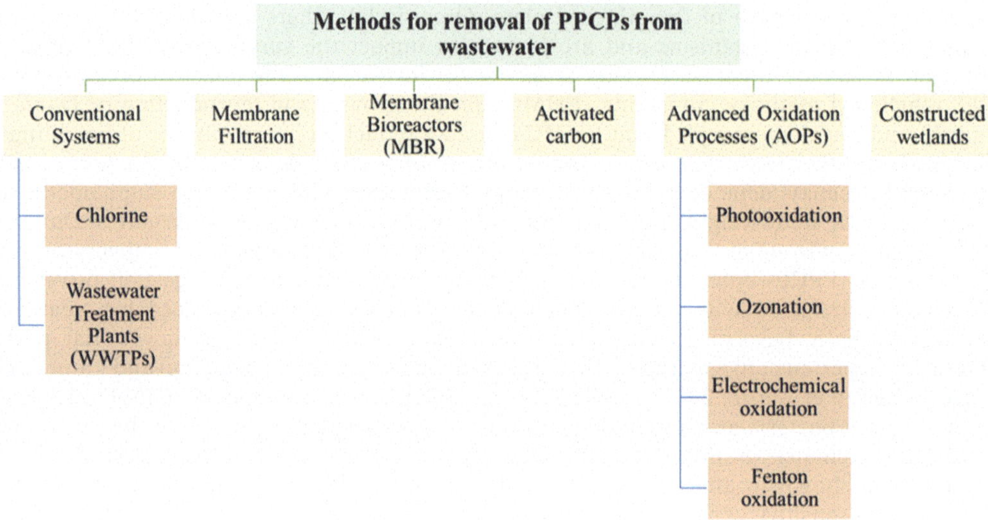

Figure 10.3 Schematic diagram showing different methods for removal of PPCPs from wastewater.

wastewater cannot be utilized by microorganisms and may even inhibit their activity or cause their bioaccumulation in the food chain. The conventional WWTPs have a limited capacity to eliminate pharmaceutical products from the wastewater. This study concluded that a significant portion of the PPCPs contained in urban wastewater cannot be entirely removed by traditional treatment techniques, as a result, they persist in effluents and pollute surface and underground waters, the primary sources of drinking water. To lessen the environmental impact and potential harm caused by effluents, more efficient and targeted treatments are needed.

10.6.1.2 Membrane filtration

It has been established that the majority of contaminants can only be partially eliminated by traditional wastewater treatment methods. Therefore, to stop the release of PPCPs into the environment, improvements in existing wastewater treatment methods and further treatment of the generated sludge are needed (Kumar *et al.*, 2023). Comparatively, more sophisticated wastewater treatment techniques, like those utilizing membrane technology, can remove drugs at higher rates. Fundamentally, membrane-based filtration techniques are classified into four basic types: microfiltration (MF), ultrafiltration (UF), nanofiltration (NF) and reverse osmosis (RO). NF and RO membranes (due to their small pore size, 0.001–0.008 μg) have been suggested to be used effectively for the removal of pharmaceutical products from wastewater (Yoon *et al.*, 2006). But a smaller-sized compound can cross these filters and fouling often occurs when cake is formed on a membrane's surface, which greatly decreases the overall separation performance. On the other side, after membrane separation operations, the disposal of concentrates becomes another challenge that has yet to be resolved (Rosman *et al.*, 2018). In an experiment conducted by Liu *et al.* (2023) the removal efficiency of RO membranes was analyzed for PPCPs like carbamazepine, ibuprofen and triclosan and was found to be 97.47%, 98.93% and 99.01%, respectively.

10.6.1.3 Membrane bioreactors (MBRs)

In comparison to the traditional activated sludge system, MBRs that combine the traditional biological method with membrane filtering technology, exhibit a number of advantages, particularly high biomass concentrations, effective solid–liquid separation and a small footprint. Long sludge retention times (SRT) can improve the elimination of some organic compounds, but it is still difficult to effectively remove developing pollutants like PPCPs. Furthermore, due to the negative effects that antibiotics have on microorganisms, a serious membrane fouling was seen in the MBR system while treating antibiotics (Chen *et al.*, 2020).

10.6.1.4 Activated carbon

Due to its adaptability and effectiveness, adsorption is a particularly intriguing method for the removal of PPCPs. Since activated carbon (AC) does not produce toxins, it can be employed for adsorption (Delgado *et al.*, 2019; Zhu *et al.*, 2022). The most widely used ACs come in two forms, granular or powdered (GAC or PAC, respectively). Both GAC and PAC can be used for the treatment of wastewater, however, PAC is typically more effective with faster adsorption rates (owing to the smaller size of the particles) and GAC has the primary benefit of regeneration/reuse post saturation (Delgado *et al.*, 2019). Activated carbon has been produced from *Albizia lebbeck* seed pods for the removal of cephalexin (Ahmed & Theydan, 2012), lotus stalk for removal of trimethoprim (Liu *et al.*, 2012), and paper mill sludge for the removal of citalopram (Calisto *et al.*, 2014). Even though ACs from different raw materials are widely available, a lot of work is still being done on producing carbons from different starting materials (like agricultural and industrial residues) to reduce production costs and encourage value-added recycling of waste (Calisto *et al.*, 2014). The primary objective is to produce a carbon with a high adsorption capacity using low-cost, ecologically acceptable methods (without utilizing external activation method) and, at the same time, to suggest a novel approach to valorizing the industrial byproducts (Calisto *et al.*, 2014).

10.6.1.5 Advanced oxidation processes (AOPs)

These processes mainly use higher concentrations of hydroxyl radicals which transform the recalcitrant PPCPs by reacting with them through chemical or photochemical reactions like photooxidation, ozonation and electrochemical oxidation (Rosman *et al.*, 2018).

10.6.1.5.1 Photooxidation

When photon energy from artificial or natural light interacts with the target molecule, it triggers a photochemical reaction that causes the target contaminant to mineralize, this process is called photolysis (Rosman *et al.*, 2018). The radiation wavelength within ultraviolet spectrum that is 200–400 nm is commonly used for photolysis. A number of pharmaceuticals can undergo degradation upon absorption of solar radiation while some others like ibuprofen, naproxen, triclosan and triclocaban are not photoactive or produce toxins upon photooxidation (Rivera-Utrilla *et al.*, 2013).

10.6.1.5.2 Ozonation

For ozonation, a number of reagents can be used to increase the oxidation reaction in the form of O_3/H_2O_2, O_3/UV, $O_3/H_2O_2/UV$ and O_3/activated carbon systems for the removal of pharmaceuticals from water. In recent years, this method has attracted a lot of interest among researchers worldwide (Rivera-Utrilla *et al.*, 2013; Rosman *et al.*, 2018). Ozone (O_3) instability encourages a spontaneous breakdown with a water matrix component to produce a hydroxyl group ($^{\cdot}OH$). This process of ozone breakdown occurs at a basic medium (i.e., at high pH) (Rosman *et al.*, 2018). According to Adams *et al.* (2002), ozonation effectively removes paracetamol from wastewater. Further, the addition of H_2O_2 accelerates the process of ozone breakdown as the reaction of O_3 and H_2O_2 results in a radical chain mechanism that produces hydroxyl radicals (Rosman *et al.*, 2018). For instance, clofibric acid and ibuprofen cannot be effectively eliminated by ozone alone but can be successfully removed by O_3/H_2O_2 (Snyder *et al.*, 2006). UV rays may be used during the ozonation process in place of peroxide as the oxidizing agent. UV works by giving the energy to chemical compounds through radiation that can be absorbed by reactant molecules, which can then move into an excited state but require a lot of time to complete the reaction (Balcioglu & Otker, 2004). For the treatment of water, several researchers have investigated O_3/UV or $O_3/H_2O_2/UV$ combinations. The most potent oxidation process when compared among O_3/H_2O_2, $O_3/H_2O_2/UV$ and O_3/UV is $O_3/H_2O_2/UV$ (Rosman *et al.*, 2018).

10.6.1.5.3 Electrochemical oxidation

Due to their compatibility with environmental conditions, simplicity and ease of automation, electrochemical oxidation process is the most widely used electrochemical method for wastewater cleanup. The electron transfer occurs between the electrode (anode) and the pollutant. The material of the electrode has a significant impact on the electrochemical process because some anodes prefer partial oxidation while others prefer the selective oxidation of organic pollutants. Comninellis (1994) classified the anodes into two types, the active anodes (Pt, IrO_2 and RuO_2) and the non-active anodes (PbO_2, SnO_2 and boron-doped diamond (BDD)). Water is oxidized in both types of anodes, designated as M, producing a physiosorbed hydroxyl radical (M(OH)). This radical interacts so intensely with the surface of 'active' anodes that it is converted to chemisorbed 'active oxygen,' or superoxide MO. As opposed to this, the surface of 'non-active' anodes only weakly interacts with M(OH), and this radical then directly reacts with organics until entire mineralization is reached. The BDD anode is considered as the most effective 'non-active' electrode available and is regarded as the best anode for treating PPCPs via advanced oxidation (Marselli *et al.*, 2003).

10.6.1.5.4 Fenton oxidation

The Fenton reaction produces extremely reactive hydroxyl radicals by decomposing hydrogen peroxide with ferrous ions (the reaction occurs at an optimum pH range of 2.8–3.0) (John *et al.*, 2022). Fenton technology, among several AOPs, has the benefit of comparatively minimal reactions, an easy technique, and a high oxidation capacity, making it an intriguing option for the management of PPCPs. However, typical Fenton technology has various limitations, including significant iron loss, a limited pH range, and the difficulty of Fe^{2+} recovery in practice (Qian *et al.*, 2021). To address these disadvantages, considerable work was invested in the conceptualization and creation of heterogeneous Fenton-like catalysts. There are methods like electro-Fenton oxidation (generation of H_2O_2 by electrolysis), bio-electro Fenton process (electricity for H_2O_2 generation is produced by electrochemically active cells i.e., microbial fuel cells) (Wang *et al.*, 2018), photo-Fenton process (Fenton oxidation in presence of artificial light sources) (Guo *et al.*, 2023) and so on. The research is still going on and new reaction mechanisms are being discovered, so, it is difficult to classify a single AOP as the best method for removing PPCPs from contaminated water.

10.6.1.6 Constructed wetlands

Constructed wetlands are artificial ecosystems constructed in a specific environment, primarily used for wastewater treatment. Macrophytes, substrate for macrophyte growth, water depth, wetland structure and proper hydraulic retention time (HRT) are just a few of the many criteria that must be carefully measured for CWs to operate well. CWs are regarded as more effective and sustainable for treating various PPCPs. The use of CW systems can remove more than half of all PPCPs under ideal circumstances. The macrophytes remove PPCPs by elimination (under favourable conditions) or by direct uptake (Kumar *et al.*, 2022). Li *et al.* (2014) reported the degradation of ibuprofen, caffeine and naproxen by using CWs. The biomass produced by macrophytes can be used as raw material in industries (paper and pulp), for biofuel production and so on. Although CWs have advantages over other methods of PPCPs removal, there are studies that showed that the presence of excess organic matter, the number and type of PPCPs, pH and temperature reduce the degradation/absorption efficiency of CWs (Kumar *et al.*, 2022).

10.7 CONCLUSION AND FUTURE PROSPECTIVES

In conclusion, the presence and persistence of PPCPs in wastewater have raised serious concerns about potential environmental risks. Humans use these chemicals heavily on a daily basis, and they can enter the ecosystem in a number of ways, including veterinary and human medications, nutraceuticals, bioactive food supplements, wastewater treatment facilities, PPCP manufacturing industries, agricultural runoff and natural cycles. When PPCPs are added to sewage or wastewater, they go through a variety of transformations. Because traditional sewage treatment (STPs) and wastewater treatment plants (WWTPs) were unable to properly process these chemicals, various advanced methods such as membrane filtration, MBRs, activated carbon and advanced oxidation processes (which appear to be promising) have been suggested. Due to their bioaccumulation and biomagnification, these substances may have a number of negative impacts on the aquatic ecosystem and disrupt the food chain which ultimately puts human health at risk. This necessitates the development of better wastewater treatment technology that specifically targets PPCPs. Exciting opportunities for sustainable management exist as a result of the future predictions for the fate and behaviour of PPCPs in wastewater. Enhancements in treatment technologies, encouraged by ongoing research and development initiatives, will increase the effectiveness of PPCP removal. The amount of PPCP released into wastewater will be reduced by the incorporation of source control measures, regulations and public awareness campaigns. Potential harm will be reduced through environmental risk evaluations and the establishment of safe concentration limits. The sustainability of resources will be greatly impacted by resource recovery and water reuse. We can ultimately create a path towards efficient PPCP management in wastewater, protecting both human and environmental health, through cooperation among researchers, policymakers and stakeholders.

REFERENCES

Adams C., Wang Y., Loftin K. and Meyer M. (2002). Removal of antibiotics from surface and distilled water in conventional water treatment processes. *Journal of Environmental Engineering*, **128**, 253–260, https://doi.org/10.1061/(ASCE)0733-9372(2002)128:3(253)

Afsa S., Hamden K., Lara Martin P. A. and Mansour H. B. (2020). Occurrence of 40 pharmaceutically active compounds in hospital and urban wastewaters and their contribution to Mahdia coastal seawater contamination. *Environmental Science and Pollution Research*, **27**, 1941 1955, https://doi.org/10.1007/s11356-019-06866-5

Agnihotri V. and Thathola P. (2019). Pharmaceutical and personal care products (PPCPs) in wastewater/freshwater sources: distribution and related health concerns. *Himalayan Ecology*, **27**, 63.

Ahmed M. J. and Theydan S. K. (2012). Adsorption of cephalexin onto activated carbons from *Albizia lebbeck* seed pods by microwave-induced KOH and K_2CO_3 activations. *Chemical Engineering Journal*, **211**, 200–207, https://doi.org/10.1016/j.cej.2012.09.089

Al-Baldawi I. A., Mohammed A. A., Mutar Z. H., Abdullah S. R. S., Jasim S. S. and Almansoory A. F. (2021). Application of phytotechnology in alleviating pharmaceuticals and personal care products (PPCPs) in wastewater: source, impacts, treatment, mechanisms, fate, and SWOT analysis. *Journal of Cleaner Production*, **319**, 128584, https://doi.org/10.1016/j.jclepro.2021.128584

Alstadhaug K. B., Ofte H. K., Müller K. I. and Andreou A. P. (2020). Sudden caffeine withdrawal triggers migraine – a randomized controlled trial. *Frontiers in Neurology*, **11**, 1002, https://doi.org/10.3389/fneur.2020.01002

Anand U., Adelodun B., Cabreros C., Kumar P., Suresh S., Dey A., Ballesteros F. and Bontempi E. (2022). Occurrence, transformation, bioaccumulation, risk and analysis of pharmaceutical and personal care products from wastewater: a review. *Environmental Chemistry Letters*, **17**, 3883–3904, https://doi.org/10.1007/s10311-022-01498-7

Antonopoulou M., Dormousoglou M., Spyrou A., Dimitroulia A. A. and Vlastos D. (2022). An overall assessment of the effects of antidepressant paroxetine on aquatic organisms and human cells. *Science of the Total Environment*, **852**, 158393, https://doi.org/10.1016/j.scitotenv.2022.158393

Bahamonde P. A., Fuzzen M. L., Bennett C. J., Tetreault G. R., McMaster M. E., Servos M. R. and Munkittrick K. R. (2015). Whole organism responses and intersex severity in rainbow darter (*Etheostoma caeruleum*) following exposures to municipal wastewater in the Grand River basin, ON, Canada. Part A. *Aquatic Toxicology*, **159**, 290–301, https://doi.org/10.1016/j.aquatox.2014.11.023

Balakrishna K., Rath A., Praveenkumarreddy Y., Guruge K. S. and Subedi B. (2017). A review of the occurrence of pharmaceuticals and personal care products in Indian water bodies. *Ecotoxicology and Environmental Safety*, **137**, 113–120, https://doi.org/10.1016/j.ecoenv.2016.11.014

Balcioglu I. A. and Otker M. (2004). Pre-treatment of antibiotic formulation wastewater by O_3, O_3/H_2O_2, and O_3/UV processes. *Turkish Journal of Engineering and Environmental Sciences*, **28**, 325–331.

Banerjee S. and Maric F. (2023). Mitigating the environmental impact of NSAIDs-physiotherapy as a contribution to one health and the SDGs. *European Journal of Physiotherapy*, **25**, 51–55, https://doi.org/10.1080/21679169.2021.1976272

Bavumiragira J. P. and Yin H. (2022). Fate and transport of pharmaceuticals in water systems: a processes review. *Science of the Total Environment*, **823**, 153635, https://doi.org/10.1016/j.scitotenv.2022.153635

Bayer A., Asner R., Schüssler W., Kopf W., Weiß K., Sengl M. and Letzel M. (2014). Behavior of sartans (antihypertensive drugs) in wastewater treatment plants, their occurrence and risk for the aquatic environment. *Environmental Science and Pollution Research*, **21**, 10830–10839, https://doi.org/10.1007/s11356-014-3060-z

Bedner M. and MacCrehan W. A. (2006a). Transformation of acetaminophen by chlorination produces the toxicants 1,4-benzoquinone and N-acetyl-p-benzoquinone imine. *Environmental Science and Technology*, **40**, 516–522, https://doi.org/10.1021/es0509073

Bedner M. and MacCrehan W.A. (2006b). Reactions of the amine-containing drugs fluoxetine and metoprolol during chlorination and dechlorination processes used in wastewater treatment. *Chemosphere*, **65**, 2130–2137, https://doi.org/10.1016/j.chemosphere.2006.06.016

Bijlsma L., Pitarch E., Fonseca E., Ibanez M., Botero A. M., Claros J. and Hernandez F. (2021). Investigation of pharmaceuticals in a conventional wastewater treatment plant: removal efficiency, seasonal variation and impact of a nearby hospital. *Journal of Environmental Chemical Engineering*, **9**, 105548, https://doi.org/10.1016/j.jece.2021.105548

Bisognin R. P., Wolff D. B., Carissimi E., Prestes O. D. and Zanella R. (2021). Occurrence and fate of pharmaceuticals in effluent and sludge from a wastewater treatment plant in Brazil. *Environmental Technology*, **42**, 2292–2303, https://doi.org/10.1080/09593330.2019.1701561

Blonç M., Lima J., Balasch J. C., Tort L., Gravato C. and Teles M. (2023). Elucidating the effects of the lipids regulators fibrates and statins on the health status of finfish species: a review. *Animals*, **13**, 792–792, https://doi.org/10.3390/ani13050792

Botero-Coy A. M., Martínez-Pachón D., Boix C., Rincón R. J., Castillo N., Arias-Marín L. P. and Hernandez F. (2018). An investigation into the occurrence and removal of pharmaceuticals in Colombian wastewater. *Science of the Total Environment*, **642**, 842–853, https://doi.org/10.1016/j.scitotenv.2018.06.088

Březinova T. D., Vymazal J., Koželuh M. and Kule L. (2018). Occurrence and removal of ibuprofen and its metabolites in full-scale constructed wetlands treating municipal wastewater. *Ecological Engineering*, **120**, 1–5, https://doi.org/10.1016/j.ecoleng.2018.05.020

Calisto V., Ferreira C. I., Santos S. M., Gil M. V., Otero M. and Esteves V. I. (2014). Production of adsorbents by pyrolysis of paper mill sludge and application on the removal of citalopram from water. *Bioresource Technology*, **166**, 335–344, https://doi.org/10.1016/j.biortech.2014.05.047

Cerveny D., Cisar P., Brodin T., McCallum E. S. and Fick J. (2022). Environmentally relevant concentration of caffeine – effect on activity and circadian rhythm in wild perch. *Environmental Science and Pollution Research*, **29**, 54264–54272, https://doi.org/10.1007/s11356-022-19583-3

Chabchoubi I. B., Bouchhima R. A., Louhichi N., Baanannou A., Masmoudi S. and Hentati O. (2023). Short-term effects of various non-steroidal anti-inflammatory drugs (NSAIDs) on *Danio rerio* embryos. *MethodsX*, **10**, 102215, https://doi.org/10.1016/j.mex.2023.102215

Chen M., Ren L., Qi K., Li Q., Lai M., Li Y., Li X. and Wang Z. (2020). Enhanced removal of pharmaceuticals and personal care products from real municipal wastewater using an electrochemical membrane bioreactor. *Bioresource Technology*, **311**, 123579, https://doi.org/10.1016/j.biortech.2020.123579

Chen Y., Wang J., Xu P., Xiang J., Xu D., Cheng P., Wang X., Wu L., Zhang N. and Chen Z. (2022). Antidepressants as emerging contaminants: occurrence in wastewater treatment plants and surface waters in Hangzhou, China. *Frontiers in Public Health*, **10**, 963257–963257, https://doi.org/10.3389/fpubh.2022.963257×

Comninellis C. (1994). Electrocatalysis in the electrochemical conversion/combustion of organic pollutants for waste water treatment. *Electrochimica Acta*, **39**, 1857–1862, https://doi.org/10.1016/0013-4686(94)85175-1

Cooney J., Lenczewski M., Leal-Bautista R. M., Tucker K., Davis M. and Rodriguez J. (2023). Analysis of sunscreens and antibiotics in groundwater during the Covid-19 pandemic in the Riviera Maya, Mexico. *Science of the Total Environment*, **894**, 164820, https://doi.org/10.1016/j.scitotenv.2023.164820

Cory W. C., Welch A. M., Ramirez J. N. and Rein L. C. (2019). Naproxen and its phototransformation products: persistence and ecotoxicity to toad tadpoles (*Anaxyrus terrestris*), individually and in mixtures. *Environmental Toxicology and Chemistry*, **38**, 2008–2019, https://doi.org/10.1002/etc.4514

Damkjaer K., Weisser J. J., Msigala S. C., Mdegela R. and Styrishave B. (2018). Occurrence, removal and risk assessment of steroid hormones in two wastewater stabilization pond systems in Morogoro, Tanzania. *Chemosphere*, **212**, 1142–1154, https://doi.org/10.1016/j.chemosphere.2018.08.053

Delgado N., Capparelli A., Navarro A. and Marino D. (2019). Pharmaceutical emerging pollutants removal from water using powdered activated carbon: study of kinetics and adsorption equilibrium. *Journal of Environmental Management*, **236**, 301–308, https://doi.org/10.1016/j.jenvman.2019.01.116

Do Q. T. T., Otaki M., Otaki Y., Tushara C. and Sanjeewa I. W. (2022). Pharmaceutical contaminants in shallow groundwater and their implication for poor sanitation facilities in low-income countries. *Environmental Toxicology and Chemistry*, **41**, 266–274, https://doi.org/10.1002/etc.5110

Ebele A.J., Oluseyi T., Drage D.S., Harrad S. and Abdallah M.A.E. (2020). Occurrence, seasonal variation and human exposure to pharmaceuticals and personal care products in surface water, groundwater and drinking water in Lagos State, Nigeria. *Emerging Contaminants*, **6**, 124–132, https://doi.org/10.1016/j.emcon.2020.02.004

Elveren M. and Osma E. (2022). Effects of pharmaceuticals and personal care products (PPCPs) in water on wheat (*Triticum aestivum* L.). *Environmental Engineering & Management Journal*, **21**, 423–430, https://doi.org/10.30638/eemj.2022.040

Falahi O. A. A., Abdullah S. R. S., Hasan H. A., Othman A. R., Ewadh H. M., Al-Baldawi I. A., Shrauddin S. S. N. K., Kurniawan S. B. and Ismail N. (2022). Elimination of mixed ibuprofen and paracetamol from spiked domestic wastewater via a pilot continuous aerated sub-surface constructed wetland system. *Journal of Water Process Engineering*, **50**, 103308 https://doi.org/10.1016/j.jwpe.2022.103308

Falfushynska H., Poznanskyi D., Kasianchuk N., Horyn O. and Bodnar O. (2022). Multimarker responses of Zebrafish to the effect of ibuprofen and gemfibrozil in environmentally relevant concentrations. *Bulletin of Environmental Contamination and Toxicology*, **109**(6), 1010–1017, https://doi.org/10.1007/s00128-022-03607-2

Gebuijs I. G. E., Metz J. R., Zethof J., Carels C. E. L., Wagener F. A. D. T. G. and Von den Hoff J. W. (2020). The anti-epileptic drug valproic acid causes malformations in the developing craniofacial skeleton of zebrafish larvae. *Mechanisms of Development*, **163**, 103632, https://doi.org/10.1016/j.mod.2020.103632

Golbaz S., Zamanzadeh M., Yaghmaeian K., Nabizadeh R., Rastkari N. and Esfahani H. (2023). Occurrence and removal of psychiatric pharmaceuticals in the Tehran South municipal wastewater treatment plant. *Environmental Science and Pollution Research*, **30**, 27041–27055, https://doi.org/10.1007/s11356-022-23667-5

Goswami P., Guruge K. S., Tanoue R., Tamamura Y. A., Jinadasa K. B. S. N., Nomiyama K., Kunisue T. and Tanabe S. (2022). Occurrence of pharmaceutically active compounds and potential ecological risks in wastewater from hospitals and receiving waters in Sri Lanka. *Environmental Toxicology and Chemistry*, **41**, 298–311, https://doi.org/10.1002/etc.5212

Guerrero-Gualan D., Valdez-Castillo E., Crisanto-Perrazo T. and Toulkeridis T. (2023). Methods of removal of hormones in wastewater. *Water*, **15**, 353, https://doi.org/10.3390/w15020353

Guo J., Zhang Y., Li J., Wu F. and Luo L. (2023). Molecular oxygen activation by citric acid boosted pyrite–photo–Fenton process for degradation of PPCPs in water. *Molecules*, **28**, 607, https://doi.org/10.3390/molecules28020607

Güzel E. (2021). Occurrence and environmental risks assessment of DEET (N, N-diethyl-m-toluamide) pesticide in Seyhan River, Turkey. *Journal of Anatolian Environmental and Animal Sciences*, **6**, 345–351, https://doi.org/10.35229/jaes.895045

Hsieh C. Y., Wu Y. C., Mudigonda S., Dahms H. U. and Wu M. C. (2023). Assessing the effects of ozonation on the concentrations of personal care products and acute toxicity in sludges of wastewater treatment plants. *Toxics*, **11**, 75, https://doi.org/10.3390/toxics11010075

Huang H., Wu J., Ye J., Ye T., Deng J., Liang Y. and Liu W. (2018). Occurrence, removal, and environmental risks of pharmaceuticals in wastewater treatment plants in south China. *Frontiers of Environmental Science and Engineering*, **12**, 1–11, https://doi.org/10.1007/s11783-018-1053-8

John J., Ramesh K. and Chellam P. V. (2022). Metal-organic frameworks (MOFs) as a catalyst for advanced oxidation processes – micropollutant removal. In: Advanced Materials for Sustainable Environmental Remediation. Elsevier. pp. 155–174.

Júnior C. A. M., da Costa Luchiari N. and Gomes P. C. F. L. (2019). Occurrence of caffeine in wastewater and sewage and applied techniques for analysis: a review. *Eclética Química*, **44**, 11–26.

Jurado A., Labad F., Scheiber L., Criollo R., Nikolenko O., Pérez S. and Ginebreda A. (2022). Occurrence of pharmaceuticals and risk assessment in urban groundwater. *Advances in Geosciences*, **59**, 1–7, https://doi.org/10.5194/adgeo-59-1-2022

Kitchen L. W., Lawrence K. L. and Coleman R. E. (2009). The role of the United States military in the development of vector control products, including insect repellents, insecticides, and bed nets. *Journal of Vector Ecology*, **34**, 50–61, https://doi.org/10.1111/j.1948-7134.2009.00007.x

Korkmaz N. E., Savun-Hekimoğlu B., Aksu A., Burak S. and Caglar N. B. (2022). Occurrence, sources and environmental risk assessment of pharmaceuticals in the Sea of Marmara, Turkey. *Science of the Total Environment*, **819**, 152996, https://doi.org/10.1016/j.scitotenv.2022.152996

Košnář Z., Mercl F., Chane A. D., Pierdonà L., Míchal P. and Tlustoš P. (2021). Occurrence of synthetic polycyclic and nitro musk compounds in sewage sludge from municipal wastewater treatment plants. *Science of The Total Environment*, **801**, 149777, https://doi.org/10.1016/j.scitotenv.2021.149777

Kumar S., Pratap B., Dubey D., Kumar A., Shukla S. and Dutta V. (2022). Constructed wetlands for the removal of pharmaceuticals and personal care products (PPCPs) from wastewater: origin, impacts, treatment methods, and SWOT analysis. *Environmental Monitoring and Assessment*, **194**, 885, https://doi.org/10.1007/s10661-022-10540-8

Kumar M., Silori R., Mazumder P. and Tauseef S. M. (2023). Screening of pharmaceutical and personal care products (PPCPs) along wastewater treatment system equipped with root zone treatment: a potential model for domestic waste leachate management. *Journal of Environmental Management*, **335**, 117494, https://doi.org/10.1016/j.jenvman.2023.117494

Langbehn R. K., Michels C. and Soares H. M. (2021). Antibiotics in wastewater: from its occurrence to the biological removal by environmentally conscious technologies. *Environmental Pollution*, **275**, 116603, https://doi.org/10.1016/j.envpol.2021.116603

Lee G., Lee S., Ha N., Kho Y., Park K., Kim P. and Choi K. (2019). Effects of gemfibrozil on sex hormones and reproduction related performances of *Oryzias latipes* following long-term (155 d) and short-term (21 d) exposure. *Ecotoxicology and Environmental Safety*, **173**, 174–181, https://doi.org/10.1016/j.ecoenv.2019.02.015

Leese J. M., McMahon J. and Colosi J. C. (2021). Effects of wastewater treatment plant effluent in a receiving stream on reproductive behavior of fathead minnows (*Pimephales promelas*). *Fishes*, **6**, 14, https://doi.org/10.3390/fishes6020014

Lei H., Yao K., Yang B., Xie L. and Ying G. (2023). Occurrence, spatial and seasonal variation, and environmental risk of pharmaceutically active compounds in the Pearl River basin, South China. *Frontiers of Environmental Science & Engineering*, **17**, 46, https://doi.org/10.1007/s11783-023-1646-8

Lenart-Boroń A., Prajsnar J., Guzik M., Boroń P., Grad B. and Żelazny M. (2022). Antibiotics in groundwater and River Water of Białka – a Pristine Mountain River. *Applied Sciences*, **12**, 12743, https://doi.org/10.3390/app122412743

Li X., Gu W., Chen B., Zhu Z., Zhang B. (2021). Functional modification of HHCB: Strategy for obtaining environmentally friendly derivatives. *Journal of Hazardous Materials*, **416**, 126116.

Li Y., Zhu G., Ng W. J. and Tan S. K. (2014). A review on removing pharmaceutical contaminants from wastewater by constructed wetlands: design, performance and mechanism. *Science of the Total Environment*, **468**, 908–932, https://doi.org/10.1016/j.scitotenv.2013.09.018

Liu H., Zhang J., Bao N., Cheng C., Ren L. and Zhang C. (2012). Textural properties and surface chemistry of lotus stalk-derived activated carbons prepared using different phosphorus oxyacids: adsorption of trimethoprim. *Journal of Hazardous Materials*, **235**, 367–375, https://doi.org/10.1016/j.jhazmat.2012.08.015

Liu L., Wu W., Zhang J., Lv P., Xu L. and Yan Y. (2018). Progress of research on the toxicology of antibiotic pollution in aquatic organisms. *Acta Ecologica Sinica*, **38**, 36–41, https://doi.org/10.1016/j.chnaes.2018.01.006

Liu N., Jin X., Fenga C., Wang Z., Wua F., Johnsond A. C., Xiaoe H., Hollerte H. and Giesyf J. P. (2020). Ecological risk assessment of fifty pharmaceuticals and personal care products (PPCPs) in Chinese surface waters: a proposed multiple-level system. *Environment International*, **136**, 105454–105454, https://doi.org/10.1016/j.envint.2019.105454

Liu Q., Feng X., Chen N., Shen F., Zhang H., Wang S., Sheng Z. and Li J. (2022). Occurrence and risk assessment of typical PPCPs and biodegradation pathway of ribavirin in wastewater treatment plants. *Environmental Science and Ecotechnology*, **11**, 100184, https://doi.org/10.1016/j.ese.2022.100184

Liu J., Duan L., Gao Q., Zhao Y. and Gao F. (2023). Removal of typical PPCPs by reverse osmosis membranes: optimization of treatment process by factorial design. *Membranes*, **13**, 355, https://doi.org/10.3390/membranes13030355

Loganathan P., Vigneswaran S., Kandasamy J., Cuprys A. K., Maletskyi Z. and Ratnaweera H. (2023). Treatment trends and combined methods in removing pharmaceuticals and personal care products from wastewater–a review. *Membranes*, **13**, 158, https://doi.org/10.3390/membranes13020158

López-Serna R., Jurado A., Vázquez-Suñé E., Carrera J., Petrović M. and Barceló D. (2013). Occurrence of 95 pharmaceuticals and transformation products in urban groundwaters underlying the metropolis of Barcelona, Spain. *Environmental Pollution*, **174**, 305–315, https://doi.org/10.1016/j.envpol.2012.11.022

Ma L. D., Li J., Li J. J., Liu M., Yan D. Z., Shi W. Y. and Xu G. (2018). Occurrence and source analysis of selected antidepressants and their metabolites in municipal wastewater and receiving surface water. *Environmental Science: Processes & Impacts*, **20**, 1020–1029, https://doi.org/10.1039/C8EM00077H

Machado G. C., Abdel-Shaheed C., Underwood M. and Day R. O. (2021). Non steroidal anti inflammatory drugs (NSAIDs) for musculoskeletal pain. *British Medical Journal*, **372**.

Madikizela L. M. and Chimuka L. (2017). Occurrence of naproxen, ibuprofen, and diclofenac residues in wastewater and river water of KwaZulu-Natal Province in South Africa. *Environmental Monitoring and Assessment*, **189**, 1–12, https://doi.org/10.1007/s10661-017-6069-1

Mao F., He Y. and Gin K. Y. H. (2019). Occurrence and fate of benzophenone-type UV filters in aquatic environments: a review. *Environmental Science: Water Research & Technology*, **5**, 209–223, https://doi.org/10.1039/C8EW00539G

Marselli B., Garcia-Gomez J., Michaud P. A., Rodrigo M. A. and Comninellis C. (2003). Electrogeneration of hydroxyl radicals on boron-doped diamond electrodes. *Journal of the Electrochemical Society*, **150**, D79–D83, https://doi.org/10.1149/1.1553790

Merel S. and Snyder S. A. (2016). Critical assessment of the ubiquitous occurrence and fate of the insect repellent N, N-diethyl-m-toluamide in water. *Environment International*. **96**, 98–117, https://doi.org/10.1016/j.envint.2016.09.004

Michalaki A. and Grintzalis K. (2023). Acute and transgenerational effects of non-steroidal anti-inflammatory drugs on *Daphnia magna*. *Toxics*, **11**, 320, https://doi.org/10.3390/toxics11040320

Musee N. (2018). Environmental risk assessment of triclosan and triclocarban from personal care products in South Africa. *Environmental Pollution*, **242**, 827–838, https://doi.org/10.1016/j.envpol.2018.06.106

Mussa Z. H., Al-Qaim F. F., Jawad A. H., Scholz M. and Yaseen Z. M. (2022). A comprehensive review for removal of non-steroidal anti-inflammatory drugs attained from wastewater observations using carbon-based anodic oxidation process. *Toxics*, **10**, 598, https://doi.org/10.3390/toxics10100598

Nozaki K., Tanoue R., Kunisue T., Tue N. M., Fujii S., Sudo N., Isobe T., Nakayama K., Sudaryanto A., Subramanian A. and Bulbule K. A. (2023). Pharmaceuticals and personal care products (PPCPs) in surface water and fish from three Asian countries: species-specific bioaccumulation and potential ecological risks. *Science of the Total Environment*, **866**, 161258, https://doi.org/10.1016/j.scitotenv.2022.161258

Ofrydopoulou A., Nannou C., Evgenidou E., Christodoulou A. and Lambropoulou D. (2022). Assessment of a wide array of organic micropollutants of emerging concern in wastewater treatment plants in Greece: occurrence, removals, mass loading and potential risks. *Science of the Total Environment*, **802**, 149860, https://doi.org/10.1016/j.scitotenv.2021.149860

Okoye C. O., Okeke E. S., Okoye K. C., Echude D., Andong F. A., Chukwudozie K. I., Okoye H. U. and Ezeonyejiaku C. D. (2022). Occurrence and fate of pharmaceuticals, personal care products (PPCPs) and pesticides in African water systems: a need for timely intervention. *Heliyon*, **8**, e09143, https://doi.org/10.1016/j.heliyon.2022.e09143

Osma E., Cigir Y., Karnjanapiboonwong A. and Anderson T. A. (2018). Evaluation of selected pharmaceuticals on plant stress markers in wheat. *International Journal of Environmental Research*, **12**, 179–188, https://doi.org/10.1007/s41742-018-0081-3

Pai C. W. and Wang G. S. (2022). Treatment of PPCPs and disinfection by-product formation in drinking water through advanced oxidation processes: comparison of UV, UV/Chlorine, and UV/H_2O_2. *Chemosphere*, **287**, 132171, https://doi.org/10.1016/j.chemosphere.2021.132171

Peng X., Ou W., Wang C., Wang Z., Huang Q., Jin J. and Tan J. (2014). Occurrence and ecological potential of pharmaceuticals and personal care products in groundwater and reservoirs in the vicinity of municipal landfills in China. *Science of the Total Environment*, **490**, 889–898, https://doi.org/10.1016/j.scitotenv.2014.05.068

Penrose M. T. and Cobb G. P. (2023). Evaluating seasonal differences in paraben transformation at two different wastewater treatment plants in Texas and comparing parent compound transformation to byproduct formation. *Water Research*, **235**, 119798, https://doi.org/10.1016/j.watres.2023.119798

Pérez D. J., Lombardero L. R. and Doucette W. J. (2023). Influence of exposure time, physicochemical properties, and plant transpiration on the uptake dynamics and translocation of pharmaceutical and personal care products in the aquatic macrophyte *Typha latifolia*. Science of the Total Environment, 165107, https://doi.org/10.1016/j.scitotenv.2023.165107

Pinkston K. E. and Sedlak D. L. (2004). Transformation of aromatic ether- and amine-containing pharmaceuticals during chlorine disinfection. *Environmental Science & Technology*, **38**, 4019–4025, https://doi.org/10.1021/es035368l

Plahuta M., Tišler T., Toman M. J. and Pintar A. (2017). Toxic and endocrine disrupting effects of wastewater treatment plant influents and effluents on a freshwater isopod *Asellus aquaticus* (Isopoda, Crustacea). *Chemosphere*, **174**, 342–353, https://doi.org/10.1016/j.chemosphere.2017.01.137

Pusceddu F. H., Guimarães M. M., Lopes L. O., Souza L. S., Cortez F. S., Pereira C. D. S. and Cesar A. (2022). Biological effects of the antihypertensive losartan under different ocean acidification scenarios. *Environmental Pollution*, **292**, 118329, https://doi.org/10.1016/j.envpol.2021.118329

Qian H., Yu G., Hou Q., Nie Y., Bai C., Bai X., Wang H., Ju M., Qian H., Yu G., Hou Q., Nie Y., Bai C., Bai X., … and Ju M. (2021). Ingenious control of adsorbed oxygen species to construct dual reaction centers ZnO@FePc photo-Fenton catalyst with high-speed electron transmission channel for PPCPs degradation. *Applied Catalysis B: Environmental*, **291**, 120064, https://doi.org/10.1016/j.apcatb.2021.120064

Ramírez-Morales D., Masís-Mora M., Montiel-Mora J. R., Cambronero-Heinrichs J. C., Briceño-Guevara S., Rojas-Sánchez C. E. and Rodríguez-Rodríguez C. E. (2020). Occurrence of pharmaceuticals, hazard assessment and ecotoxicological evaluation of wastewater treatment plants in Costa Rica. *Science of the Total Environment*, **746**, 141200, https://doi.org/10.1016/j.scitotenv.2020.141200

Rastogi A., Tiwari M. K. and Ghangrekar M. M. (2021). A review on environmental occurrence, toxicity and microbial degradation of non-steroidal anti-inflammatory drugs (NSAIDs). *Journal of Environmental Management*, **300**, 113694–113694, https://doi.org/10.1016/j.jenvman.2021.113694

Ravichandran M. K. and Philip L. (2022). Fate of carbamazepine and its effect on physiological characteristics of wetland plant species in the hydroponic system. *Science of the Total Environment*, **846**, 157337, https://doi.org/10.1016/j.scitotenv.2022.157337

Ricky R. and Shanthakumar S. (2022). Phycoremediation integrated approach for the removal of pharmaceuticals and personal care products from wastewater – a review. *Journal of Environmental Management*, **302** (Pt A), 113998–113998, https://doi.org/10.1016/j.jenvman.2021.113998

Rivera-Utrilla J., Sánchez-Polo M., Ferro-García M. Á., Prados-Joya G. and Ocampo-Pérez R. (2013). Pharmaceuticals as emerging contaminants and their removal from water. A review. *Chemosphere*, **93**, 1268–1287, https://doi.org/10.1016/j.chemosphere.2013.07.059

Rodriguez-Narvaez O. M., Peralta-Hernandez J. M., Goonetilleke A. and Bandala E. R. (2017). Treatment technologies for emerging contaminants in water: a review. *Chemical Engineering Journal*, **323**, 361–380, https://doi.org/10.1016/j.cej.2017.04.106

Rosal R., Rodea-Palomares I., Boltes K., Fernández-Piñas F., Leganés F., Gonzalo S. and Petre A. (2010). Ecotoxicity assessment of lipid regulators in water and biologically treated wastewater using three aquatic organisms. *Environmental Science and Pollution Research*, **17**, 135–144, https://doi.org/10.1007/s11356-009-0137-1

Rosman N., Salleh W. N. W., Mohamed M. A., Jaafar J., Ismail A. F. and Harun Z. (2018). Hybrid membrane filtration-advanced oxidation processes for removal of pharmaceutical residue. *Journal of Colloid and Interface Science*, **532**, 236–260, https://doi.org/10.1016/j.jcis.2018.07.118

Salahinejad A., Meuthen D., Attaran A., Chivers D. P. and Ferrari M. C. (2023). Effects of common antiepileptic drugs on teleost fishes. *Science of the Total Environment*, **866**, 161324, https://doi.org/10.1016/j.scitotenv.2022.161324

Schlüsener M. P., Hardenbicker P., Nilson E., Schulz M., Viergutz C. and Ternes T. A. (2015). Occurrence of venlafaxine, other antidepressants and selected metabolites in the Rhine catchment in the face of climate change. *Environmental Pollution*, **196**, 247–256, https://doi.org/10.1016/j.envpol.2014.09.019

Seethalakshmi P. S., Charity O. J., Giakoumis T., Kiran G. S., Sriskandan S., Voulvoulis N. and Selvin J. (2022). Delineating the impact of COVID-19 on antimicrobial resistance: an Indian perspective. *Science of the Total Environment*, **818**, 151702, https://doi.org/10.1016/j.scitotenv.2021.151702

Senta I., Kostanjevecki P., Krizman-Matasic I., Terzic S. and Ahel M. (2019). Occurrence and behavior of macrolide antibiotics in municipal wastewater treatment: possible importance of metabolites, synthesis byproducts, and transformation products. *Environmental Science & Technology*, **53**, 7463–7472, https://doi.org/10.1021/acs.est.9b01420

Sharma B. M., Bečanová J., Scheringer M., Sharma A., Bharat G. K., Whitehead P. G., Klánová J. and Nizzetto L. (2019). Health and ecological risk assessment of emerging contaminants (pharmaceuticals, personal care products, and artificial sweeteners) in surface and groundwater (drinking water) in the Ganges River Basin, India. *Science of the Total Environment*, **646**, 1459–1467, https://doi.org/10.1016/j.scitotenv.2018.07.235

Silori R. and Tauseef S. M. (2022). A review of the occurrence of pharmaceutical compounds as emerging contaminants in treated wastewater and aquatic environments. *Current Pharmaceutical Analysis* **18**, 345–379, https://doi.org/10.2174/157341291 8666211119142030

Singh V. and Suthar S. (2021). Occurrence, seasonal variations, and ecological risk of pharmaceuticals and personal care products in River Ganges at two holy cities of India. *Chemosphere*, **268**, 129331–129331, https://doi.org/10.1016/j.chemosphere.2020.129331

Snyder S. A., Wert E. C., Rexing D. J., Zegers R. E. and Drury D. D. (2006). Ozone oxidation of endocrine disruptors and pharmaceuticals in surface water and wastewater. *Ozone: Science and Engineering*, **28**, 445–460, https://doi.org/10.1080/01919510601039726

Sodhi K. K., Kumar M. and Singh D. K. (2021). Insight into the amoxicillin resistance, ecotoxicity, and remediation strategies. *Journal of Water Process Engineering*, **39**, 101858, https://doi.org/10.1016/j.jwpe.2020.101858

Subedi B. and Kannan K. (2015). Occurrence and fate of select psychoactive pharmaceuticals and antihypertensives in two wastewater treatment plants in New York State, USA. *Science of the Total Environment*, **514**, 273–280, https://doi.org/10.1016/j.scitotenv.2015.01.098

Subedi B., Balakrishna K., Joshua D. I. and Kannan K. (2017). Mass loading and removal of pharmaceuticals and personal care products including psychoactives, antihypertensives, and antibiotics in two sewage treatment plants in southern India. *Chemosphere*, **167**, 429–437, https://doi.org/10.1016/j.chemosphere.2016.10.026

Sui Q., Cao X., Lu S., Zhao W., Qiu Z. and Yu G. (2015). Occurrence, sources and fate of pharmaceuticals and personal care products in the groundwater: a review. *Emerging Contaminants*, **1**, 14–24, https://doi.org/10.1016/j.emcon.2015.07.001

Sun C., Dudley S., Trumble J. and Gan J. (2018). Pharmaceutical and personal care products-induced stress symptoms and detoxification mechanisms in cucumber plants. *Environmental Pollution*, **234**, 39–47, https://doi.org/10.1016/j.envpol.2017.11.041

Świacka K., Maculewicz J., Świeżak J., Caban M. and Smolarz K. (2022). A multi-biomarker approach to assess toxicity of diclofenac and 4-OH diclofenac in *Mytilus trossulus* mussels – first evidence of diclofenac metabolite impact on molluscs. *Environmental Pollution*, **315**, 120384, https://doi.org/10.1016/j.envpol.2022.120384

Tandiono E. J. and Budiyanti E. (2023). The effect of coffee consumption on acute increased blood pressure in normotensive teens. *Journal of Urban Health Research*, **1**, 79–84.

Tasselli S. and Guzzella L. (2020). Polycyclic musk fragrances (PMFs) in wastewater and activated sludge: analytical protocol and application to a real case study. *Environmental Science and Pollution Research*, **27**, 30977–30986 https://doi.org/10.1007/s11356-019-06767-7

Tasselli S., Valenti E. and Guzzella L. (2021). Polycyclic musk fragrance (PMF) removal, adsorption and biodegradation in a conventional activated sludge wastewater treatment plant in Northern Italy. *Environmental Science and Pollution Research*, **28**, 38054–38064, https://doi.org/10.1007/s11356-021-13433-4

Thalla A. K. and Vannarath A. S. (2020). Occurrence and environmental risks of nonsteroidal anti-inflammatory drugs in urban wastewater in the southwest monsoon region of India. *Environmental Monitoring and Assessment*, **192**, 1–13, https://doi.org/10.1007/s10661-019-7904-3

Tran N. H., Li J., Hu J. and Ong S. L. (2014). Occurrence and suitability of pharmaceuticals and personal care products as molecular markers for raw wastewater contamination in surface water and groundwater. *Environmental Science and Pollution Research*, **21**, 4727–4740, https://doi.org/10.1007/s11356-013-2428-9

Tran H. T., Dang B. T., Thuy L. T. T., Hoang H. G., Bui X. T., Le V. G., Lin C., Nguyen M. K., Nguyen K. Q., Nguyen P. T., Binh Q. A. and Bui T. P. T. (2022). Advanced treatment technologies for the removal of organic chemical sunscreens from wastewater: a review. *Current Pollution Reports*, **8**, 288–302, https://doi.org/10.1007/s40726-022-00221-y

Trapp S. and Legind C. N. (2011). Uptake of organic contaminants from soil into vegetables and fruits. In: Dealing with Contaminated Sites: from Theory towards Practical Application, F. Swartjes (ed.), Springer, Dordrecht, pp. 369–408.

Tsui M. M., Leung H. W., Lam P. K. and Murphy M. B. (2014). Seasonal occurrence, removal efficiencies and preliminary risk assessment of multiple classes of organic UV filters in wastewater treatment plants. *Water Research*, **53**, 58–67, https://doi.org/10.1016/j.watres.2014.01.014

Ulvi A., Aydın S. and Aydın M. E. (2022). Fate of selected pharmaceuticals in hospital and municipal wastewater effluent: occurrence, removal, and environmental risk assessment. *Environmental Science and Pollution Research*, **29**, 75609–75625, https://doi.org/10.1007/s11356-022-21131-y

Wang J. and Wang S. (2016). Removal of pharmaceuticals and personal care products (PPCPs) from wastewater: a review. *Journal of Environmental Management*, **182**, 620–640, https://doi.org/10.1016/j.jenvman.2016.07.049

Wang W., Lu Y., Luo H., Liu G., Zhang R. and Jin S. (2018). A microbial electro-fenton cell for removing carbamazepine in wastewater with electricity output. *Water research*, **139**, 58–65, https://doi.org/10.1016/j.watres.2018.03.066

Wang S., Huo Z., Gu J. and Xu G. (2021). Benzophenones and synthetic progestin in wastewater and sediment from farms, WWTPs and receiving surface water: distribution, sources, and ecological risks. *RSC Advances*, **11**, 31766–31775, https://doi.org/10.1039/D1RA05333G

Xu M., Yan S., Sun S., Ni Z., Wu W. and Sun J. (2022). N, N-diethyl-m-toluamide (DEET) degradation by• OH and SO4•−-assisted AOPs in wastewater treatment: theoretical studies into mechanisms, kinetics, and toxicity. *Journal of Environmental Chemical Engineering*, **10**, 108435, https://doi.org/10.1016/j.jece.2022.108435

Yazdan M. M. S., Kumar R. and Leung S. W. (2022). The environmental and health impacts of steroids and hormones in wastewater effluent, as well as existing removal technologies: a review. *Ecologies*, **3**, 206–224, https://doi.org/10.3390/ecologies3020016

Yilmaz B., Terekeci H., Sandal S. and Kelestimur F. (2020). Endocrine disrupting chemicals: exposure, effects on human health, mechanism of action, models for testing and strategies for prevention. *Reviews in Endocrine and Metabolic Disorders*, **21**, 127–147, https://doi.org/10.1007/s11154-019-09521-z

Yoon Y., Westerhoff P., Snyder S. A. and Wert E. C. (2006). Nanofiltration and ultrafiltration of endocrine disrupting compounds, pharmaceuticals and personal care products. *Journal of Membrane Science*, **270**, 88–100, https://doi.org/10.1016/j.memsci.2005.06.045

Yu X., Yu F., Li Z., Shi T., Xia Z. and Li G. (2023). Occurrence, distribution, and ecological risk assessment of artificial sweeteners in surface and ground waters of the middle and lower reaches of the Yellow River (Henan section, China). *Environmental Science and Pollution Research*, **30**, 52609–52623, https://doi.org/10.1007/s11356-023-26073-7

Zemann M., Wolf L., Grimmeisen F., Tiehm A., Klinger J., Hötzl H. and Goldscheider N. (2015). Tracking changing X-ray contrast media application to an urban-influenced karst aquifer in the Wadi Shueib, Jordan. *Environmental Pollution*, **198**, 133–143, https://doi.org/10.1016/j.envpol.2014.11.033

Zeng Y., Zhang Y., Zhang H., Wang J., Lian K. and Ai L. (2022). Uptake and transport of different concentrations of PPCPs by vegetables. *International Journal of Environmental Research and Public Health*, **19**, 15840, https://doi.org/10.3390/ijerph192315840

Zhang Y., Wang B., Cagnetta G., Duan L., Yang J., Deng S. and Yu G. (2018). Typical pharmaceuticals in major WWTPs in Beijing, China: occurrence, load pattern and calculation reliability. *Water Research*, **140**, 291–300, https://doi.org/10.1016/j.watres.2018.04.056

Zhang M., Shen J., Zhong Y., Ding T., Dissanayake P. D., Yang Y. and Tsang Y. F. (2022). Sorption of pharmaceuticals and personal care products (PPCPs) from water and wastewater by carbonaceous materials: a review. *Critical Reviews in Environmental Science and Technology*, **52**, 727–766, https://doi.org/10.1080/10643389.2020.1835436

Zhou R., Lu G., Yan Z., Jiang R., Shen J. and Bao X. (2019). Parental transfer of ethylhexyl methoxy cinnamate and induced biochemical responses in zebrafish. *Aquatic Toxicology*, **206**, 24–32, https://doi.org/10.1016/j.aquatox.2018.11.001

Zhou R., Lu G., Yan Z., Jiang R., Bao X. and Lu P. (2020). A review of the influences of microplastics on toxicity and transgenerational effects of pharmaceutical and personal care products in aquatic environment. *Science of the Total Environment*, **732**, 139222–139222, https://doi.org/10.1016/j.scitotenv.2020.139222

Zhu X., He M., Sun Y., Xu Z., Wan Z., Hou D., Alessi D. S. and Tsang D. C. (2022). Insights into the adsorption of pharmaceuticals and personal care products (PPCPs) on biochar and activated carbon with the aid of machine learning. *Journal of Hazardous Materials*, **423**, 127060, https://doi.org/10.1016/j.jhazmat.2021.127060

doi: 10.2166/9781789063752_0203

Chapter 11

A review of occurrence of emerging contaminants and the advanced analytical techniques used for detection and removal of these pollutants in wastewater

Masixole Sihlahla[1,2,4] *and Sihle Mngadi*[1,2,3*]

[1]*Department of Chemical Sciences, University of Johannesburg, Doornfontein Campus, P.O. Box 17011, Johannesburg 2028, South Africa*
[2]*Department of Science and Innovation-National Research Foundation South African Research Chair Initiative (DSI-NRF SARChI) in Nanotechnology for Water, University of Johannesburg, Doornfontein 2028, South Africa*
[3]*Scientific Services, Laboratories, Chemical Sciences, uMngeni-uThukela Water, Pietermaritzburg, South Africa*
[4]*Department of Chemistry, College of Science and Engineering and Technology, University of South Africa, Johannesburg, South Africa*
**Corresponding author: sihlemngadi1966@gmail.com*

ABSTRACT

This chapter focuses on reviewing the literature that provides reliable and quantitative information on emerging contaminants (ECs) in wastewater, focusing on their occurrence, detection and removal efficiency using advanced analytical techniques. In addition, providing knowledge on areas of ECs that are non-regulated since the environmental legislations and policies are being developed. Some of the classes of ECs include pharmaceuticals, nanomaterials, herbicides, personal care products and microplastics and many more. ECs have been identified as an environmental problem globally and are a result of different compounds ranging from inorganic to organic compounds which are released into the environment. ECs are commonly found in aquatic environments and the main source of ECs are municipal wastewater, domestic discharge, hospital effluents, industrial wastewater and agricultural run-off. The presence of ECs poses health problems and ecological impacts associated with them. The elevated concentration of ECs in wastewater has necessitated a need to research their varying detection techniques and different ways of removal. Water contamination by ECs is also attributed to an increase in urbanization, industrialization and agricultural activities. Current wastewater treatment plants are inefficient in the removal of ECs as they were not initially designed for the treatment and removal of ECs, these may result in the transformation of EC products that are undetected and unregulated. These products exhibit similar toxicity as their parental ECs, while some of the ECs have been recognized as endocrine-disrupting chemicals. Due to new ECs being introduced, there is a gap in knowledge of their detection and treatment techniques for their removal, thus demonstrating a need for integrated analytical approaches that compliments the screening and removal of target and non-targeted ECs with biological assays. Development advancement of analytical techniques has enabled the detection, identification and treatment of ECs in trace concentration (mg to ng/L), and the development and advancement of hybrid treatment systems are emerging promising treatment solutions. The use of nanomaterials and phytoremediation approaches are new approaches widely studied as a good potential process for remediating the ECs in wastewater.

Keywords: emerging contaminants, wastewater, advanced analytical techniques, occurrence, detection and removal

11.1 INTRODUCTION

Emerging contaminants (ECs) consist of extensive and broader groups of human-induced/anthropogenic compounds frequently present within wastewater but currently have been recognized as major pollutants for terrestrial and aquatic environments (Ifon *et al.*, 2023; Pal *et al.*, 2023). ECs are defined as a wide range group of synthetic or naturally occurring chemical compounds, these compounds have adverse effects on humans, animals and the environment (Kadac-Czapska *et al.*, 2023). ECs include a diverse amount of substantially utilized compounds and products such as artificial sweeteners, pharmaceutical and veterinary drugs, pesticides and herbicides among others (Figure 11.1). ECs are commonly distributed throughout the environment but are predominantly present in wastewater effluents (Hu *et al.*, 2023; Kadac-Czapska *et al.*, 2023).

The monitoring and control of ECs increase within the environmental compartments has proven to be a very strenuous task because these compounds are present in our essential products such as pharmaceuticals and personal care products (PCPs) that we use daily. Apart from the increasing use of ECs in daily essential life activities and a surge in environmental pollution through these contaminants, the disadvantageous health effects caused by bioaccumulation and biomagnification of ECs within the environment cannot be disregarded (Jaffari *et al.*, 2023). However, the occurrence of ECs in effluent discharge from wastewater treatment plants (WWTPs) is fundamentally impacted by population density, geolocations and usage patterns of ECs containing products and materials (Parida *et al.*, 2021). The industrial and urbanization developments have led to an increase in new chemicals containing ECs being produced and used in daily activities which represents a concern for authorities, researchers, the general population and the environment (Ren *et al.*, 2023a, 2023b). The inherent hazard they pose to human health has attracted public interest and the adverse environmental impact is due to their continuous discharge and environmental pollution without proper monitoring protocols being implemented (Bellas & León 2023; Majumder *et al.* 2023; Ullah *et al.*, 2023).

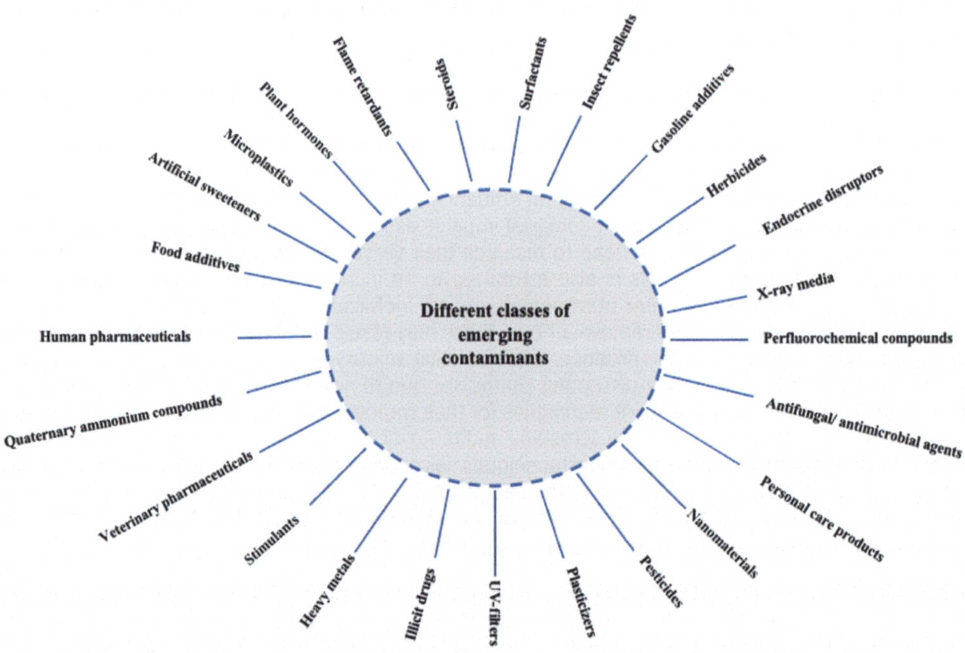

Figure 11.1 Different classes of emerging contaminants.

The technological advancement of analytical methodologies and instrumentation have empowered researchers in identifying, detecting and quantification of a wide range of ECs that are found in low concentrations within aquatic environments such as wastewater, drinking water, surface and groundwater analysis studies (Hanna *et al.*, 2023; Hawash *et al.*, 2023; Milanović *et al.*, 2023). Research work conducted on wastewater and WWTPs has gained attention in environmental assessments due to a variety of recently identified and quantified compounds/pollutants that are present because of anthropogenic activities (Martínez-Huitle *et al.*, 2023). Many studies recently documented in literature have reported on the occurrence of EC pollutants within the environment, which are present in a wide range of concentrations from micrograms per litre (μg/L) to nanograms per litre (ng/L). These pollutants are commonly classified as ECs and do not have any legislative regulations or guidelines. These pollutants are a concern to environmentalists and healthcare workers as they pose adverse health effects and are undetectable in conventional WWTPs and effluent discharge and are not regulated (Ofrydopoulou *et al.*, 2022).

The detection of various new ECs compounds in various water bodies has highlighted the health and safety concerns as these pollutants are potentially hazardous to human and animal consumption and have been deemed to be unsafe (Barbosa *et al.*, 2023). The occurrence of ECs within the environment results in human and animal health implications and environmental endangerment (Ghasemi *et al.*, 2023; Kumar *et al.*, 2022). The presence of ECs has been identified in various water bodies such as surface and groundwater, as well as wastewater that is discharged from treatment plants (Puri *et al.*, 2023). The existence of these pollutants within the environment may pose various health risks to humans and animals such as cancers, endocrine disruptions, neurotoxicity, reproduction problems, bacterial resistance and feminization in aquatic species among other health problems (Shanmuganathan *et al.*, 2023). The detection and removal of ECs in the environment has become an extreme environmental concern and has proven to be greatly imperative in expanding the efficiency of wastewater treatment methodologies in combating and removal of EC pollutants. Although concentration levels of EC compounds are vastly different globally when in comparison between continents, countries and also at a regional scale (Martínez-Huitle *et al.*, 2023). A major source of EC contamination has been identified as untreated wastewater and effluents discharged from WWTPs. The majority of current WWTPs were not intentionally designed to remove ECs and transformative by-products that are subsequently discharged into various aquatic bodies. In addition, the current knowledge available is inadequate to enable the development of new innovative analytical methodologies that can be used to monitor and control the release of ECs from WWTPs into the environment (Krishnan *et al.*, 2023). It has been reported that the majority of ECs are non-biodegradable and persistent and to overcome this disadvantage, newly developed advanced WWTPs have to be equipped with treatment methods to improve the effluent concentration of EC and wastewater management which will help to comply with new strict/ rigorous discharge regulations and laws applied. Therefore, to preserve and safeguard water bodies from such pollutants, the following alternatives must be scrutinized:

- Identifying the source of ECs and inspecting the wastewater quality
- Evaluation and identification of hazardous effects of ECs on the environment, humans and animal health, and these would be possible by assessing the nature and toxicity of these compounds and their transformative by-products produced during various stages of water treatment procedures
- Develop or improve WWTPs technologies for wastewater treatment processes that will reduce the discharge of ECs and environmental impacts caused by ECs
- Conduct environmental assessments to monitor the discharge.

Other sources of ECs include the fertilizer industry, adhesives, food and drinks packaging and fire retardants among others (Dubey *et al.*, 2023; Faisal *et al.*, 2023; Sathya *et al.*, 2023). Conventional WWTPs removal of ECs via common processes is regularly time-consuming. These common processes are facilitated by various characteristics such as the type of target ECs and the accessibility of microorganisms that can degrade the ECs present. Consequently, there is a heightened requirement to investigate and apply efficient methods for the removal of ECs in wastewater (Ali *et al.*, 2023).

ECs enter the food chain in diverse routes and result in the biomagnification of these compounds. Consequently, an increment of ECs concentrations in living organisms at sequential trophic levels can be observed in the food chain animals (Almazrouei *et al.*, 2023). Extended exposures to ECs induce adverse effects on the aquatic ecosystems, and at times modify the hormonal and metabolic mechanisms of humans and animals (Gunathilaka *et al.*, 2023). The toxicity of EC compounds is driven by various factors; namely their chemical structure, the body's ability to absorb ECs and detoxification ability of the body. The chemical, physical and ecotoxicity profile of EC compounds are associated with its distinctive molecular structure, and the toxicity is assessed based on the functional groups attached to the compound, their reactivity and by-products (Korzeniowski *et al.*, 2023; Parida *et al.*, 2021). The primary pathway for human exposure to ECs is via the consumption of foods, animal products and drinks that are associated with contaminated water, soil, plants, microorganisms and animals (Chaturvedi *et al.*, 2023; Interdonato *et al.*, 2023) (Figure 11.2). This can be exhibited as bioaccumulation of ECs, especially for the species placed on top of the food chain (du Plessis *et al.*, 2023; Thacharodi *et al.*, 2023).

Immediately when ECs or their transformative by-products reach environmental partition, they can experience different mechanisms such as absorption, adsorption, hydrolysis, dilution, biodegradation, complexation and chemical oxidation among other processes. Each of these processes possesses the ability to result in the degradation, transformation or persistence of ECs within the environment (Lofrano *et al.*, 2020).

Various review studies have focused on the occurrence of ECs in different environmental matrices especially in freshwater bodies and wastewater (Foglia *et al.*, 2023). Some of the review studies have also analysed the present instrumentation and methodologies used for the detection of ECs and their

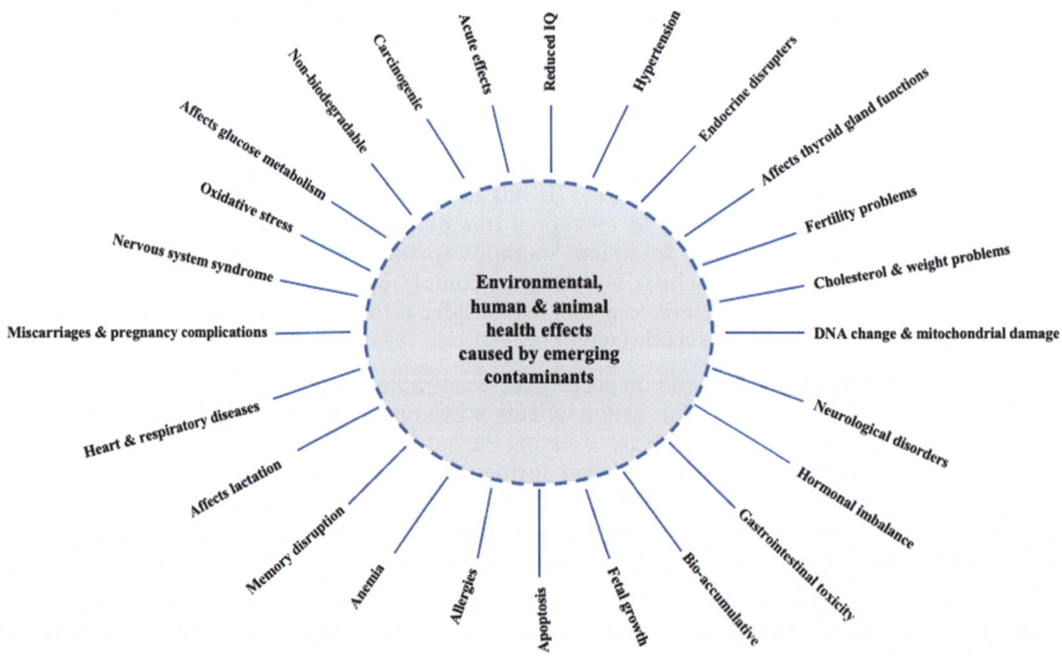

Figure 11.2 Adverse effects of emerging contaminants on animal and human health as well as on the environment.

removal and remediation procedures. A handful of the current review studies have attentively focused on the available current removal and treatment techniques such as advanced oxidation process (AOPs), coagulation, adsorption, membrane-based and flocculation among others for removal of ECs in wastewater. Other studies have focused on the application of biological techniques for removal of ECs such as microalgal removal of ECs in wastewater (Ramesh *et al.*, 2023). Some studies have focused on bioremediation and biotransformation of ECs and converting them into less/ non-toxic by-products in wastewater, while other studies have highlighted the environmental impact, risks, ecotoxicity and health concerns of ECs and their removal (Puri *et al.*, 2023). Present-day studies have investigated the application of biochar and activated carbon (AC) among others as adsorbent materials to remove ECs from wastewater (Alyasiri *et al.*, 2023).

Nevertheless, discussing and understanding the recent progress researchers have made on ECs, feasibility, advantages and limitations in the development and application of novel techniques for the detection and removal of ECs are important to implement and integrate using a robust low-cost approach within conventional WWTPs. Therefore, according to the best knowledge of the authors, a broad study on occurrence, detection, removal techniques from wastewater and a summary of ECs health and environmental impacts have not been efficiently structured in one study present in literature studies.

Therefore, this current book chapter critically examines the recent literature research work and innovative developments and advancements in the detection and removal of ECs from wastewater, with in-depth research on the types of ECs, health concerns, sources and environmental occurrence. In addition, the economic feasibility of implementation of these removal techniques, the advantages and disadvantages associated with different removal techniques have been discussed. The review chapter also imparts awareness of the future perspectives for research work and legislative policies and strategies that can be implemented to monitor and minimize the release of ECs.

11.1.1 Review methodology

The most relevant literature review was obtained from established trustworthy databases such as Google Scholar, Scopus, Science Direct and among other reliable information sources. The following terms were used as keywords during the search process: Emerging contaminants (ECs); toxicity of emerging contaminants; Occurrence of ECs; detection of ECs; physiochemical, biological, chemical and removal and remediation process of ECs; emerging hybrid removal and remediation process of ECs. The current review chapter aims to analytically provide critical information on existing literature on emerging contaminates, newly developed remediation techniques of ECs from wastewater and highlights the knowledge on removal techniques and implementation of hybrid technologies for removal of ECs and wastewater remediation. This review chapter also outlines the existing data in highlighting key areas for future research which would assist in addressing gaps in existing knowledge/literature. Especially regarding upscaling and implementation of new ECs removal techniques and application of hybrid techniques. In addition, this review chapter highlights the commercial feasibility of the current techniques and recommends a handful of propitious innovative policy-based perspectives. This chapter investigates the challenges associated with ECs ranging from occurrence, health and environmental impacts, and innovative treatment techniques. The interest in research studies based on ECs has evidently escalated over the past decade, therefore, this work provides state-of-the-art information on ECs with a specific focus on recent developments and advancements in the occurrence, detection and removal techniques of ECs.

11.2 OCCURRENCE

Various factors influence the occurrence of ECs in the environment, the major factors being the physio-chemical properties of the environment and the wastewater. In addition, the type of source for the ECs influences the exposure degree and the properties of ECs (Majumder *et al.*, 2023). Most studies present in the literature have identified municipal wastewater discharge as a major source of ECs. Other sources (non-point and point) such as stormwater, household and hospital, industrial wastewater, and agricultural run-off have also been identified as sources of ECs (Coxon & Eaton, 2023).

Current research has pivoted more on the occurrence and detection of ECs in influents and effluents (Dubey *et al.*, 2023). On the other hand, the particulate phase, which includes the sludge matrix, has not been investigated in detail due to a lack of instrumentation techniques that can be used to analyse such a sample matrix. However, the detection of these contaminants in the particulate phase is important to assess the destiny and hazardous adverse influence on the environment and human and animal health (Dubey *et al.*, 2023).

Most of the ECs are discharged into the environment via different mechanism routes such as urine and faecal matter in the case of drugs consumed by humans and animals, agricultural application of wastewater, sewage sludge and manure, landfill leachate, manufacturing, industrial wastewater, domestic use and release of products containing ECs. The major contribution of ECs to the environment stems from anthropogenic activities such as conventional WWTPs effluent release, whereby ECs are collected and accumulated from industry and urban discharge and are incompletely removed (Sewwandi *et al.*, 2023). Most conventional WWTPs comprise physical, chemical and biological treatment systems which are not suitable in design aspects to efficiently remove ECs due to their complex molecular structure, low concentration and non-biodegradable in wastewater (Figure 11.3). The effluents generated from such WWTPs have been the most significant point source of ECs. Despite advanced technologies implanted in urban WWTPs being able to remove ECs, their removal efficiencies are limited well developed areas. For underdeveloped places with economic and technical limitations, the use of advanced WWTPs is hindered. ECs will not easily be removed or bio-transformed into less toxic compounds in wastewater, which leads to inefficient removal by current WWTPs systems. This indicates that ECs can be effortlessly released from WWTPs and be discharged into the environment

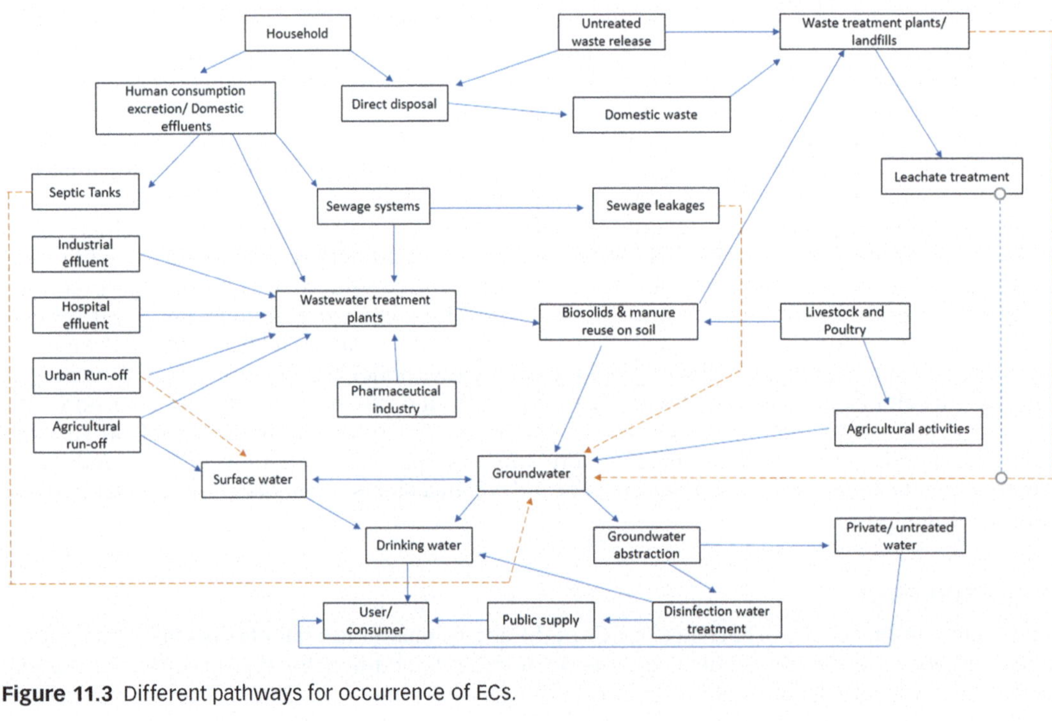

Figure 11.3 Different pathways for occurrence of ECs.

compartments via direct disposal of wastewater. In addition, studies have been documented in the literature that report low removal efficiency of ECs in conventional WWTPs that still use conventional methods such as treatment of sludge, flocculation and sedimentation (Puri *et al.*, 2023).

Various EC compounds go through microbially mediated reactions during the wastewater treatment process. Biodegradation is often identified as the dominant pathway for the remediation of ECs from wastewater, however, the use of biodegradation methods generally results in the transformation of ECs into new compounds that are likely to be me hazardous and get discharged from WWTPs without any detection or removal action (Douna & Yousefi 2023; Zambrano *et al.*, 2023).

Sources of ECs within the environmental matrices are present in two forms namely the point source and diffuse source (Onnis *et al.*, 2023). Point source refers to contaminant pollution that emanates from distinct locations whose pollution inputs into the environment can certainly identified in a spatial distinct manner. Important examples of point sources include domestic, hospital and industrial effluents, WWTPs effluents, sewage storm-water overflow, waste disposal sites, resource extraction and buried septic tanks among others (Ulucan-Altuntas *et al.*, 2023). Diffuse source refers to contamination pollution that originates from poorly distinct locations that usually occur over a wide broad geographical scale. Examples of such source points include stormwater and urban run-off, leakage from sewage systems and diffuse aerial deposition, and agriculture run-offs from the application of chemicals such as bio-solids, pesticides and manure among others (Niu *et al.*, 2022; Onnis *et al.*, 2023).

Insufficient information on toxicity, impact and concentration levels of ECs present within various environmental compartments results in problems for government and environmental authorities to manage their application and also control their discharge levels. There are no laws in practice regulating the upper permitted concentration levels of ECs in wastewater discharge, ECs present in potable water or in the environment (Reid *et al.*, 2019).

11.3 DETECTION

In the past couple of decades, tremendous progress has been made in the detection of ECs in water treatment and industrial effluents and the development of new analytical methods for the characterization, identification and quantification of rapidly evolving ECs (Hemida *et al.*, 2023; Ieda & Hashimoto 2023; van den Hurk *et al.*, 2023). Various types of ECs are continuously discharged into water surroundings either purposely or unintentionally, with little or no legislation in place or minimal care thus posing a health risk to humans and animals (van den Hurk *et al.*, 2023).

With the requirement to investigate the occurrence, transportation, and fate of ECs, it is important to unequivocally identify the new ECs and determine their concentrations (Angelakis *et al.*, 2023). Majority of the EC compounds are easily soluble in water due to their chemical structure, therefore posing a potential harm to the human, animal and aquatic life through the water cycle. There has been an increase in the number of ECs detected in the environment (drinking, ground, surface and waste) comprising of parents and their derivatives. The identification and assessment of ECs in water have proven to be an important task scientifically, which requires highly sensitive analytical techniques capable of reaching the nanogram per litre (ng/L) scale. The analysis of ECs must be conducted using sensitive, selective, robust and automated techniques that apply to a wide range of compounds that are present in wastewater. Therefore, there is a need for the development of methods that are fast, responsive and efficient in the detection and determination of the wide range of ECs which in turn can be used for monitoring purposes (Ghosh *et al.*, 2023; Kumar *et al.*, 2022). There are different analytical methods have been investigated by researchers globally for the analysis of ECs in wastewater (Kumar *et al.*, 2023). In WWTPs, different water samples such as influent and effluent are the common matrices that are mostly used in the development stages of analytical techniques that can be used for the efficient qualitative and quantitative target analysis of ECs (Ghosha & Biswasb, 2023). Mass spectrometry coupled liquid chromatography (LC–MS) or tandem MS has been recently upgraded for analysing ECs in different matrices like solid sludge and water at low extreme concentrations (ng/L or μg/L). In the field of environmental analytical chemistry, these advanced techniques have been widely used for the detection and quantification of over 3000 active chemicals (Angelakis *et al.*, 2023).

The analysis of ECs essentially needs a method with a low limit of detection (LOD) and high selectivity. Analytical instrumental analysis of ECs in wastewater is generally conducted by chromatographic and spectrochemical techniques which are coupled to mass spectrometry (MS) (Li *et al.*, 2023). ECs are mostly polar compounds, and their analysis by gas chromatography (GC) is hindered by limitations of volatility and/or thermal stability. These above-mentioned limitations can be overcome by the application of derivatization processes such as silylation, acylation and alkylation among others. The application of the GC for the detection of ECs has proven to be an inexpensive technique that could be implemented in different laboratories globally. Most GC methods reported in the literature are coupled to MS detection in both tandem and single modality (Angelakis *et al.*, 2023). GC commonly demands the addition of a derivatization step which would help improve the chromatographic behaviour of analytes while also improving selectivity, sensitivity and peak resolution (Li *et al.*, 2023).

Despite the advantages provided by GC, the liquid chromatography (LC) based methods are the most commonly used for ECs detection in wastewater (Khurana *et al.*, 2022). LC provides a higher versatility in the analysis as it covers a wide range of compounds that can be detected without the need for derivatization. The choice of detection preferred with most LC methods is the MS. Within LC techniques, high-performance liquid chromatography (HPLC) has emerged as an improved modality along with ultra-high version (UHPLC) (Chaturvedi *et al.*, 2023). Most of the ECs are usually found at ppb to ppt levels in different environment matrices, therefore, liquid chromatography with tandem mass spectroscopy (LC-MS/MS) is used. The major advantage of the LC-MS/MS is the high selectivity and sensitivity (Khurana *et al.*, 2022). Volatile organic compounds are usually analysed using GC, while LC is utilized for detecting polar and less volatile compounds. The MS techniques have shown outstanding results in precise analysis of ECs that are present within complex matrix samples of wastewater (Sewwandi *et al.*, 2023). GC–MS and LC–MS have been extensively used for analytical analysis of ECs including pharmaceuticals and metabolites, endocrine disrupting compounds, UV filters and flame retardants among others.

Inductively coupled plasma mass spectrometry (ICP-MS) has been used for the analysis of nanomaterials detection in wastewater. This technique provides some advantages such as low LOD, high precision, low cost and simultaneous analysis of multi-elements and isotopes within a few minutes (less analysis time required). The coupling of ICP with various detectors has also been used for the analysis of nanomaterials. Improved detection limits of the ECs to low concentration are achieved by these analytical instruments (Gumbi *et al.*, 2022; Inarmal *et al.*, 2023).

Recently, most studies focusing on the detection of the ECs have been successful through using sophisticated instrumentation which includes, gas chromatography/mass spectrometry (GC/MS), GC/MS/MS, ultra-high-performance liquid chromatography–tandem mass spectrometry (UHPLC/MS/MS), triple quadrupole mass spectrometer (TQ-MS), high-performance liquid chromatography (HPLC) and LC–electrospray tandem MS (LC–ES/MS/MS) (Table 11.1). The application of these techniques ensures low concentration up to ppt levels with good precision and accuracy.

Currently, there is insufficient data in global comparison studies on the detection, analysis and occurrence of ECs in wastewater and other water bodies. A study conducted by Nikolopoulou *et al.* (2023), on the investigation of ECs examines water samples obtained from three separate WWTPs in Lagos, Nigeria. The detection and identification approach were executed using ultra performance liquid chromatography mass spectrometry (UPLC-QToF-MS), 250 compounds were identified in the samples analysed from the WWTPs. A total of 182 compounds were quantified from the 250 detected, and 78 of those compounds had a high significant environmental risk score index. The majority of the compounds detected at high concentrations were pharmaceuticals and were from hospital WWTP, with salicylic acid having the highest concentration of 72.4 mg/kg followed by ciprofloxacin and ofloxacin at 24.4 and 28.4 mg/kg, respectively.

Table 11.1 Different detection techniques of ECs.

Source	Country	Detection Technique	EC Analysed	Concentration	References
River	China	LC/MS/MS	Antibiotics	0.1 and 74 ng L^{-1}	Chung *et al.* (2016)
River	USA	LC–MS/MS	Sulfamethazine, tylosin, and atrazine	1.87, 0.30, and 754.2 ng/L	Albero *et al.* (2018)
Surface water	South Africa and Botswana	LC–MS	Antiretroviral drug ritonavir; ibuprofen	64.52 μg/L; 1097	Selwe *et al.* (2022)
Waste and surface	South Africa	Triple TOF coupled LCHPLC	Pharmaceuticals, drugs, and metabolites., personal care products, pesticides and food additives	ng.L^{-1} to μg.L^{-1}	Abafe *et al.* (2023)
Freshwater, marine and terrestrial apex predators	United Kingdom, Germany, Netherlands and Sweden	LC–HRM	Pharmaceuticals and antibiotics	ng.L^{-1} to μg.L^{-1}	Gkotsis *et al.* (2022)
Rivers and wastewater	China	(UPLC–MS/MS)	Antidepressants	0.6 and 87 ng/L	Karlsson *et al.* (n.d.)
Surface water	India	LC/Q-TOF-MS	Pharmaceuticals and agrochemicals	ng.L^{-1} to μg.L^{-1}	Richards *et al.* (2023)
Surface water	Brazil	LC–MS/MS with (ESI) and HPLC	Pharmaceutical products and herbicides	1.6 to 14.5 μg L^{-1}	Gomes *et al.* (2022)
Wastewater	South Africa	SPE and (LC–MS)	Sulfamethoxazole hydroxylamine, sulfamethoxazole, prednisolone and ivermectin	0.05215 0.979 mg/L	Inarmal and Moodley (2023)

Ng *et al.* (2023), conducted a study using state-of-the-art target screening approach for 2362 ECs compounds and their transformation by-products. The analysed ECs included three major categories: industrial chemicals, plant protection products (PPPs) and (PPCPs). A total of 586 in the samples which consisted of 158 PPP's, 71 industrial chemicals, 348 PPCPs and 9 other chemicals. The sample variety used in the study consisted of influent and affluent wastewater, groundwater and river water. Gas chromatography high resolution mass spectrometry (GC–HRSM) and LC–MS/MS were used for the detection and analysis of ECs compounds respectively.

11.4 REMOVAL

11.4.1 Physiochemical methods

Physicochemical mechanism is one of the widely used approaches used to remove the ECs from wastewater and surface waters through sorption onto biomass during wastewater treatment (Tholozan *et al.*, 2023). The removal of pathogens, odour and reduction of turbidity is generally assisted by the physio-chemical treatment process. These processes provide the advantage of decreasing the pollutants in drinking water, however, their removal efficiencies have been proven insufficient for ECs. Coagulation–flocculation is the standard physio-chemical process that is usually necessary for water treatment processes. The removal of ECs using physiochemical methods is divided into two, that is membrane and adsorption techniques (Jaffari *et al.*, 2023).

11.4.1.1 Adsorption methods

The adsorption treatment process is common and important in water treatment due to its simplicity in design and it yields by-products and has proven to be insensitive to toxic substances. Adsorption is defined as a phenomenon of accumulation of analyte(s) material onto the adsorbent material. The use of adsorption techniques for the removal of EC' from wastewater has been explored and recorded in the literature (Kordbacheh & Heidari, 2023).

The adsorbent materials adsorb the target analyte(s) on their surface comprised of porous networks and consequently eliminate the pollutants from the wastewater. Adsorbent materials generally have features of large surface area and high porosity. The adsorption process has been established to be an excellent method aimed at the separation and elimination of dilute contaminants and provides benefits such as recovery, recycling and reuse of the adsorbent material (Rathi & Kumar, 2021).

A widely used adsorbent material is activated carbon which could be differentiated into two groups namely powdered activated carbon and granular activated carbon (Figure 11.4). The main disadvantage of commercial activated carbon is the high production and regeneration costs. Even after activated carbon adsorbent has been used, it can also be regenerated for future use. However, the regeneration process can result in carbon loss thus resulting in the production of adsorbent with lower adsorption efficiency when compared to freshly prepared activated carbon. Currently, there's a developing interest in alternative low-cost adsorbent materials derived from waste materials or by-products from agricultural or industry processes. The removal efficiency of these adsorbent materials is based on their properties such as porosity, surface area, pore diameter, functional groups and the chemical nature of the specific target ECs. In addition, the existence of dissolved organic matter may affect the efficiency of the adsorption process. The occurrence of dissolved organic matter within might result in competition for the available adsorption site with the target analyte(s) and thus resulting in reduced adsorption capacity for ECs (Kurniawan *et al.*, 2023).

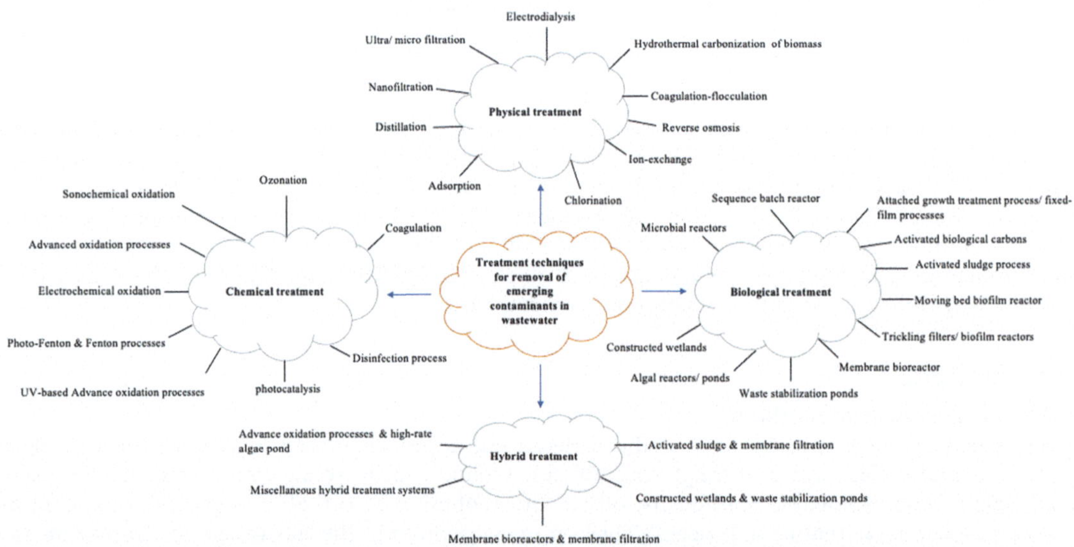

Figure 11.4 Treatment techniques commonly used for removal of emerging contaminants in wastewater.

Filtration processes using membrane methods such as reverse osmosis, micro, nano and ultrafiltration depending on the pore size and separation mechanisms are generally used in WWTPs to remove ECs. The advantage of the membrane filtration technique is the high value of eluent discharged without the need for the addition of other purification chemicals. Membrane filtration methods are now commonly applied to WWTPs for wastewater treatment reclamation and drinking water as they have been proven to efficiently remove the majority of organic and inorganic pollutant compounds (Nikolopoulou *et al.*, 2023). The implementation of reverse osmosis and nanofiltration techniques have been proven as capable alternatives for removing ECs in wastewater processes, occurring in three steps as follows:

Step 1: The adsorbate material (analyte (s) is transferred to the exterior surface of the adsorbent material, and this process is driven by film diffusion which is also known as the external diffusion mechanism.

Step 2: This is a movement of the analyte(s) from the adsorbent surface into the adsorbent pores which is also known as the porous diffusion mechanism.

Step 3: The analyte(s) material is fixed/retained on the adsorbent pores and this is known as the adsorbate surface reaction mechanism.

The adsorption treatment process is identified as user friendly and reliable wastewater treatment process owing to the versatility of usage as it provides its lack of sensitivity to unsafe materials, and its ease of operation (Raninga *et al.*, 2023).

Several adsorbents are utilized in wastewater treatment plants to remove emerging contaminants, and these include carbon materials, polymers, activated carbon, activated carbon, metal–organic complexes, carbon nanotubes, zeolites, organic carbon complexes, mineral substances, carbon nanotubes, biochar and amongst others. These adsorbents may originate from several materials, and these include plants, carbon nanotubes, fly ash, ion exchange resin, some minerals, organic resins and so on. Some of the requirements of these adsorbents are that they must be efficient, effective, cost-effective, greener to the environment and have good regeneration capacity (Sellaoui *et al.*, 2023)

Based on the literature that was recently published, the adsorption process is regarded as the greener approach for the removal of ECs from wastewater resources. This is because the process is capable of removing a high percentage of ECs with lesser secondary sludge generation. Furthermore, this process can be combined with other systems, and thus increases the removal efficiency (Coxon & Eaton, 2023).

11.4.1.2 Membrane technology

This treatment technology involves the filtration of the solution by retaining the analyte(s) on a membrane material and the target analyte(s) that are removed are analysed by different filtration characteristics such as surface charge, pore size and hydrophobicity which are based on the membrane properties (Ghasemi *et al.*, 2023). To combat the growing concern about ECs, high pressure membrane methods such as nanofiltration and reverse osmosis were developed and fully utilized for the removal of ECs from wastewater, surface and drinking water while also removing any additional contaminants present (Onnis *et al.*, 2023). Other membrane tools that have been developed and used for the elimination of ECs comprise distillation, forward osmosis and electrodialysis. Ionized ECs are hampered by electrodialysis reversal, while electrostatic, repulsion and sieving are three mechanisms used in nanofiltration for the removal of ECs. Physiochemical treatment approaches, for example, adsorption and membrane technology have been identified to have a fair chance of removal of ECs in wastewater, although more research studies are required for the removal of a large spectrum of ECs as they need further evaluation and refabrication (Amalina *et al.*, 2023).

The global scarcity of water along with an increase in population over the years has necessitated the use of wastewater treatment as an alternative source of drinking water (Shehata *et al.*, 2023). This process has a high potential for high concentration of the ECs hence, their removal is key if water is planned for usage in industrial applications and irrigation in farming (du Plessis *et al.*, 2023).

Some of the methods that have been applied to remove the ECs are capable of totally removing them while others are not effective enough, but this depends on the compound of interest and WWTP. Several methods have been established for the degradation, reduction, and removal of ECs that would help mitigate their negative influences the environment (Ng *et al.*, 2023; Tholozan *et al.*, 2023). Basic wastewater treatment processes usually include the mediation and extraction of soluble and insoluble contaminants. Several extraction methods such as biological processes, adsorption, membrane-enhanced oxidation technology and construction of wetlands are ways used to efficiently remove ECs in wastewater (Ng *et al.*, 2023).

The efficient and cost-effective removal of ECs in different water matrices has proven to be a difficult challenge, especially in wastewater treatment. In recent years, several techniques have been established to remove the ECs from wastewater. From an analytical perspective, wastewater treatment has proven to be difficult due to the multifaceted nature of its matrix, additionally, to that, its properties vary depending on the inputs to the WWTPs (Coxon & Eaton, 2023). The conventional WWTPs consist of two treatment stages, namely the primary treatment (application of physiochemical properties) and secondary treatment (application of activated sludge biological reactor). The efficient removal of ECs from wastewater generally relies on their nature: physiochemical properties, biological and chemical form and application process settings.

Traditional municipal WWTPs are not efficiently intended to remove the ECs that exist at low concentrations such as micrograms/litre (μg/L) and nanogram/litre (ng/L). Advanced technologies that are affiliated with non-conventional wastewater treatment have been upgraded due to the advance of new methods. Substantial wastewater treatment expertise may be characterized as advanced oxidation, biological and physical treatments and phase-changing practices. Conventional techniques reported for ECs removals generally involved the use of bacteria, both aerobic and anaerobic, but these systems have proven to be energy-consuming, expensive and have lower removal efficiencies (Sellaoui *et al.*, 2023).

Sedimentation is a primary treatment process widely used to remove ECs in general conventional WTTP's however, their removal is restricted due to the hydrophilic nature of a number of ECs. On the other hand, the biological process (e.g., ASP) is regarded as the secondary treatment process and is good at removing the PPCPs through biodegradation partition, biotransformation and adsorption. Parameters such as sludge age and its adsorption capacity, the way the reactor is designed, nature of the ECs have a huge impact on effectively eliminating the PPCPs in ASP. Different ECs belonging to one category can display significant inconsistency in their biodegradability. When the ECs are not entirely removed in the secondary treatment process this is likely to be attributed to the alteration of ECs into by-products or metabolites thereby yielding low elimination effectiveness (Jaffari *et al.*, 2023).

Advanced technologies implemented in treatment plants include the following: AOPs, constructed wetlands (CW), adsorption techniques, hydrolysis processes, chlorination and membrane filtration. These various treatment processes can be implemented due to the WWTP and the effluent quality requirements and depending on the specific end-uses. Different AOPs have been applied for the elimination and degradation of ECs from wastewater, that is, photocatalysis, photo-Fenton, photolysis, ozonation, ultrasonication, Fenton's oxidation and solar-driven processes among others (Sellaoui *et al.*, 2023).

During the sludge treatment process and sludge treatment process, high concentrations of the ECs present in the particulate and liquid phases are removed via the biodegradation mechanism. The compound biodegradation mechanism is affected by several factors such as microbial diversity, redox, toxicity of ECs, hydraulic retention time, temperature, pH, molecular features, potentials, availability of the ECs to microorganisms and physicochemical properties, amongst others (van den Hurk *et al.*, 2023). The toxicity of various ECs is mostly affected by the catalytic activity of particular enzymes subject to their genetic capacities that are degraded by the microorganism. The ECs that do not pose any threat to the environment are not likely to have an impact on microbial activity. Newly developed wastewater treatment technologies for the removal of ECs have been categorized as physiochemical and biological techniques (Jaffari *et al.*, 2023).

11.4.2 Biological methods

Biological treatment that involves the removal or elimination of ECs through biodegradation mechanisms is extensively used. During the biodegradation process, large ECs molecular compounds are broken down into small compounds or biomineralized into inorganic compounds such as water and carbon dioxide using microorganisms such as fungi, bacteria and microalgae among others (Raninga *et al.*, 2023). The efficiency rate for the removal or degradation of ECs is affected by different influences such as the treatment method utilized, physiochemical properties of the target EC species, their biological persistence and WWTPs operational settings. The common biological treatment techniques used are generally distributed into two categories, namely conventional and non-conventional. The categorization of these processes is based on the properties of the wastewater being treated, removal efficiency, process operation, maintenance and treatment challenges (Coccia & Bontempi, 2023).

One of the important issues that is faced by the wastewater challenges in algal-bioremediation strategies. The nature and ecotoxicity of wastewater depend on two factors; the type of wastewater and the source of waste. Procedures such as direct toxicity and water quality assessment, in-vitro and in-vivo bioassays are utilized to regulate the quality and toxicity of wastewater. Biological methods are normally applied for the elimination of ECs in wastewater. Biological treatments are generally differentiated into two groups, namely aerobic and anaerobic processes (Bellas & León, 2023). Aerobic processes include the application of aerobic bioreactor, membrane bioreactor, sequencing batch reactor and trickling filter. Whereas anaerobic process examples include the application of sludge reactors and anaerobic film reactors. The biological treatment process involves the use of microorganisms for the degradation of the specific target ECs in wastewater into smaller less toxic molecules or biomineralized into simple organic molecules. The main advantages of biological methods over physical methods include operational costs and complexities provided that the target analyte(s) are ready to be biodegradable/oxidized by the microorganisms, the main driver for this technique is that it effectively destroys the ECs rather than concentrates them. Generally, less information is documented regarding the biodegradation mechanisms of ECs in environmental conditions (Jatoi *et al.*, 2023). Implementation of biological methods for the removal of ECs has been proven to be ineffective in some instances due to numerous non-biodegradable ECs. In addition, most of the ECs compounds are potentially harmful and can hinder microbial growth which may consequently hinder the biodegradation process. Hence, wastewater characteristics are crucially important when it comes to the selection of biological wastewater treatment (Amalina *et al.*, 2023). Biological treatment processes are time-consuming because of the slow growth rate of the microorganisms used and the exhausting mineralization periods necessary particularly when fresh microorganism culture is used. In addition, the application of this treatment method is time-consuming as there is a need for the cultivation of microorganisms before the degradation process (Ren *et al.*, 2023a, 2023b).

11.4.3 Chemical treatments

Chemical treatment methods possess the capability to attain high elimination efficiency of a targeted wide variety of ECs as a result of the chemical characteristics of the wastewater along with the operating conditions of WWTPs (Barbosa *et al.*, 2023). Newly developed advanced treatment techniques such as chemical treatment methods are more efficient and are referred to as chemical oxidation methods (Coxon & Eaton, 2023).

Treatment techniques based on chemical interaction aim to modify or convert contaminating pollutants into minimal detrimental effects or biodegradable compounds, and this is achieved by mineralizing them or transforming them into inorganic compounds that are less hazardous such as nitrogen, water and carbon dioxide (Zhou *et al.*, 2023).

These treatment techniques utilize methods such as hydrogen peroxide, ozone, chlorine, oxide-based-metals and metal-based catalysis coupled with sources like sun, gamma, UV radiation, electric current and ultrasound among others. There are two types of chemical treatment methods frequently used and these are conventional oxidation and AOPs (Folorunsho *et al.*, 2023).

11.4.3.1 Conventional oxidation methods

The oxidation treatment process involves the use of oxidizing agents including chlorine and ozone as the essential methods for the removal of ECs in wastewater (Figure 11.4). During the treatment process, reactions based on chemical interactions in wastewater can be reactive resulting in the development of by-products (Kurniawan *et al.*, 2023).

Photolysis, ozonation, the Fenton process and chlorination are dome of some examples of conventional oxidation methods. Therefore, a careful selection of oxidants to be used is required before selecting this treatment technique. The utilization of less reactive species such as bromine and chlorine has been demonstrated to remove the majoring of ECs, studies conducted revealed high removal efficiencies for ECs such as diclofenac (100%), naproxen (95%), while other ketoprofen, triclosan, bisphenol A, ibuprofen removal ranged from 38% to 84% (Kurniawan *et al.*, 2023).

Ozone is extensively used in the wastewater treatment process for colour removal, decontamination properties, reduction of organic pollutants, taste and odour control in drinking water. Ozone is reactive towards organic pollutants directly or indirectly through interaction with ozone molecules and the free radicals (made up of hydroxyl OH radicals) that are formed through ozone decomposition (Sewwandi *et al.*, 2023). The hydroxyl radical is produced during the Fenton reaction process during the reaction of iron and hydrogen peroxide. Fenton reaction has been advocated for as a viable treatment method for effluents because of the availability and non-toxicity of iron, but the major concern is the poor removal efficiency of EC when compared with other oxidation methods and also the production of $Fe(OH)_3$ sludge (Folorunsho *et al.*, 2023). The application of the catalyst can help enhance the removal effectiveness of ECs in Fenton treatment. The compound structures of ECs are damaged/broken down during the photolysis process and this is caused by the radiation or light during the treatment process. Two photolysis methods are applied for the degradation of ECs (degradation through photosynthesizers such as hydrogen peroxide) from wastewater and these are direct and indirect photolysis (Folorunsho *et al.*, 2023).

Ozonation is the most commonly applied method in WWTPs to help enhance the biodegradability efficiency of ECs. The ozonation method has been proven to efficiently remove all forms of EC compounds (90–100%) due to its dominant oxidant that reacts with aromatic rings and results in degrading the EC compound structures. However, the treatment method requires large amounts of energy, application of ozonation is high-costly and requires immense energy operation (Sellaoui *et al.*, 2023). Moreover, studies have shown that the activation of free radicals and the formation of oxidative metabolites pose a major hindrance in this method, and more research work is required to solve these problems (Abafe *et al.*, 2023). The advantages of using the ionization radiation technique comprise of good penetration range in water matrix samples such as wastewater, and no requirement for additional chemicals during the treatment process. Furthermore, it is insensitive to colour and suspended particles present within wastewater, the uncooperative ECs compounds present within wastewater can be degraded in situ by reactive species that are created in the course of radiolysis of water. Disadvantages of this technique include the costly price of radioisotopes used and safety concerns regarding the use of isotopes (Almazrouei *et al.*, 2023).

11.4.3.2 Advance oxidation processes

Advanced oxidation processes (AOPs) are newly developed removal methods of ECs in wastewater and provide several advantages over conventional chemical treatment techniques. Research conducted has reported that the application of improved oxidation techniques resulted in high degradation efficiencies of ECs and this has been achieved by either using the method alone or coupling it with other removal methods (Bhattu & Singh, 2023).

Even though some of the chemicals that are used in biological oxidation techniques have proven challenging to remove, the use of biological oxidation processes for wastewater treatment provides several advantages. This method can be applied before or subsequently to the biological treatment process. Removal efficiencies ranging between 80% and 90% were achieved utilizing this technique, and in some instances above 90% removal efficiencies were achieved by coupling different removal methods such as coagulation, flocculation with innovative oxidation methods like sonolysis, ozonation, UV radiation and electrocatalytic oxidation among others (Ren *et al.*, 2023a, 2023b).

WWTPs and other wastewater management systems that apply AOPs have been described to be highly effective in the elimination of the ECs. The hydroxyl radicals have been proven to be more potent oxidizing agents in organic compounds but they are not regarded as catalysts. This chemical property has been critical in wastewater treatment processes (Ifon *et al.*, 2023).

WWTPs that use AOPs coupled with ozone has been reported to successfully remove ECs. Various AOPs were recently studied to efficiently remove the ECs from aquatic environments. Researchers have studied AOPs as an alternative method to effectively remove and degrade the ECs in comparison with conventional wastewater treatment techniques. The disadvantages emanating from conventional treatment techniques are now conquered by the expansion of new AOPs like; ultrasonic, photocatalytic, iron treatment and Fenton processes. To conquer these disadvantages, the use of the electro-Fenton process has been explored in which hydrogen peroxide is induced in situ via an electrochemical process under supervised conditions. AOPs have displayed high-efficiency removal of ECs compounds such as tetracycline, triclosan and acetaminophen among others present in products such as PCPs and pharmaceuticals which are present in complex wastewater matrices at low concentrations levels present within the effluent discharge. A major disadvantage in the application of Fenton processes is high operational and maintenance costs (Bikiaris *et al.*, 2023).

11.4.4 Emerging and hybrid treatment technology

As highlighted, the main aim of the current work is to address and discuss the novelty in the treatment process that can be used to close the gap of the shortcoming faced by the conventional methods used to remove the ECs. Some of the disadvantages of the conventional methods; are physical process including problems of disposing of large amounts of waste after the process, high operational costs, fouling of the membrane filters, and formation of large amounts of sludge. On the other hand, chemical process shortcomings include the challenge of removing the microplastics, the formation of sludge in large quantities, issue of separating the photocatalytic particles in a suspension (Zhou *et al.*, 2023). The microbiological process faces issues of high costs related to maintenance, precipitates forming during the process, and failure to function at high levels during cold weather conditions. Based on the shortcomings mentioned, this work looks at bringing solutions on the potential solution to remove the ECs in wastewater. Recent approaches focus on coupling the processes, for example, (physical + chemical processes), (physical + biological processes) and (chemical + biological processes). These approaches are mostly efficient, effective, greener, and environmentally friendly, with low costs for the removal of ECs in wastewater (Ren *et al.*, 2023a, 2023b). Biodegradation with photocatalysis combined with biological treatment is an approach that uses a photo catalyst instead of a porous membrane that can change the pollutants into degradable forms (Richards *et al.*, 2023). Other researchers have started using the electrochemical approach and there is a need to focus more on these studies and a potential way to remediate the ECs. For the adsorption approach, the use of nanomaterials has found high interest. Nanotechnology is a growing field to address the current conventional approaches which have several drawbacks (Bhattu & Singh, 2023). Some of the nanomaterials; graphene, carbon nanotubes, and nanomembrane, amongst others are currently being investigated for the remedial options for ECs in wastewater treatment plants. Recent studies conducted by Bikiaris *et al.* (2023) revealed that nano biochar (dendor) is an efficient nanomaterial to remediate ECs in wastewater. Furthermore, the occurrence of microplastics, pharmaceuticals and personal care products remains a huge global issue, hence, different approaches investigated include the use of magnetic materials since they are known to be flexible and can be easily modified to enhance separation and removal. Generally, during sample preparations including SPE, magnetic materials can be added forming magnetic solid phase extraction (MSPE). Research shows that more work needs to be done on utilizing the magnetic materials coupling with different sample preparations as an alternative approach to remove the ECs in wastewater (Hawash *et al.*, 2023; Karlsson *et al.*, n.d.).

The occurrence of emerging contaminants (metal ions, radio nuclei and organics) is a global challenge and different nanomaterials are utilized and these include the advanced porous nanomaterials, for example, porous aromatic framework, covalent organic framework and metal-organic frameworks (MOFs). These nanomaterials are easily modified making them excellent adsorbents to remove the ECs (Zhou *et al.*, 2023). There is a promising approach to synthesizing and fabricating advanced porous materials and reviewing the adsorption tools between porous materials and environmental toxins (Krishnan *et al.*, 2023).

The biodegradation process of ECs by photocatalytic process combined with biological treatment techniques provides numerous benefits such as ecologically friendly, low-cost operation and maintenance and it is a renewable treatment technique (Gondi *et al.*, 2022). The presence of ECs in wastewater can also be controlled by applying molecular structured impressed and non-imprinted polymers. Most EC compounds are degraded by anaerobic digestion processes. The anaerobic membrane reactor is the current anaerobic digesting technology that possesses high device stability and offers a large microbial community. In comparison with the conventional anaerobic digestion approach, the anaerobic bioreactor membrane has better decomposition capabilities for ECs as well as biogas production (Ulacan *et al.*, 2023).

In summary, the development of new technologies for removing emerging contaminants is essential for safeguarding water resources and protecting human health. Metal-organic frameworks, nanotechnology technologies, amongst others, show promise in effectively eliminating a widespread variety of emerging contaminants from water sources. Additionally, integrating freshwater ecology with ecotoxicology provides valuable perceptions of the environmental impacts of these pollutants and aids in the design of effective mitigation strategies.

11.5 CONCLUSION

The treatment of wastewater and reclamation processes has become an important goal in global water security and sustenance processes, especially in countries that experience severe drought climate conditions and lack proper WWTPs, while also experiencing expeditious population growth, therefore driving the demand for safe and clean water even more. Based on the aforementioned reasons, the monitoring of ECs within water bodies that are used for potable and non-potable activities should be prioritized to ensure the protection and safety of freshwater ecosystems, which have been proven to be susceptible to pollution through anthropogenic activities especially taking into consideration the enormous amounts of the ECs detected in aquatic ecosystems that are not regulated. Additionally, no ecological and human risk factors associated with these ECs have been established.

Studies reported in research literature have shown that the occurrence of ECs in wastewater and other water bodies such as pharmaceutical products, nanomaterials, personal care products, pesticides and herbicides among many more are a global threat to human and animal health as well the environment as a whole. The presence and transportation pathways of these ECs into aquatic environments can be explained by various activities such as domestic, hospital, industrial, municipal and agricultural discharges and by the use of conventional treatment methods used in WWTPs which have been proven to be inefficient in complete removal of these ECs. Such pollutants are pervasive in aquatic environments and are seldom detected in drinking water in various countries globally.

In addition, studies reported also identify that the majority of conventional treatment techniques for ECs used in WWTPs are inefficient in lowering the concentrations of ECs in effluent discharge to be below the health and safety required guideline values. Therefore, this highlights the requirement for modification of existing treatment methods and the development of new innovative methods that would help improve removal efficiencies of ECs in WWTPs while also reducing their environmental impact. The presence of ECs at a low concentration within environmental compartments has resulted in adverse effects and this is due to their toxicity in nature along with frequent occurrence which impacts human and animal health, especially the aquatic species. Consequently, newly developed treatment techniques must be incorporated into existing WWTPs and processing systems to try and combat the ECs pandemic.

The major problems associated with ECs are linked to their adverse effects and the health risks they pose to human health, animals and the environment. A few of the publications in the literature associated some of the health effects in humans and animals with the presence of ECs in the environment they occupy, water and food consumed. The presence of these persistent ECs such as endocrine disruptors and pharmaceutical compounds highlights the concerns about the reusability and recyclability of wastewater. The nature, behaviour and complexity of the ECs matrices have been proven to be more likely to pose negative impacts than single contaminants. This highlights the significance of further research studies to evaluate the prospective harmful exposure of humans, animals and the environment to ECs by assessing their toxicity. In addition, the detection, identification and quantification of the transformative by-products formed during the treatment process has proven to be another problem. Moreover, some of these bio-transformed products may be more reactive as compared to parent compounds therefore posing higher toxicity. Proper analytical methods along with newly developed detection techniques are required for the identification, quantification, and removal of EC' and their transformation products. As reported key issues such as the efficiency of treatment methods used in WWTPs and environmental health adversities can be addressed through monitoring the process of WWTPs and affluent discharge. Advances in treatment technologies have proven to be favourable methods for the elimination of ECs from wastewater. Significant research studies have been devoted to developing and advancing treatment methodologies for the removal of ECs in wastewater. Nonetheless, insufficient important knowledge regarding the removal mechanisms of ECs still exists which indicates a major obstacle in ensuring the safety of reused water from WWTPs.

Chromatographic methods such as HPLC and GC are frequently used techniques for detection, identification and quantification of ECs present in the environment especially in wastewater at trace levels. To achieve enhanced sensitivity, accuracy and precision which would help in the quantification of ECs in trace complex analyte sources such as wastewater, various advanced analytical techniques have been developed, tested and optimized to ensure high ECs removal efficiency is achieved. In addition, newly developed fast analysis analytical methods coupled with various sample preparation techniques such as liquid–liquid microextraction and solid phase microextraction have been developed to help reduce sample preparation steps and time while also reducing analysis time, eliminating sample contamination, reducing solvent consumption and costs. The application of these newly developed techniques for the detection and quantification of ECs in wastewater enabled the possibility of toxicological evaluations of ECs. Measures to treat and remove ECs in wastewater are still not yet fully acknowledged, therefore it is important to study and develop new removal techniques. Constructed wetlands have been identified as low-cost systems that are economically viable and researchers have been exploring in removing ECs.

Currently, ECs are not regulated as this is demonstrated by the non-existence of threshold limits discharge limits are set by authorities and governing bodies. Continuous discharge of ECs into the various environmental bodies at trace levels poses a risk to human and animal health, and the environment. Until now, there is a paucity in information concerning the source, transportation, detection, removal, reactivity and adverse effects of ECs and this has contributed heavily to the limitation of wastewater quality legislation towards the discharge of ECs in water bodies. ECs have been demonstrated to react differently in natural environments revealing contrasting ecotoxicological effects that are inadequately assessed by traditional tests in labs. A holistic approach is necessary and must also be applied when environmental assessment studies are conducted, and this should include identifying the impact of ECs within the environment throughout their complete life cycle and their fate.

Interest in recently developed advanced treatment techniques for ECs such as membrane filtration, AOPs, adsorption and application of hybrid systems has demonstrated to be more efficient in the removal of ECs in wastewater. However, various application factors such as performance, costs and reliability from one process to another. Therefore, there is a requirement for assessment of the adverse effects of pollution on the removal efficiency of the technique, stability of the treatment mechanism and microbial organisms used for biological processes. The assessment of these components could help to produce a standard base from which newly developed treatment techniques can be easily integrated into existing treatment methods for the removal of ECs in wastewater and water purification processes.

11.6 FUTURE PERSPECTIVE

When developing new methodologies that can be applied in the detection and monitoring of ECs in water bodies particularly wastewater, sustainability should be prioritized as a necessary characteristic. However, due to new ECs structures and their metabolites in wastewater at trace level, this results in obstacles to developing accurate and quick scientific methodologies. Moreover, additional research is required to refine composition interpretation and the accuracy and sensitivity of the methods. It is critical to upscale the research pilot-plant to scale real WWTPs that consist of various ECs that are different and vary in concentrations to clarify the common ramifications of ECs compounds on their removal rates rather than focusing on experimental work systems that contain one compound used as a model ECs. A huge amount of newly discovered ECs arise from pollutant compounds that are undetected during wastewater treatment due to a lack of knowledge and legislation and therefore are never reported. Therefore, it is imperative to develop new methods for the elimination of ECs and new legislation for ecotoxicity tests and evaluate various health effects by applying different techniques that have appropriate termination points for ECs. More exhaustive research work is required which will fill the knowledge gaps about ECs in conventional WWTPs and newly developed advanced treatment methods. In addition, research work should incorporate emphasis on WWTPs removal techniques and the fate of ECs in waste biomass. Integrating nanotechnology science and engineering techniques, while also improving EC integrity are all crucial components of achieving sufficient elimination of various ECs in wastewater. Newly developed EC removal studies and techniques should be inclusive of parameters that affect efficiency such as operational parameters, EC degradation mechanisms, reaction kinetics and reactor design among others. Evaluation of different treatment methods in real-world environments and operational scenarios should be implemented rather than focusing on laboratory batch studies.

In addition, future investigative research assessment studies should focus on clarification of the predominantly breakdown mechanisms of ECs and the developments of their by-products during their removal and degradation process. The potential toxicity risk of these by-products towards the environment should be assessed as some turned out to be far more toxic in comparison with parent compounds. Furthermore, new and more stringent regulations and policies are required to warrant the protection, health and safety of our environments.

The application of hybrid systems in WWTPs has gained attention from researchers and policymakers due to the high removal efficiency of ECs in conducted field studies, although common biological systems are used mostly in numerous countries globally. The integration of chemical processes such as AOPs with biological systems still requires more in-depth research, and this could enhance the suitability and application of hybrid removal technologies. The economic feasibility of hybrid techniques is a major concern for policymakers, along with the design and operational parameters for these advanced systems.

Despite all the research progress and technological advancement on removal techniques for ECs in wastewater, there are still significant limitations that obstruct the progress. The focal point of future research work should be the following:

- Monitoring of ECs and the transformative by-products because these contaminants are not assessed due to a paucity of knowledge on their toxicity and persistence within the environment, therefore, necessitates the development of new protocols and procedures that will focus on removal and assessment of ECs and their transformative by-products in wastewater.
- Concentration reduction of ECs from point source-point must be accentuated according to stricter legislation, public awareness programmes and commanding limited release of ECs to the environment.
- Explore the application of sustainable 'green' removal techniques such as filtration, adsorption and application of nanotechnology among others at an industrial scale as an efficient and cost-effective alternative for reduction and removal of ECs concentration from various point sources.
- In-depth research is essential for the incorporation of existing studies on wastewater treatment techniques with new innovative physical, chemical and biological strategies such as the application of ultrasound with adsorption and UV radiation for ECs removal.
- Application of effective strategies for selecting suitable treatment methodologies and operation costs. The selection criteria should include various factors such as water quality and source, reliability, removal efficiency of targeted ECs, flexibility, environmental compatibility, maintenance, and operating costs among others.
- Advancement in analytical instrumentation provides a possibility to identify and measure emerging classes of ECs, while new methodologies will help remove these ECs from wastewater and the environment as a whole.

REFERENCES

Abafe O. A., Lawal M. A. and Chokwe T. B. (2023). Non-targeted screening of emerging contaminants in South African surface and wastewater. *Emerging Contaminants*, **9**(4), 100246, https://doi.org/10.1016/j.emcon.2023.100246

Albero B., Tadeo J. L., Escario M., Miguel E. and Pérez R. A. (2018). Persistence and availability of veterinary antibiotics in soil and soil-manure systems. *Science of the Total Environment*, **643**, 1562–1570. https://doi.org/10.1016/j.scitotenv.2018.06.314

Ali A., Khalid Z. and Ajarem J. S. (2023). Wastewater treatment by using microalgae: insights into fate, transport, and associated challenges. *Chemosphere*, **338**, 139501, https://doi.org/10.1016/j.chemosphere.2023.139501

Almazrouei B., Islayem D., Alskafi F., Catacutan M. K., Amna R., Nasrat S., Sizirici B. and Yildiz I. (2023). Steroid hormones in wastewater: sources, treatments, environmental risks, and regulations. *Emerging Contaminants*, **9**(2), 100210, https://doi.org/10.1016/j.emcon.2023.100210

Alyasiri H., Rushdi S. and Al Sharify Z. T. 2023. July. Recent advances in the application of activated carbon for the removal of pharmaceutical contaminants from wastewater: a review. In: AIP Conference Proceedings (Vol. **2787**, No. 1). AIP Publishing, https://doi.org/10.1063/5.0150157

Amalina F., Abd Razak A. S., Krishnan S., Zularisam A. W. and Nasrullah M. (2023). The synthesization of activated carbon from electrocoagulated palm oil mill effluent sludge for wastewater treatment. *Materials Today: Proceedings*, https://doi.org/10.1016/j.matpr.2023.03.514

Angelakis A. N., Tzanakakis V. A., Capodaglio A. G. and Dercas N. (2023). A critical review of water reuse: lessons from prehistoric Greece for present and future challenges. *Water*, **15**(13), 2385, https://doi.org/10.3390/w15132385

Barbosa Jr. F., Rocha B. A., Souza M. C., Bocato M. Z., Azevedo L. F., Adeyemi J. A., Santana A. and Campiglia A. D. (2023). Polycyclic aromatic hydrocarbons (PAHs): updated aspects of their determination, kinetics in the human body, and toxicity. *Journal of Toxicology and Environmental Health, Part B*, **26**(1), 28–65, https://doi.org/10.1080/10937404.2022.2164390

Bellas J. and León V. M. (2023). Future trends and challenges in relation to contaminants of emerging concern. *Contaminants of Emerging Concern in the Marine Environment*, **13**, 465–473, https://doi.org/10.1016/B978-0-323-90297-7.00013-5

Bhattu M. and Singh J. (2023). Recent advances in nanomaterials based sustainable approaches for mitigation of emerging organic pollutants. *Chemosphere*, **321**, 138072, https://doi.org/10.1016/j.chemosphere.2023.138072

Bikiaris N. D., Koumentakou I., Samiotaki C., Meimaroglou D., Varytimidou D., Karatza A., Kalantzis Z., Roussou M., Bikiaris R. D. and Papageorgiou G. Z. (2023). Recent advances in the investigation of poly (lactic acid) (PLA) nanocomposites: incorporation of various nanofillers and their properties and applications. *Polymers*, **15**(5), 1196, https://doi.org/10.3390/polym15051196

Chaturvedi M., Joy S., Gupta R. D., Pandey S. and Sharma S. (2023). Endocrine disrupting chemicals (EDCs): chemical fate, distribution, analytical methods and promising remediation strategies – a critical review. *Environmental Technology Reviews*, **12**(1), 286–315, https://doi.org/10.1080/21622515.2023.2205026

Chung H. S., Choi J. H., Abd El-Aty A. M., Lee Y. J., Lee H. S., Kim S., Jung H. J., Kang T. W., Shin H. C. and Shim J. H. (2016). Simultaneous determination of seven multiclass veterinary antibiotics in surface water samples in the Republic of Korea using liquid chromatography with tandem mass spectrometry. *Journal of Separation Science*, **39**(24), 4688–4699. https://doi.org/10.1002/jssc.201600968

Coccia M. and Bontempi E. (2023). New trajectories of technologies for the removal of pollutants and emerging contaminants in the environment. *Environmental Research*, **229**, 115938, https://doi.org/10.1016/j.envres.2023.115938

Coxon S. and Eaton C. (2023). Review of contaminants of potential human health concern in wastewater and stormwater.

Douna B. K. and Yousefi H. (2023). Removal of PFAS by biological methods. *Asian Pacific Journal of Environment and Cancer*, **6**(1), 53–68, https://doi.org/10.31557/APJEC.2023.6.1.53

Dubey M., Vellanki B. P. and Kazmi A. A. (2023). Removal of emerging contaminants in conventional and advanced biological wastewater treatment plants in India – a comparison of treatment technologies. *Environmental Research*, **218**, 115012, https://doi.org/10.1016/j.envres.2022.115012

du Plessis M., Fourie C., Stone W. and Engelbrecht A. M. (2023). The impact of endocrine disrupting compounds and carcinogens in wastewater: implications for breast cancer. *Biochimie*, **209**, 103–115, https://doi.org/10.1016/j.biochi.2023.02.006

Faisal A. A., Taha D. S., Hassan W. H., Lakhera S. K., Ansar S. and Pradhan S. (2023). Subsurface flow constructed wetlands for treating of simulated cadmium ions-wastewater with presence of *Canna indica* and *Typha domingensis*. *Chemosphere*, **338**, 139469, https://doi.org/10.1016/j.chemosphere.2023.139469

Foglia A., González-Camejo J., Radini S., Sgroi M., Li K., Eusebi A. L. and Fatone F. (2023). Transforming wastewater treatment plants into reclaimed water facilities in water-unbalanced regions. An overview of possibilities and recommendations focusing on the Italian case. *Journal of Cleaner Production*, **410**, 137264, https://doi.org/10.1016/j.jclepro.2023.137264

Folorunsho O., Bogush A. and Kourtchev I. (2023). A new on-line SPE LC-HRMS method for simultaneous analysis of selected emerging contaminants in surface waters. *Analytical Methods*, **15**(3), 284–296, https://doi.org/10.1039/D2AY01574A

Ghasemi F., Fahimi-Kashani N., Bigdeli A., Alshatteri A. H., Abbasi-Moayed S., Al-Jaf S. H., Merry M. Y., Omer K. M. and Hormozi-Nezhad M. R. (2023). Based optical nanosensors – a review. *Analytica Chimica Acta*, **1238**, 340640, https://doi.org/10.1016/j.aca.2022.340640

Ghosh S., Falyouna O., Onyeaka H., Malloum A., Bornman C., AlKafaas S. S., Al-Sharify Z. T., Ahmadi S., Dehghani M. H., Mahvi A. H. and Nasseri S. (2023). Recent progress on the remediation of metronidazole antibiotic as emerging contaminant from water environments using sustainable adsorbents: a review. *Journal of Water Process Engineering*, **51**, 103405, https://doi.org/10.1016/j.jwpe.2022.103405

Ghosha S. and Biswasb A. (2023). Emerging contaminants in. In: Current Developments in Biotechnology and Bioengineering: Bioremediation of Endocrine Disrupting Pollutants in Industrial Wastewater, I. Haq, A. S. Kalamdhad and A. Pandey (eds), Elsevier, Amsterdam, Netherlands, Oxford, UK, Cambridge, US p. 153.

Gkotsis G., Nika M. C., Nikolopoulou V., Alygizakis N., Bizani E., Aalizadeh R., Badry A., Chadwick E., Cincinelli A., Claßen D., Danielsson S., Dekker R., Duke G., Drost W., Glowacka N., Göckener B., Jansman H. A. H., Juergens M., Knopf B., … Thomaidis N. S. (2022). Assessment of contaminants of emerging concern in European apex predators and their prey by LC-QToF MS wide-scope target analysis. *Environment International*, **170**. https://doi.org/10.1016/j.envint.2022.107623

Gomes M. P., Brito J. C. M., Vieira F., Kitamura R. S. A. and Juneau P. (2022). Emerging contaminants in streams of Doce River watershed, Minas Gerais, Brazil. *Frontiers in Environmental Science*, **9**, 1–11, https://doi.org/10.3389/fenvs.2021.801599

Gumbi B. P., Moodley B., Birungi G. and Ndungu P. G. (2022). Risk assessment of personal care products, pharmaceuticals, and stimulants in Mgeni and Msunduzi Rivers, KwaZulu-Natal, South Africa. *Frontiers in Water*, **4**, 867201, https://doi.org/10.3389/frwa.2022.867201

Gunathilaka M. L., Bao S., Liu X., Li Y. and Pan Y. (2023). Antibiotic pollution of planktonic ecosystems: a review focused on community analysis and the causal chain linking individual- and community-level responses. *Environmental Science & Technology*, **57**(3), 1199–1213, https://doi.org/10.1021/acs.est.2c06787

Gondi R., Kavitha S., Yukesh Kannah R., Parthiba Karthikeyan O., Kumar G., Kumar Tyagi V. and Rajesh Banu J. (2022). Algal-based system for removal of emerging pollutants from wastewater: A review. *In Bioresource Technology*, (Vol. 344), Elsevier Ltd., https://doi.org/10.1016/j.biortech.2021.126245

Hanna N., Tamhankar A. J. and Lundborg C. S. (2023). Antibiotic concentrations and antibiotic resistance in aquatic environments of the WHO Western Pacific and South-East Asia regions: a systematic review and probabilistic environmental hazard assessment. *The Lancet Planetary Health*, **7**(1), e45–e54, https://doi.org/10.1016/S2542-5196(22)00254-6

Hawash H. B., Moneer A. A., Galhoum A. A., Elgarahy A. M., Mohamed W. A., Samy M., El-Seedi H. R., Gaballah M. S., Mubarak M. F. and Attia N. F. (2023). Occurrence and spatial distribution of pharmaceuticals and personal care products (PPCPs) in the aquatic environment, their characteristics, and adopted legislations. *Journal of Water Process Engineering*, **52**, 103490, https://doi.org/10.1016/j.jwpe.2023.103490

Hemida M., Ghiasvand A., Macka M., Gupta V., Haddad P. R. and Paull B. (2023). Recent advances in miniaturization of portable liquid chromatography with emphasis on detection. *Journal of Separation Science*, **46**, 2300283, https://doi.org/10.1002/jssc.202300283

Hu Y., Cheng H. and Tao S. (2023). Environmental and human health impacts of geothermal exploitation in China and mitigation strategies. *Critical Reviews in Environmental Science and Technology*, **53**(11), 1173–1196, https://doi.org/10.1080/10643389.2022.2128236

Ieda T. and Hashimoto S. (2023). GC × GC and computational strategies for detecting and analyzing environmental contaminants. *TrAC Trends in Analytical Chemistry*, **165**, 117118, https://doi.org/10.1016/j.trac.2023.117118

Ifon B. E., Adyari B., Hou L., Ohore O. E., Rashid A., Yu C. P. and Anyi H. (2023). Urbanization influenced the interactions between dissolved organic matter and bacterial communities in rivers. *Journal of Environmental Management*, **341**, 117986, https://doi.org/10.1016/j.jenvman.2023.117986

Inarmal N. and Moodley B. (2023). Selected pharmaceutical analysis in a wastewater treatment plant during COVID-19 infection waves in South Africa. *Environmental Science: Water Research and Technology*, **9**, 1566–1576, https://doi.org/10.1039/d3ew00059a

Interdonato L., Siracusa R., Fusco R., Cuzzocrea S. and Di Paola R. (2023). Endocrine disruptor compounds in environment: focus on women's reproductive health and endometriosis. *International Journal of Molecular Sciences*, **24**(6), 5682, https://doi.org/10.3390/ijms24065682

Jaffari Z. H., Jeong H., Shin J., Kwak J., Son C., Lee Y. G., Kim S., Chon K. and Cho K. H. (2023). Machine-learning-based prediction and optimization of emerging contaminants' adsorption capacity on biochar materials. *Chemical Engineering Journal*, **466**(2023), 143073, https://doi.org/10.1016/j.cej.2023.143073

Jatoi A. S., Ahmed J., Akhter F., Sultan S. H., Chandio G. S., Ahmed S., Hashmi Z., Usto M. A., Shaikh M. S., Siddique M. and Maitlo G. (2023). Recent advances and treatment of emerging contaminants through the bio-assisted method: a comprehensive review. *Water, Air, & Soil Pollution*, **234**(1), 49, https://doi.org/10.1007/s11270-022-06037-2

Kadac-Czapska K., Knez E., Gierszewska M., Olewnik-Kruszkowska E. and Grembecka M. (2023). Microplastics derived from food packaging waste – their origin and health risks. *Materials*, **16**(2), 674, https://doi.org/10.3390/ma16020674

Karlsson O., Zheng H., Chen Y., Wang J., Xu P., Xiang J., Xu D., Cheng P., Wang X., Wu L., Zhang N. and Chen Z. (n.d.). Antidepressants as emerging contaminants: occurrence in wastewater treatment plants and surface waters in Hangzhou, China. *Frontiers in Public Health*, **10**, 963257, https://doi.org/10.3389/fpubh.2022.963257

Khurana P., Pulicharla R. and Brar S. K. (2022). Analytical challenges of antibiotic–metal complexes in wastewaters: a mini-review. *Environmental Nanotechnology, Monitoring & Management*, **18**, 100747, https://doi.org/10.1016/j.enmm.2022.100747

Kordbacheh F. and Heidari G. (2023). Water pollutants and approaches for their removal. *Materials Chemistry Horizons*, **2**(2), 139–153, https://doi.org/10.22128/MCH.2023.684.1039

Korzeniowski S. H., Buck R. C., Newkold R. M., Kassmi A. E., Laganis E., Matsuoka Y., Dinelli B., Beauchet S., Adamsky F., Weilandt K. and Soni V. K. (2023). A critical review of the application of polymer of low concern regulatory criteria to fluoropolymers II: fluoroplastics and fluoroelastomers. *Integrated Environmental Assessment and Management*, **19**(2), 326–354, https://doi.org/10.1002/ieam.4646

Krishnan R. Y., Manikandan S., Subbaiya R., Karmegam N., Kim W. and Govarthanan M. (2023). Recent approaches and advanced wastewater treatment technologies for mitigating emerging microplastics contamination – a critical review. *Science of the Total Environment*, **858**, 159681, https://doi.org/10.1016/j.scitotenv.2022.159681

Kumar R., Qureshi M., Vishwakarma D. K., Al-Ansari N., Kuriqi A., Elbeltagi A. and Saraswat A. (2022). A review on emerging water contaminants and the application of sustainable removal technologies. *Case Studies in Chemical and Environmental Engineering*, **6**, 100219, https://doi.org/10.1016/j.cscee.2022.100219

Kumar S., Yadav S., Kataria N., Chauhan A. K., Joshi S., Gupta R., Kumar P., Chong J. W. R., Khoo K. S. and Show P. L. (2023). Recent advancement in nanotechnology for the treatment of pharmaceutical wastewater: sources, toxicity, and remediation technology. *Current Pollution Reports*, **9**, 1–33, https://doi.org/10.1007/s40726-023-00251-0

Kurniawan T. A., Lo W. H., Liang X., Goh H. H., Othman M. H. D., Chong K. K., Mohyuddin A., Kern A. O. and Chew K. W. (2023). Heavy metal removal from aqueous solutions using biomaterials and/or functional composites: recent advances and the way forward in wastewater treatment using digitalization. *Journal of Composites Science*, **7**(2), 84, https://doi.org/10.3390/jcs7020084

Li D., Huang W. and Huang R. (2023). Analysis of environmental pollutants using ion chromatography coupled with mass spectrometry: a review. *Journal of Hazardous Materials*, **458**(2023), 131952, https://doi.org/10.1016/j.jhazmat.2023.131952

Lofrano G., Sacco O., Venditto V., Carotenuto M., Libralato G., Guida M., Meric S. and Vaiano V. (2020). Occurrence and potential risks of emerging contaminants in water. In: Visible Light Active Structured Photocatalysts for the Removal of Emerging Contaminants: Science and Engineering. Elsevier, Amsterdam, Netherlands, Oxford, UK, Cambridge, US pp. 1–25, https://doi.org/10.1016/B978-0-12-818334-2.00001-8

Majumder S., Sharma P., Singh S. P. and Nadda A. K. (2023). Engineered biochar for the effective sorption and remediation of emerging pollutants in the environment. *Journal of Environmental Chemical Engineering*, **11**(2), 109590, https://doi.10.1016/j.jece.2023.109590

Martínez-Huitle C. A., Rodrigo M. A., Sirés I. and Scialdone O. (2023). A critical review on latest innovations and future challenges of electrochemical technology for the abatement of organics in water. *Applied Catalysis B: Environmental*, **11**, 122430, https://doi.org/10.1016/j.jece.2023.109590

Milanović M., Đurić L., Milošević N. and Milić N. (2023). Comprehensive insight into triclosan – from widespread occurrence to health outcomes. *Environmental Science and Pollution Research*, **30**(10), 25119–25140, https://doi.org/10.1007/s11356-021-17273-0

Ng N., Wan Ibrahim W. A. and Sutirman Z. A. (2023). Magnetic nanomaterials for preconcentration and removal of emerging contaminants in the water environment. *Nanotechnology for Environmental Engineering*, **8**, 297–315, https://doi.org/10.1007/s41204-022-00296

Nikolopoulou V., Ajibola A. S., Aalizadeh R. and Thomaidis N. S. (2023). Wide-scope target and suspect screening of emerging contaminants in sewage sludge from Nigerian WWTPs by UPLC-qToF-MS. *Science of the Total Environment*, **857**, 159529, https://doi.org/10.1016/j.scitotenv.2022.159529

Niu L., Liu W., Juhasz A., Chen J. and Ma L. (2022). Emerging contaminants antibiotic resistance genes and microplastics in the environment: introduction to 21 review articles published in CREST during 2018–2022. *Critical Reviews in Environmental Science and Technology*, **52**(23), 4135–4146, https://doi.org/10.1080.10643389.2022.2117847

Ofrydopoulou A., Nannou C., Evgenidou E., Christodoulou A. and Lambropoulou D. (2022). Assessment of a wide array of organic micropollutants of emerging concern in wastewater treatment plants in Greece: occurrence, removals, mass loading and potential risks. *Science of the Total Environment*, **802**, 149860, https://doi.org/10.1016/j.scitotenv.2021.149860

Onnis P., Byrne P., Hudson-Edwards K. A., Frau I., Stott T., Williams T., Edwards P. and Hunt C. O. (2023). Source apportionment of mine contamination across streamflows. *Applied Geochemistry*, **151**, 105623, https://doi.org/10.1016/j.apgeochem.2023.105623

Pal S. K., Masum M. M. H., Salauddin M., Hossen M. A., Ruva I. J. and Akhie A. A. (2023). Appraisal of stormwater-induced runoff quality influenced by site-specific land use patterns in the south-eastern region of Bangladesh. *Environmental Science and Pollution Research*, **30**(13), 36112–36126, https://doi.org/10.1007/s11356-022-24806-8

Parida V. K., Saidulu D., Majumder A., Srivastava A., Gupta B. and Gupta A. K. (2021). Emerging contaminants in wastewater: a critical review on occurrence, existing legislations, risk assessment, and sustainable treatment alternatives. *Journal of Environmental Chemical Engineering*, **9**(5), 105966, https://doi.org/10.1016/j.jece.2021.105966

Puri M., Gandhi K. and Kumar M. S. (2023). Emerging environmental contaminants: a global perspective on policies and regulations. *Journal of Environmental Management*, **332**, 117344, https://doi.org/10.1016/j.jenvman.2023.117344

Ramesh B., Saravanan A., Kumar P. S., Yaashikaa P. R., Thamarai P., Shaji A. and Rangasamy G. (2023). A review on algae biosorption for the removal of hazardous pollutants from wastewater: limiting factors, prospects and recommendations. *Environmental Pollution*, **327**, 121572, https://doi.org/10.1016/j.envpol.2023.121572

Raninga M., Mudgal A., Patel V. K., Patel J. and Sinha M. K. (2023). Modification of activated carbon-based adsorbent for removal of industrial dyes and heavy metals: a review. *Materials Today: Proceedings*, **77**, 286–294, https://doi.org/10.1016/j.matpr.2022.11.358

Rathi B. S. and Kumar P. S. (2021). Application of adsorption process for effective removal of emerging contaminants from water and wastewater. *Environmental Pollution*, **280**, 116995, https://doi.org/10.1016/j.envpol.2021.116995

Reid A. J., Carlson A. K., Creed I. F., Eliason E. J., Gell P. A., Johnson P. T. J., Kidd K. A., MacCormack T. J., Olden J. D., Ormerod S. J., Smol J. P., Taylor W. W., Tockner K., Vermaire J. C., Dudgeon D. and Cooke S. J. (2019). Emerging threats and persistent conservation challenges for freshwater biodiversity. *Biol Rev*, **94**, 849–873, https://doi.org/10.1111/brv.12480

Ren C., Bai R., Chen W., Li J., Zhou X., Tian X. and Zhao F. (2023a). Advances in nanomaterial-microbe coupling system for removal of emerging contaminants. *Chemical Research in Chinese Universities*, **39**, 1–6, https://doi.org/10.1007/s40242-023-3053-x

Ren C., Yu C. W. and Cao S. J. (2023b). Development of urban air environmental control policies and measures. *Indoor and Built Environment*, **32**(2), 299–304, https://doi.org/10.1177/1420326X221120380

Richards L. A., Guo S., Lapworth D. J., White D., Civil W., Wilson G. J. L., Lu C., Kumar A., Ghosh A., Khamis K., Krause S., Polya D. A. and Gooddy D. C. (2023). Emerging organic contaminants in the river Ganga and key tributaries in the middle Gangetic Plain, India: characterization, distribution & controls. *Environmental Pollution*, **327**, 121626, https://doi.org/10.1016/j.envpol.2023.121626

Sathya R., Arasu M. V., Al-Dhabi N. A., Vijayaraghavan P., Ilavenil S. and Rejiniemon T. S. (2023). Towards sustainable wastewater treatment by biological methods – challenges and advantages of recent technologies. *Urban Climate*, **47**, 101378, https://doi.org/10.1016/j.uclim.2022.101378

Sellaoui L., Gómez-Avilés A., Dhaouadi F., Bedia J., Bonilla-Petriciolet A., Rtimi S. and Belver C. (2023). Adsorption of emerging pollutants on lignin-based activated carbon: analysis of adsorption mechanism via characterization, kinetics and equilibrium studies. *Chemical Engineering Journal*, **452**, 139399, https://doi.org/10.1016/j.cej.2022.139399

Selwe K. P., Thorn J. P. R., Desrousseaux A. O. S., Dessent C. E. H. and Sallach J. B. (2022). Emerging contaminant exposure to aquatic systems in the Southern African Development Community. *Environmental Toxicology and Chemistry*, **41**(2), 382–395. https://doi.org/10.1002/etc.5284

Sewwandi M., Wijesekara H., Rajapaksha A. U., Soysa S. and Vithanage M. (2023). Microplastics and plastics-associated contaminants in food and beverages; global trends, concentrations, and human exposure. *Environmental Pollution*, **317**, 120747, https://doi.org/10.1016/j.envpol.2022.120747

Shanmuganathan R., Kadri M. S., Mathimani T., Le Q. H. and Pugazhendhi A. (2023). Recent innovations and challenges in the eradication of emerging contaminants from aquatic systems. *Chemosphere*, **332**, 138812, https://doi.org/10.1016/j.chemosphere.2023.138812

Shehata N., Egirani D., Olabi A. G., Inayat A., Abdelkareem M. A., Chae K. J. and Sayed E. T. (2023). Membrane-based water and wastewater treatment technologies: issues, current trends, challenges, and role in achieving sustainable development goals, and circular economy. *Chemosphere*, **320**, 137993, https://doi.org/10.1016/j.chemosphere.2023.137993

Thacharodi A., Hassan S., Hegde T. A., Thacharodi D. D., Brindhadevi K. and Pugazhendhi A. (2023). Water a major source of endocrine-disrupting chemicals: an overview on the occurrence, implications on human health and bioremediation strategies. *Environmental Research*, **231**, 116097, https://doi.org/10.1016/j.envres.2023.116097

Tholozan L. V., Valério Filho A., Maron G. K., Carreno N. L. V., da Rocha C. M., Bordin J. and da Rosa G. S. (2023). Sphagnum perichaetiale Hampe biomass as a novel, green, and low-cost biosorbent in the adsorption of toxic crystal violet dye. *Environmental Science and Pollution Research*, **30**(18), 52472–52484, https://doi.org/10.1007/s11356-023-26068-4

Ullah H., Lun L., Rashid A., Zada N., Chen B., Shahab A., Li P., Ali M. U., Lin S. and Wong M. H. (2023). A critical analysis of sources, pollution, and remediation of selenium, an emerging contaminant. *Environmental Geochemistry and Health*, **45**, 1359–1389, https://doi.org/10.1007/s10653-022-01354-1

Ulucan-Altuntas K., Manav-Demir N., Ilhan F., Gelgor H. B., Huddersman K., Tiwary A. and Debik E. (2023). Emerging pollutants removal in full-scale biological treatment plants: a case study. *Journal of Water Process Engineering*, **51**, 103336, https://doi.org/10.1016/j.jwpe.2022.103336

van den Hurk R. S., Pursch M., Stoll D. R. and Pirok B. W. (2023). Recent trends in two-dimensional liquid chromatography. *TrAC Trends in Analytical Chemistry*, **166**, 117166, https://doi.org/10.1016/j.trac.2023.117166

Zambrano J., García-Encina P. A., Hernández F., Botero-Coy A. M., Jiménez J. J. and Irusta-Mata R. (2023). Kinetics of the removal mechanisms of veterinary antibiotics in synthetic wastewater using microalgae–bacteria consortia. *Environmental Technology & Innovation*, **29**, 103031, https://doi.org/10.1016/j.eti.2023.10303

Zhou Q., Yu C., Meng L., Ji W., Liu S., Pan C., Lan T., Wang L. and Qu B. (2023). Research progress of applications for nano-materials in improved QuEChERS method. *Critical Reviews in Food Science and Nutrition*, 1–20, https://doi.org/10.1080/10408398.2023.2225613

doi: 10.2166/9781789063752_0227

Chapter 12

Abatement of pharmaceutical compounds in wastewater using green nanomaterials: an eco-friendly alternative to conventional nanomaterials

*Akshay Botle[1], Sayli Salgaonkar[1], Gayatri Barabde[1,2] and Mihir Herlekar[1]**

[1]Department of Environmental Science, The Institute of Science, Dr. Homi Bhabha State University, Mumbai 400032, India
[2]Department of Analytical Chemistry, The Institute of Science, Dr. Homi Bhabha State University, Mumbai 400032, India
**Corresponding author: tevs.mihir@iscm.ac.in; mihirherlekar1@gmail.com*

ABSTRACT

Pharmaceuticals and their remnants have been acknowledged for their ability to save many lives, but they have also developed a new set of emerging pollutants due to the difficulty in treating them in wastewater worldwide. Increased consumption of drugs has led to adverse impacts on aquatic ecosystems. Even at low levels, these contaminants cause various problems because of their persistent nature and long-lasting negative effects. Therefore, various conventional methods such as activated sludge process, chemical precipitation, membrane filtration, ozonation, adsorption, and photocatalysis have been proposed for their removal. These are limited by high costs, inefficient removal, the production of toxic materials, and the need for significant investment. Nanotechnology has begun to explore various effective strategies for treating wastewater with the help of various nanomaterials. Nanomaterials have been inspected for their potential to eradicate water impurities and improve the effectiveness of conventional technologies. However, the conventional methods of producing nanomaterials involve the usage of hazardous and toxic substances, which create additional pollution. Green nanomaterials present plenty of promising avenues for wastewater treatment and have been recognized to be efficient in providing clean and affordable removal of pharmaceuticals, with features such as increased surface area, higher reactivity, target specificity, low energy and cost consumption, sustainability, improved physical and chemical properties, and effective regeneration. This has led to the development of innovative trends for creating novel, environmentally friendly nanomaterials for the removal and degradation of pharmaceutical substances. This paper focuses on these new trends in the development of greener nanomaterials globally and evaluates their performance for the abatement of pharmaceuticals from wastewater. The paper concludes with the beneficial aspects of green nanomaterials over conventional technologies and the future scope of research.

Keywords: wastewater treatment, pharmaceutical compounds, conventional technologies, nanotechnology, green nanomaterials

12.1 INTRODUCTION TO EMERGING CONTAMINANTS IN WASTEWATER

12.1.1 Background and significance of the topic

Life originates through water on Earth where there are 2.5% freshwater resources, of which only 1% is accessible. Of the 1% of available freshwater found in the form of rivers, ponds, lakes, and so on, 30% is underground water and 69% is trapped in glaciers and ice caps. Considering the growing demand, approximately 80 million people per year add to global freshwater consumption, leading to an increase of 64 billion cubic meters per year (Elgarahy *et al.*, 2021).

In many industrial sectors, wastewater discharges (effluents) enter the aquatic environment directly or indirectly. The annual release of water pollutants is estimated to be about 300–400 million tons (Elgarahy *et al.*, 2021). Thus, the unaltered release of several newly identified compounds in the aquatic environment, regardless of their origin, has become a significant concern worldwide, where typical concentration ranges of these organic contaminants are in parts per billion to parts per trillion. Such composites are categorized as 'emerging contaminants (ECs).' In recent years, scientists, engineers, and the public have been worried about 'emerging pollutants' and their negative impacts on the living ecosystem (Rout *et al.*, 2021). The United States Department of Defense and the United States Environmental Protection Agency classify them as potential, likely, or definite health and environmental concerns (USDoD, 2011; US EPA, 2012). In water bodies, there are several types of emerging pollutants: pharmaceuticals, surfactants, endocrine disruptive compounds, plasticizers (Kumar *et al.*, 2022), and others. All these are illustrated in Figure 12.1.

Pharmaceuticals are probably the most concerning emerging contaminants in modern use (Fernández-López *et al.*, 2016). It is estimated that around 200,000 tons of antibiotics and other pharmaceutical compounds are produced annually for veterinary and human usage (Khalil *et al.*, 2021). Antibiotics are extensively practiced in the healthcare division as they possess powerful antimicrobial and pathogenic actions (Varma *et al.*, 2020). Some of the most regularly found pharmaceutical-based pollutants are carbamazepine, propranolol, tetracycline, phenytoin, ibuprofen, estradiol, X-ray contrast, and fenofibric acid. Many of these pharmaceutical compounds found in wastewater treatment plants (WWTPs) with their uses and disastrous impacts are mentioned in Table 12.1. Water contaminated with these contaminants is hazardous to humans and the ecosystem. Moreover, because of their chronic properties, they are difficult to deteriorate (Aguilar-Pérez *et al.*,

Figure 12.1 Different types of emerging pollutants.

Table 12.1 Common types of pharmaceuticals found in WWTPs, their uses, and their side effects.

Sr. No.	Pharmaceutical Compound	Structure	Uses	Side Effects
1	Amphetamine	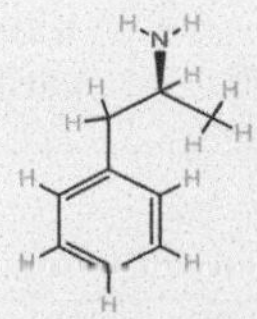	Positive mood, recreational drug	Insomnia, agitation, psychotic symptoms
2	Atenolol	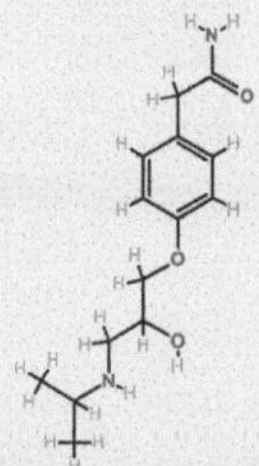	Angina pectoris, acute myocardial infarction, hypertension	Headache, confusion, diarrhea, heart failure, constipation
3	Azithromycin	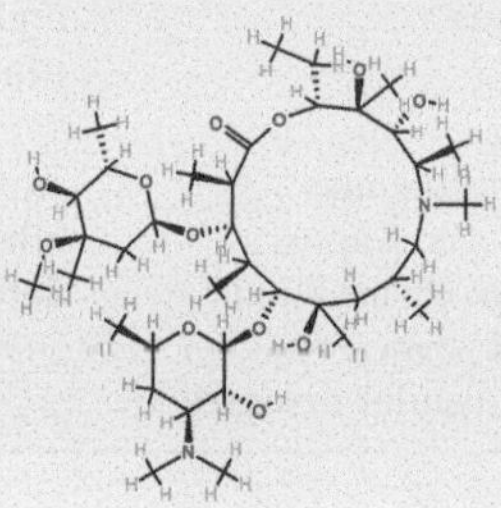	Enteric and urinary tract infections, respiratory tract infections	Dizziness, gastrointestinal upset
4	Ciprofloxacin	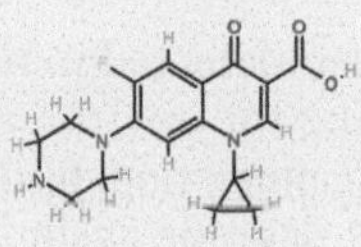	Sexually transmitted diseases, prostatitis, biliary tract infections	Vomiting, diarrhea, nausea
5	Diclofenac	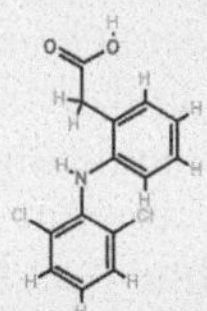	Rheumatic problems, acute joint, and mild-to-moderate pain treatment.	Gastrointestinal disorders, aplastic anemia, disturbed renal function
6	Erythromycin	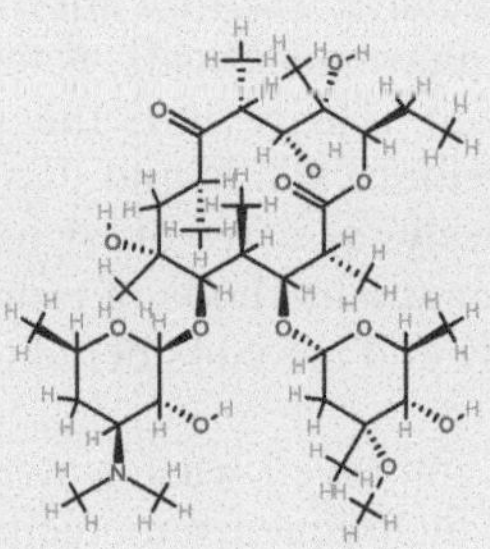	Skin infections, rheumatic fever, syphilis, intestinal amebiasis	Abdominal pain, diarrhea, rash, allergic reaction

(Continued)

Table 12.1 Common types of pharmaceuticals found in WWTPs, their uses, and their side effects (*Continued*).

Sr. No.	Pharmaceutical Compound	Structure	Uses	Side Effects
7	Ibuprofen	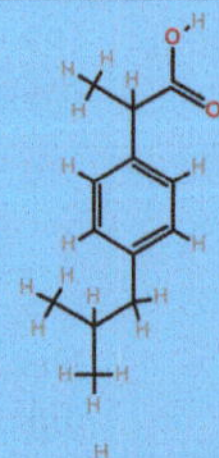	Gout, dysmenorrhea, arthritis, osteoarthritis	Kidney diseases, cardiovascular risks
8	Metronidazole	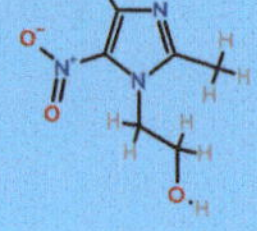	Trichomoniasis treatment, liver abscess	Neurotoxicity, optic neuropathy, peripheral neuropathy, encephalopathy
9	Norfloxacin	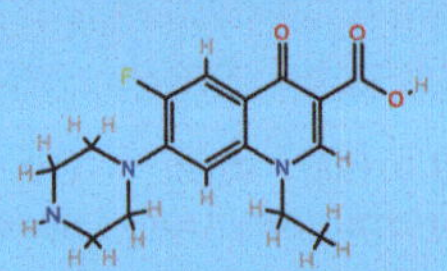	Respiratory, urinary, and gastrointestinal tract infections	Hallucinations, insomnia, central nervous system, and gastrointestinal tract illness
10	Tetracycline	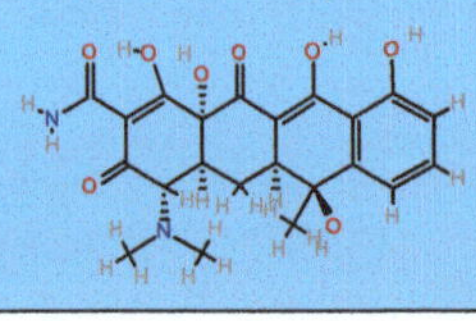	Acne, cutaneous sarcoidosis, Kaposi's sarcoma, rheumatoid arthritis, cancer, cardiovascular diseases	Antibiotic resistance, allergic reactions, liver, and dental damage

2020). Thus, a rapid increase in antibiotic consumption has been detected in nations including South Africa, Russia, China, Brazil, and India. In addition to polluting the environment, antibiotic use has led to the development of resistant microbes to such drugs. Also, many nonsteroidal anti-inflammatory medicines (NSAIDs) are extremely soluble in water and do not degrade easily; these drugs have been testified to have an adverse impact on aquatic bodies and humans (Varma *et al.*, 2020).

A large proportion of pharmaceuticals get expelled with feces and urine, entering municipal WWTPs. During sewage treatment, few composites are removed chemically or biologically while the rest are adsorbed in solid phases and degraded, but a significant amount of micropollutants are not eradicated by traditional treatment processes and further treatment is necessary to avert the discharge of these substances into the environment. Nowadays, a significant route of pharmaceuticals in habitats is urban wastewater. Clearance of unused pharmaceuticals, veterinary drugs, and feed additives directly into domestic waste can add to their release into the ecosystem (Gracia-Lor *et al.*, 2012). In most cases, WWTPs are designed to remove solids and dissolved organics but are unable to eradicate pharmaceutical compounds effectively; as a result, significant residues remain in WWTPs (Ramírez-Morales *et al.*, 2020). At present, many methods of remediating pollutants are costly, nonreplenishable, and environmentally unfriendly and are considered to be causing derivative pollutants (Azeez *et al.*, 2022).

Advances in nanoscience and nanotechnology have improved wastewater treatment procedures in recent decades. Compared with traditional treatment techniques, nanotechnology-based pathways are more effective (Malik *et al.*, 2022). Several fields of research now incorporate nanotechnology as a front-runner in innovation, especially those that involve creating and modifying bulk materials into nanoscale (Azeez *et al.*, 2022). Using nanostructured materials to scavenge and eliminate hazardous water pollutants is becoming more and more crucial because of their unique properties such as unique size, enhanced surface properties, compressibility, exceptional constancy, less intraparticle diffusion distances, recycling, and remarkable reusability. The effectiveness of nanoparticles (NPs) to remove pollutants can be increased through fabrication and functionalization. In addition, the use of flammable substances and harsh compounds for nanomaterial extraction and subsequent processes poses detrimental concerns. Thus, in light of environmental stewardship and sustainability concerns, researchers have been inspired to develop environmentally friendly approaches to synthesizing highly efficient NPs to treat and remove an array of contaminants from the environment (Gautam *et al.*, 2019).

12.1.2 Objectives of the study

This chapter aims to understand the emerging contaminants that are introduced into the environment, primarily focusing on pharmaceutical compounds and their introduction into wastewater. Furthermore, it discusses the sources, types, and toxicology of these compounds, their influence on human health and the environment, and the conventional approaches to treating these pharmaceutical compounds in WWTPs. This chapter also provides information on nanomaterials and green nanomaterials used for treating wastewater, focusing on the use of green nanomaterials for treating pharmaceutical waste in wastewater.

12.2 PHARMACEUTICAL COMPOUNDS IN WASTEWATER

12.2.1 Sources, composition, types, and toxicology of pharmaceutical compounds in wastewater

Pharmaceutical products are viewed as an emerging source of pollution by the scientific community as they have attracted worldwide attention when their production and use have increased in recent decades. There are about 3000 permitted pharmaceutically active compounds (PhACs) for human medicines in the European Union. However, global environmental researchers are studying their probable effects on the environment as they are less understood (Kermia *et al.*, 2016).

In addition to human excretion, unutilized drugs can be flushed directly in toilets, which makes WWTPs the primary source of these contaminants in aquatic resources (Pereira *et al.*, 2020). The main sources of pharmaceuticals entering various environmental systems are WWTPs, sites of manufacturing activities, sewage treatment plants, individual households, large farms, and landfills (Al-Baldawi *et al.*, 2021). Global data from 71 countries were evaluated from the year 1996 to 2020 and adopted from the German Environment Agency–Umweltbundesamt (2023) receiving pharmaceutical contaminants in WWTPs has been represented in Figure 12.2. Nevertheless, pharmaceuticals are being used and dispensed by personnel, clinics, and pharmaceutical and agrarian industries, leading to more continuous environmental entry (Agunbiade & Moodley, 2016). Insignificant bulk drug manufacturing sites, where wastewater is released, are potential sources of pharmaceuticals (Pereira *et al.*, 2020). Disposing of outdated medications in garbage, drains, and lavatories is another route. Medical antibiotics and synthesized hormones used to manage the development and proliferation of fish breeding and animal rearing are also key contributors to veterinarian medications. Moreover, routine washing releases pharmaceuticals straight into sewage systems (Al-Baldawi *et al.*, 2021). Surface and wastewater include sulfonamides, macrolides, and fluoroquinolone antibiotics. Therapeutic hormones, synthetic versions of plant/animal hormones, alter the endocrine mechanism and health of animals and humans. Analgesics are often used for pain and inflammation. Analgesics, notably meprobamate, ibuprofen, acetaminophen, diclofenac, and naproxen, survive in ground and surface water, and thus, are considered serious environmental contaminants (Tiwari *et al.*, 2017). Various sources of pharmaceuticals received in WWTPs are illustrated in Figure 12.3.

Effluents from these substances that reach WWTPs are classified as hazardous, nonhazardous, or chemo waste. There are two sorts of hazardous waste: designated wastes and distinctive wastes. Pharmaceuticals are further classified into 'P' or 'U.' These wastes are regulated due to their corrosivity, sensitivity, flammability, and toxicity (Gupta *et al.*, 2019). Thus, management of such pharmaceutical

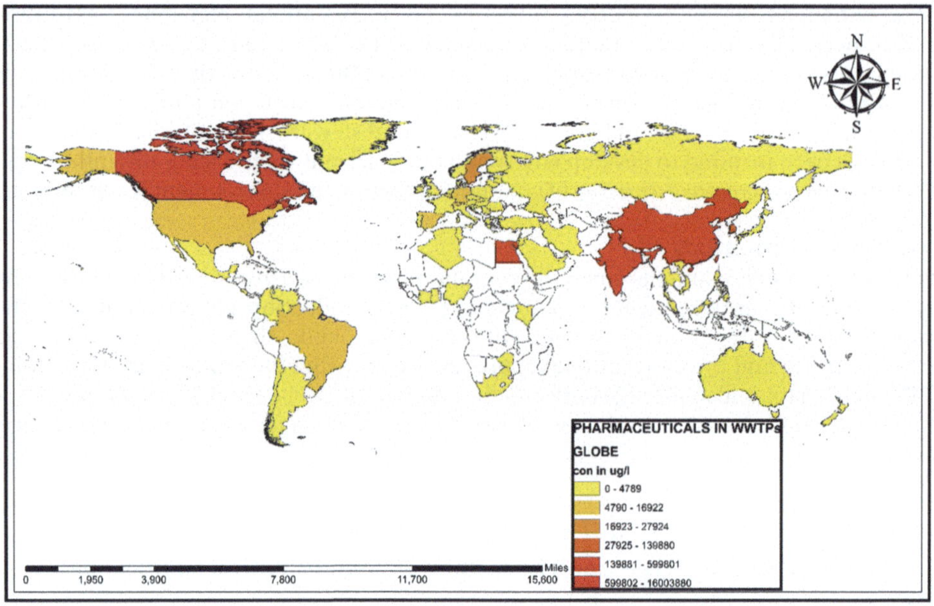

Figure 12.2 Pharmaceuticals detected in various WWTPs globally from 1996 to 2020.

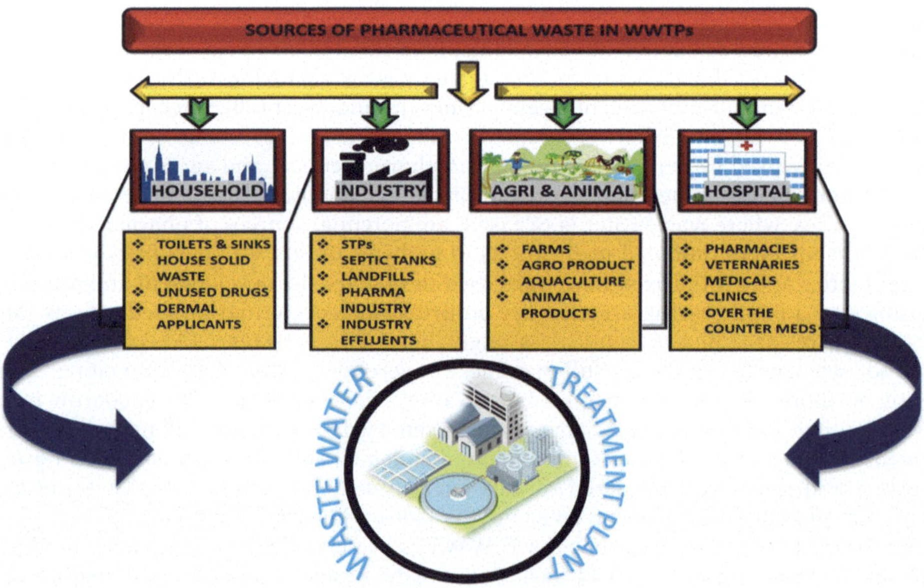

Figure 12.3 Sources of pharmaceuticals in WWTPs.

waste derivatives is necessary as they may cause detrimental toxic effects; well-known examples are the feminization of male fish, the prevention of crustacean molting, and changes in fish behavior (Huerta *et al.*, 2016).

12.2.2 Impact of pharmaceutical compounds on human health and the environment

Pharmaceuticals may mix and interact in ecosystems, even though most medications are in low quantities and may not cause damage as separate substances. Pharmaceutical combinations are more poisonous and ecologically harmful than single substances. Bacteria infect 2 million people annually, 50–70% of which bacteria are antibiotic resistant, out of which antibiotic failure causes 14,000 deaths (Zhang *et al.*, 2020a).

Pharmaceuticals differ from other pollutants. There are 4000 veterinary and human medications worldwide, 600 of which have propagated to terrestrial and aquatic ecosystems globally. Pharmaceuticals may bioaccumulate via trophic chains after entering animals via gills, cuticles, and epidermis. Plants support food webs and provide net primary production (Néstor & Mariana, 2019). Riaz *et al.* (2017) tested levofloxacin, enrofloxacin, ciprofloxacin, and their combination on *Triticum aestivum* (edible wheat). The results indicated that antibiotic-treated seedlings have smaller roots and branches. Islas-Flores *et al.* (2017) tested the lethality of diclofenac and ibuprofen on *Cyprinus carpio*, a major farmed teleost fish. Due to its resistance and ease of care, the fish is often used as a bioindicator in aquatic habitats.

Pharmaceuticals that stay intact after the process develop resistance to deterioration in the environment. Pharmaceuticals may have solitary, synergistic, or antagonistic effects, including cancerous or teratogenic, endocrine-disrupting, and antibiotic resistance effects, as well as long-term harm to living beings (Al-Baldawi *et al.*, 2021). Research discoveries considering the harmful effects of PhACs and endocrine disruptors have resulted in certain efforts at legislation in the Union of Europe (Huerta *et al.*, 2016), including diclofenac or the artificial EE2 hormone, which are added to the list of prioritized drugs by 'Water Framework Directive' for 'the specific objective of assisting in the identification of suitable procedures to confront the threat posed by such drugs' (European Commission, 2013). The US Drinking Water Contaminant Candidate List includes antibiotics and hormones as PhACs and EDCs (Environmental Protection Agency U.S., 2012). The Global Water Research Coalition considers atenolol, bezafibrate, diclofenac, erythromycin, carbamazepine, ibuprofen, naproxen, gemfibrozil, and sulfamethoxazole as paramount drugs to the water cycle (Global Water Research Coalition, 2008). Thus, due to their health impacts, including nerve and reproductive toxicity, and intrusion in metabolites, these pollutants have developed a major concern for drinking water security (Xu *et al.*, 2019).

12.2.3 Conventional methods for treating pharmaceutical compounds in wastewater

Traditional WWTP procedures remove macropollutants such as suspended particles, organic pollutants, and pathogens but not micropollutants such as stubborn pharmaceutically active chemicals (Rout *et al.*, 2021). Pharmaceutical residues have been altered by chemical, physical, and biological methods for years (Ahmed *et al.*, 2017). In coagulation, chemical agents are quickly mixed into wastewater to disperse pollutants and transform persistent pollutants into completely unstable and precipitable particulates (Thapa *et al.*, 2022). Coagulation and flocculation remove turbidity and organic materials from wastewater. Hydrolytic aluminum and iron salt coagulates are the most common (Zinicovscaia, 2016). Secondary effluents may be cleaned by sedimentation and flotation. Injecting air into wastewater creates numerous small bubbles, creating floated floc with a lesser density than wastewater (Thapa *et al.*, 2022). Adsorption moves compounds from aqueous to nonaqueous (solid phase-adsorbent) phase (Ghazal *et al.*, 2022). Physical and chemical adsorption are major forms of adsorption. Physical adsorption is reversible and nonselective. Desorbing adsorbate-saturated on activated carbon (AC) is easy (Thapa *et al.*, 2022). In ozonation, ozone is added to water by bubbling it in a tank via a nozzle. Ozone directly or indirectly reacts via radical reactions to oxidize toxic compounds (Ghazal *et al.*, 2022). Catalytic ozonation degrades wastewater organic pollutants. Advanced catalytic ozonation may improve ozone utilization and organic pollutant mineralization (Thapa *et al.*, 2022). Most chemical oxidation procedures efficiently degrade pollutants in wastewater systems to biodegradable and less hazardous chemicals. Sometimes less reactive species such as chlorine and bromine are used in WWTPs (Ahmed *et al.*, 2017).

Pharmaceutical wastewater is treated biologically, and the treatment techniques are aerobic and anaerobic. Microorganisms such as bacteria, microalgae, and fungi biodegrade big molecular compound pollutants into smaller compounds and even biomineralize compounds into water and carbon dioxide. Wastewater treatment systems remove ECs by biodegradation. Membrane bioreactor, sequence batch reactor, and activated sludge are aerobic techniques. Anaerobic techniques include film reactors, anaerobic digestors, upflow anaerobic sludge blankets, anaerobic filters, and anaerobic baffled reactors. Although the fraction of contaminants removed by chemical precipitation, primary settling, sludge absorption, and aerating volatilization, is modest, the bulk of pollutants in wastewater is eliminated by biological degradation. Using a biological floc with air, this activated sludge procedure treats sewage and industrial wastewater with bacteria and protozoa. These bacteria are capable of decomposing organic materials into water, carbon dioxide, and other inorganic chemicals. It is less expensive to build than advanced oxidation technologies and is more ecologically favorable than chlorination (Ahmed *et al.*, 2017).

12.2.4 Limitations of conventional methods

The primary drawback of conventional methods is the chemical sludge production during coagulation. Aluminum-based coagulants can increase residual aluminum in filtered water (Zinicovscaia, 2016). Temperature, pH, coagulant type, amount, and so on, affect coagulation effectiveness (Gogoi *et al.*, 2018). AC is costly to produce and regenerate, and its frequent regeneration, replacement, and discharge are environmental concerns. Nevertheless, the significant amount of organic carbon and other oxidizable chemicals makes ozonation in WWTPs difficult to deploy as more ozone is needed to fully treat ordinary sewage. To run an ozonation system, a pharmaceutical facility must purchase a power supply and transformers. Ozone generators may also minimize residual ozone and boost energy efficiency by operating intermittently. Intermittent ozonation may limit pharmacological component elimination. Pharmaceutical wastewater ozonation produces byproducts that are often more hazardous than the original substances (Ghazal *et al.*, 2022). Free chlorine seems to progressively degrade certain antibiotics, whereas others seem to be more resistant (Gogoi *et al.*, 2018). Chlorination treatment produces hazardous byproducts because chlorine interacts with organic substances. As most highly polar pharmaceuticals cannot be digested by organisms as a source of carbon and may even hinder their function, activated sludge process-based treatment plants with reduced retention times rarely eliminate these compounds (Ahmed *et al.*, 2017). Similarly, the major drawbacks of using anaerobic methods are that they are still uneconomical due to limited flow, membrane fouling, expensive capital, and operating expenses (Chernicharo *et al.*, 2015).

12.3 NANOMATERIALS FOR WASTEWATER TREATMENT

Micro-engineering, or nanotechnology, manipulates particles under 100 nm. Nanotechnology is sometimes called the 'Next Industrial Revolution' because it will cut power usage, contamination, and manufacturing costs in industrialized nations. Nanotechnology's capacity to reduce pollutants may lead to the most dramatic breakthroughs in environmental protection (Sinha *et al.*, 2020). Adsorption, photocatalysis, advanced oxidation processes (AOPs), and filtering may eradicate organic pollutants and pharmacologically active substances using nanomaterials (Nasrollahzadeh *et al.*, 2021).

Treatment of wastewater and cleanup may benefit from nanomaterials' mechanical qualities, cost-effectiveness, chemical reactivity, large surface area, and energy efficiency. These well-defined materials and controlled nanostructures of adequate porosity and size may be effective adsorbents (Nasrollahzadeh *et al.*, 2021). This section gives information about nanomaterials comprising different

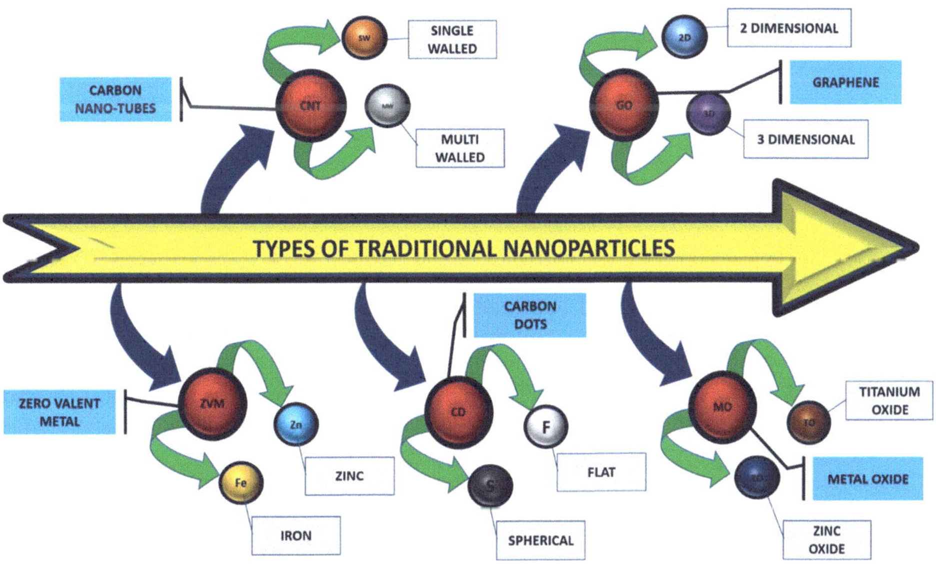

Figure 12.4 Types of traditional nanoparticles.

types, their applications in wastewater treatments, and their advantages over the conventional process currently being used for the remediation of pharmaceuticals. Thus, numerous categories of nanomaterials used in treating pharmaceutical wastes are represented in Figure 12.4.

12.3.1 Types of nanomaterials

12.3.1.1 Carbon nanotubes

Carbon nanostructures are extensively used as wastewater nano adsorbents because of their abundance, cost effectiveness, strong thermal and chemical reliabilities, large active surface areas, outstanding adsorption capabilities, and nontoxicity (Jain *et al.*, 2021). Carbon nanotubes (CNTs) are excellent adsorbents with well-defined tube-shaped geometries, enhanced physicochemical couplings, elevated aspect fractions, greater superficial area, significant sorptive capacities, hydrophobic sides, and readily changeable facades (Nasrollahzadeh *et al.*, 2021). These CNTs outperform other adsorbents due to their customizable surface chemistry, chemical inertness, hollowed morphology, greater superficial area, lighter density, high porosity, and robust engagement with pollutants (Jain *et al.*, 2021).

These qualities make them ideal for wastewater treatment (Jain *et al.*, 2021). CNTs are allotropes of carbon and may be made from a solitary graphene sheet with a roll-up or from numerous graphene sheets wrapped up. CNTs outperform activated carbon in sorption due to their elevated aspect fractions and regulated arrangement of pore size (Cerro-Lopez & Méndez-Rojas, 2019). Kariim *et al.* (2020) produced multiwalled CNT adsorbent with nickel-ferrites and activated carbonized from wood sawdust to sorb levofloxacin and metronidazole from pharmaceutical effluent. The generated multiwalled CNTs have strong levofloxacin and metronidazole adsorption capabilities.

12.3.1.2 Graphene

Graphene represents one of the most successful upcoming filtering membranes owing to its excellent selectivity, permeability, cost efficiency, and surface area. Graphene's simple form and features make it a popular wastewater treatment pollutant removal material (Ahmed *et al.*, 2022). Graphene oxide (GO) is a monomolecular graphite film comprising hydroxyl, carboxyl, carbonyl, and epoxide sets (Jain *et al.*, 2021). Despite being costly and difficult in manufacturing, GO is a superior option to graphene because of its chemical characteristics, which enable it to behave as a barrier with prolonged durability, which is necessary for a broader variety of processes needing chemical and mechanical capabilities. Moreover, GO's efficacy for treating wastewater has already been proved where 90% of ibuprofen was eradicated in just 180 min using N-dropped 3D Graphene aerogel (Ponnusamy *et al.*, 2021). Antibiotics may also be successfully removed from wastewater using graphene. It is possible to use graphene to treat wastewater despite the fact that there are several underlying challenges (Ahmed *et al.*, 2022).

12.3.1.3 Carbon and graphene dots

Carbon quantum dots (Cdots) are sub-ten-nanometer nanoparticles (NPs) having flat or quasi-spherical forms, sp2/sp3 hybridized, consisting of heterogeneous functional groups on the surface. Graphene quantum dots have (<10 nm) dimensions much bigger than their height (<2 nm) (Nasrollahzadeh *et al.*, 2021). Cdots can detect heavy metals, remove inorganic and organic contaminants, and photocatalyze contaminants owing to photo-luminescence, semiconductor, antibacterial, and photo-induced electron transfer properties. Due to their many functional groups and polar moieties, Cdots may remove hazardous chemicals from wastewater (Ahmed *et al.*, 2022).

Two-dimensional graphene is converted into zero-dimensional graphene quantum dots (GQDs). GQDs vary from carbon dots as they incorporate graphene lattices less than 100 nm in size and 10 layers thick. GQDs dissolve faster than CNTs. GQDs' broad edge effect may be adjusted by functional groups, unlike CNTs' one-dimensionality (Tian *et al.*, 2018).

12.3.1.4 Zero-valent metals nanoparticles

12.3.1.4.1 Silver nanoparticles

Silver NPs (AgNPs) can cure wastewater because of their high adsorption capacity, antibacterial characteristics, and sustainable synthesis (Ahmed *et al.*, 2022). Antimicrobial AgNPs disinfect as they penetrate microorganisms by changing their membrane structure. It generates cell-damaging free radicals (Kumar *et al.*, 2021). Junejo *et al.* (2014) showed amoxicillin-derived AgNPs with catalytic properties for the degradation of doxycycline, cefditoren, cefixime, ceftriaxone, sodium, and cefdinir in wastewater.

12.3.1.4.2 Iron nanomaterials

Iron NPs adsorb well. They oxidize and precipitate well but reduce poorly (Kumar *et al.*, 2021). Bacteria show that zero-valent iron particles are antimicrobial. *Staphylococcus aureus, Bacillus subtilus, Pseudomonas aeruginosa*, and *Escherichia coli* were tested for antiseptic activity (Sadek *et al.*, 2021). Nanoscale zero-valent iron particles inhibited bacterial development because Fe^{2+} and Fe^{3+} ions damage bacteria's cell membranes. Zero-valent iron NPs' growth inhibition boosts their wastewater treatment capability (Ahmed *et al.*, 2022). Zero-valent iron immobilized in chitosan NPs removed bisphenol A from pharmaceutical WWTPs (Dehghani *et al.*, 2020).

12.3.1.4.3 Zinc nanoparticles

Zero-valent zinc (ZVZ) has better biodegradability than zero-valent iron; however, wastewater treatment data are scarce (Ahmed *et al.*, 2022). Zinc has a higher characteristic reduction potential than iron, thus it removes the pollutant faster (Bokare *et al.*, 2013). ZVZ only abates halogenated organic molecules such as carbon tetrachloride (Lu *et al.*, 2016). ZVZ's excellent reductibility and possible integration with other chemicals may improve its removal efficiency (Ahmed *et al.*, 2022).

12.3.1.5 Metal oxide nanoparticles

12.3.1.5.1 Titanium oxides

Titanium oxide (TiO_2) does have a greater bandgap. Consequently, ultraviolet (UV) radiations produce hydroxyl radicals to ignite titanium oxide NPs. As these hydroxyl radicals injure cells, they impair bacteria, fungi, and algae cell structure and function (Kumar *et al.*, 2021). Das *et al.* (2014) found 68.14% chlorhexidine breakdown in a controlled slurry batch reactor containing TiO_2 NPs.

12.3.1.5.2 Iron oxides

Magnetite and maghemite iron oxides have magnetic characteristics that may be used to separate NPs (Kumar *et al.*, 2021). Magnetic fields may separate absorbent iron NPs. It uses no extra energy, thus rendering it energy efficient (Ahmed *et al.*, 2022). Yoon *et al.* (2017) used methacrylic acid-coated magnetite NPs to remove carbamazepine and diatrizoate from synthetic wastewater. The amount of carbamazepine removed was 68.3%, and 61.91% of diatrizoate was removed.

12.3.2 Applications of nanomaterials in wastewater treatment

Nanomaterials are attractive owing to their distinctive chemical and physical characteristics. Pharmaceutical remediation in wastewater treatment using NPs is recent research. Pharmaceuticals in wastewater pose a consequential threat to the ecosystem and human health. NPs, nanotubes, oxides, and so on remove pharmaceuticals from wastewater. These materials may selectively adsorb or break down pharmaceuticals, making wastewater pharmaceutical removal efficient and effective. Various applications of using nanomaterials have been discussed in Table 12.2.

12.3.3 Advantages and disadvantages of nanomaterials

Nanomaterials have received a lot of interest during past decades owing to their practical benefits in various disciplines. These qualities have resulted in the creation of new and better solutions. They offer various benefits, but they also have certain problems that are mentioned in Table 12.3.

Table 12.2 Applications of nanomaterials in wastewater treatment.

Sr. No.	Nanomaterial Used	Application	References
1	Magnetic MWCNTs	Used for remediation of tetracycline	Zhao *et al.* (2021)
2	TiO_2 nanoparticles	Used for remediation of meloxicam	Nadim *et al.* (2015)
3	Silver-modified TiO_2 (Ag/TiO_2) nanoparticles	Used for remediation of chloramphenicol and tartrazine	Nino-Martinez *et al.* (2008)
4	$Sn/Zn/TiO_2$ photocatalyst	Used for remediation of amoxicillin trihydrate	Mohammadi *et al.* (2015)
5	$MWCNT/TiO_2/SiO_2$ composite	Used for remediation of carbamazepine and bisphenol A	Czech and Buda (2015)
6	Multiwalled carbon nanotube	Used for remediation of carbamazepine	Ding *et al.* (2019)

Table 12.3 Advantages and disadvantages of nanomaterials.

Advantages	Disadvantages
• Advanced efficacy • Enhanced kinetics • Specific affinities for a particular contaminant • Improved photocatalysis • Remarkable microbial activity • Cost-efficient and eco-friendly • Large surface-to-volume and functionality ratio • Active catalytic sites	• Less stable • Membrane fouling and membrane clogging • Difficult to implement on a large scale • Toxic effects • Formation of harmful derivatives on degradation

12.4 GREEN NANOMATERIALS FOR WASTEWATER TREATMENT

12.4.1 Definition and characteristics of green nanomaterials

Green nanotechnology determines sustainable health and environmental safety. Green nanotechnology aims to synthesize unique nanomaterials for use in photocatalysis, semiconductors, membranes, and nanosensors, for pollutant removal and ecological cleanup, and for treatment of wastewater. Greener nanomaterial synthesis establishes a standard for cleaner, safer, and more ecological nanoproducts. Green nanomaterials are made from plants, microbes, and other natural resources. Green nanoparticle synthesis is simpler, cost-effective, more efficient, and ecologically benign than chemical synthesis (Elgarahy *et al.*, 2021).

Owing to its tiny size, a lot of NPs immediately interact with the medium, altering its reactivity. Quantum effects in NPs need lower activation energy for chemical reactions. Surface plasmon resonance in NPs helps detect environmental toxins (Sinha *et al.*, 2020).

12.4.2 Types of green nanomaterials

12.4.2.1 Synthesis of green nanomaterials

Nanoparticle synthesis may be top-down or bottom-up. Top-down scaling implies milling, crushing, and thermal/laser ablation of thin films or bulk materials. Bottom-up approaches employ NPs, molecules, atoms, and so on to consciously organize nuclei into superstructures of increasing complexity (Pandit & Gayatri, 2020).

The initial steps imply the mechanical grinding of a compact material into NPs and stabilizing them to the desired size. This strategy is hard to narrow. The second technique uses chemical procedures, including sol–gel, thermolysis, hydrothermal, hydrolysis, and gas phase, to make nanoscale material from atomic-size material. Laser ablation, aerosol technologies, UV irradiation, and photochemical reduction produce NPs. Yet, they are costly and produce harmful pollutants. These approaches also make controlling nanoparticle size, structure, and surface chemistry problematic. Nonetheless, a bottom-up approach is desirable to manufacture NPs as it initiates with smaller clusters, particles, and NPs, giving better control over particle shape and size (Devatha & Thalla, 2018). Green synthesis is the process of creating nanomaterials from plants and plant extracts, microorganisms, fungi, algae, and so on.

12.4.2.2 Plant and plant extract

Nanomaterials from plant or food leftovers are the most fascinating and ecologically benign kind of green synthesis. Plant extracts include flavonoids, terpenoids, and phenols, but proteins, glucosides, and polysaccharides also contribute to nanoparticle production. These bioactive chemicals reduce and stabilize nanoparticle precursors. This nanoparticle production approach also removes hazardous substances. The use of heated water extracts bioactive compounds without high-energy methods. Thus, most manufactured NPs may be used in biomedical applications without harsh reagents. Gold, silver, and selenium may also produce NPs. Many plant extracts containing reducing, stabilizing, and capping agents may synthesize AgNPs (Huston *et al.*, 2021). Moulton *et al.* (2010) produced colloidal AgNPs using polyphenol-rich tea leaves. Fardsadegh and Jafarizadeh-Malmiri (2019) synthesized antifungal and antibacterial selenium NPs using aloe vera.

12.4.2.3 Microorganisms

Bacteria used in green nanomaterial manufacturing are with a structured nucleus, unicellular, possessing cell walls but no organelles. Few bacteria are hazardous but many are harmless and exist naturally in the body. *Escherichia coli and B. subtilis*, for instance, are straightforward to nurture and alter genetically. These traits make bacterial nanoparticle production possible (Huston *et al.*, 2021).

Huston *et al.* (2021) generated nanomaterials made of sulfides of zinc and lead and oxides of iron and selenium, using bacterial systems. Tyrosine and tryptophan amino acids in the cell wall and cytoplasm of bacteria decrease NPs and stabilize them. Furthermore, ketose and aldose act as stabilizing/reducing agents. Since amino acids create a protective capping layer in cell walls and inside cells, they are nonhazardous to mammalian cells (Huston *et al.*, 2021). Organisms such as *B. licheniformis, P. stutzeri, Acinetobacter calcoaceuticus, B. megatherium, B. amyloliquefaciens, E. coli*, and *lactobacillus* have been widely used to produce nanomaterials (Prasad *et al.*, 2013). For instance, Mukherjee (2017) used *Microbacterium marinilacus* for generating silver, copper, and magnetic iron oxide and Tripathi *et al.* (2014) prepared spherical zinc oxide NPs by using *Rhodococcus pyridinivorans*.

12.4.2.4 Fungi

Bioreduction and bioaccumulation make fungi better nanoparticle producers. Fungi's enzymes and proteins reduce pollutants by easy hydrolysis. Hydrogenase, nitrate-dependent reductase, and others help the fungi to bioreduce several contaminants. Fungi produce organic acids, enzymes, proteins, and polysaccharides that form nanocrystals. *Fusarium solani, Coriolus vesicolous, Aspergillus niger*, and *A. fumigates* mediated bioreduction to create gold and silver NPs. Fungi may absorb metals via their strong wall binding capacity and intracellular absorption. Fungi decompose extracellular materials and release enzymes that hydrolyze bulk components to nanostructures. Fungi-mediated nanoparticle production is cheaper, easier to manage, safer, and easier to handle biomass (Aarthye & Sureshkumar, 2021). Mukherjee *et al.* (2001) found that *Verticillium*, which causes verticillium wilt, can generate silver NPs on the cell wall by minimizing aqueous silver nitrate. Ahmad *et al.* (2005) and Gericke and Pinches (2006) found that nanospheres and nanorods of gold can be created using *Verticillium luteoalbum* and *Trichothecium* enzymes.

12.4.2.5 Algae

Algae are eukaryotic, photosynthetic, non-plant species. Algae, sometimes termed green biofactories are cost-effective, critical eukaryotic creatures used in nanotechnology for their low toxicity and substantial metal bioaccumulation (Huston *et al.*, 2021). Salem *et al.* (2019) produced AgNPs from the red algae *Portieria hornemannii*. Singaravelu *et al.* (2007) revealed that *Sargassum wightii* can make gold-stable NPs.

12.4.3 Advantages of green nanomaterials over conventional nanomaterials

Traditional treatment techniques transport pollutants without decomposing the contaminant to the environment-benign product. They demand substantial investment, space, and administration (De Kwaadsteniet *et al.*, 2011; Naushad *et al.*, 2013). An alternative to this is a greener wastewater treatment method. Green synthesis offers cost-effective manufacturing, energy competence, benign procedures and byproducts, decreased surplus, and higher pharma and medical use (Hassaan & Hosny, 2018; Ijaz *et al.*, 2020). Due to their green manufacturing and capacity to reduce environmental toxins, biogenic NPs are promising materials (Gautam *et al.*, 2019). Bioprepared NPs may absorb contaminants from aqueous watercourses or degrade organics to benign categories. Biogenic NPs are biorenewable, sustainable, cheap, and energy-efficient, making them ideal for consumption and manufacturing wastewater purification methods (Gautam *et al.*, 2019). Thus, the use of natural raw materials, no toxic chemical involvements, and reduced energy requirements make these green nanomaterials more efficient than the traditionally available nanomaterials.

12.4.4 Recent research on green nanomaterials for wastewater treatment

Green nanomaterials with their unique properties have spurred an exploration into novel adsorbents. For environmental applications, a variety of new-generation nanomaterials have been created, including TiO_2, ZnO, oxides of iron, zero-valent metals, and carbon nanotubes (Abouzeid *et al.*, 2018). Instances of current trends are highlighted in Table 12.4.

Table 12.4 Recent research on green nanomaterials for wastewater treatment.

Sr. No.	Nanomaterial	Biological Component	Application in Wastewater	References
1	nZVI-Cu	Pomegranate extract	Remediation of tetracycline	Gopal *et al.* (2020)
2	Cu	Tilia extract residues	Remediation of ibuprofen, diclofenac, and naproxen	Husein *et al.* (2019)
3	Fe_3O_4	Plant extracts of lemon, black grapes, and cucumber	Removal of sulfamethoxazole, piperacillin, tazobactam, ampicillin tetracycline, erythromycin, and trimethoprim	Stan *et al.* (2017)
4	Activated carbon	Olive stones	Removal of paracetamol	García-Mateos *et al.* (2015)
5	Manganese nanoparticles	Green tea extract	Removal of mitoxantrone	He *et al.* (2021)
6	Silver oxide (Ag_2O) nanoparticles	Green leaf extract of *Punica granatum*	Removal of sulfamethoxazole	El Messaoudi *et al.* (2022)
7	Magnetic Fe_3O_4 nanoparticles	*Excoecaria cochinchinensis* extract	Rifampicin	Cai *et al.* (2019)

12.5 ABATEMENT OF PHARMACEUTICAL COMPOUNDS IN WASTEWATER USING GREEN NANOMATERIALS

12.5.1 Mechanisms of abatement using green nanomaterials

Eco-friendly NPs have been created for effective cleanup and removal of harmful pollutants (Shukla *et al.*, 2021). Electrostatic interaction adsorption removed sulfamethazine from water using modified bur cucumber biochar (Letsoalo *et al.*, 2022). Paradis-Tanguay *et al.* (2019) and Zhang *et al.* (2020b) showed high-capacity ibuprofen adsorption employing electrospun ethylene oxide (chitosan) nanofibers compound and polyacrylic-based $Fe_3O_4^-$ anion exchange resin, where physisorption removed diclofenac and ibuprofen in these experiments. Hamadeen and Elkhatib (2022) used peels of pomegranate to prepare activated biochar to remove ciprofloxacin efficiently. Interactions such as hydrophobic, π–π, electrostatic, and hydrogen bonding dominate ciprofloxacin adsorption by nanostructured activated biochar. The removal of ciprofloxacin was 89.94% and 84.74% using batch reactor and packed-bed reactor, respectively.

12.5.2 Factors affecting the efficiency of green nanomaterials in abating pharmaceutical compounds

Several variables impact nanoparticle creation, characterization, and use. The pH of the state, temperature, amount of extracts, amount of raw materials, size, and methods for nanoparticle formation are other essential aspects (Patra & Baek, 2015). Due to surface functional group protonation, the adsorbent's surface is positively charged when the solution's pH is less than the nanocomposite's pH. When the nano adsorbent's surface is negatively charged, the electrostatic interactions between it and contaminants change, affecting its adsorption capability (Liao *et al.*, 2022). For example, ciprofloxacin adsorption depends on separating media pH (i.e., adsorbent). Nevertheless, increasing pH from 2 to 10 improved ciprofloxacin adsorption with tea waste (Seedher & Sidhu, 2007). Beltrame *et al.* (2018) found that ciprofloxacin adsorption on pineapple plant leaf-activated carbon increased to pH 7, then reduced from pH 8 to pH 9.

Many amoxicillin adsorption processes have been documented, including pH (Anastopoulos *et al.*, 2020). Amoxicillin adsorption onto many adsorbents increased with pH up to 6 (Jafari *et al.*, 2018). If the solution pH rises over 6, the adsorption range stays near constant (Xing *et al.*, 2013) or declines (Belhachemi & Djelaila, 2017). With higher pH, amoxicillin's twofold negative charge and repulsive forces decrease its adsorption (Anastopoulos *et al.*, 2020). Noirod *et al.* (2016) found a similar pH impact on triclosan adsorption onto chitosan. Triclosan adsorption increased with pH and peaked at pH 3.

Temperature impacts nano adsorbent performance. Adsorption capacity rises with temperature in endothermic processes and decreases in exothermic processes (Liao *et al.*, 2022). The pyrolytically generated water hyacinth biochar as adsorbent might decrease ciprofloxacin retention efficiency from 86.05% to 59.75% when the temperature rises from 298 to 338 K (Ngeno *et al.*, 2016). Noirod *et al.* (2016) found that adsorption on chitosan for triclosan improved with temperature in the range of 5–65°C owing to increased molecular mobility, sorbent particle kinetic energy, and sorbate–sorbent collision frequency.

Site and capacity determine the adsorbent dose (Liao *et al.*, 2022). A copper alginate-carbon nanotube membrane removes tetracycline in 2000 min with 120 mg/g equilibrium adsorption (Zhang *et al.*, 2022). Adsorbent dosage enhances amoxicillin elimination (Ali *et al.*, 2020). Adsorbent dose increased tartaric acid-modified wheat grains' amoxicillin adsorption capability (Boukhelkhal *et al.*, 2016). Nevertheless, ashes of almond shells, activated carbon from *Azolla filiculoides*, activated carbon, and an ultrasound-synthesized magnetic adsorbent from olive kernel had a negative impact (Balarak *et al.*, 2017; Homem *et al.*, 2010; Jafari *et al.*, 2018; Mahmood & Abdulmajeed, 2017). Thus, the negative effect could be probably because of particle aggregation that reduces overall surface area and adsorption efficacy at larger doses, reducing adsorption capacity (Mahmood & Abdulmajeed, 2017).

12.5.3 Comparison of the effectiveness of green over conventional nanomaterials in abating pharmaceutical compounds

Nevertheless, chemical and physical techniques of extraction have major disadvantages such as low output rate, deprived surface development, high cost, elevated energy requirements, and hazardous reducing agents. Hence, developing an eco-friendly nanoparticle production process is crucial. The biological synthesis technique is clean and eco-friendly, uses active biological substances such as enzymes as capping and reducing agents, can be scaled up, and uses less energy (Karunakaran *et al.*, 2018). Green nanomaterials have various benefits over conventional NPs.

For example, Husein *et al.* (2019) removed ibuprofen, naproxen, and diclofenac from tilia leaves using green-synthesized copper NPs. The amount of ibuprofen, naproxen, and diclofenac removed was 74.4%, 86.9%, and 91.4%, respectively. Kim *et al.* (2020) found that maple leaf biochar removed tetracycline at 407.3 mg/g, making it a more effective adsorbent than conventional ones. Weng *et al.* (2018) made green synthetic Fe_3O_4 magnetite NPs from *Euphorbia cochinchinensis* extract to remove doxorubicin hydrochloride, an anti-cancer medication. Green NPs at 303 K removed 80.2% doxorubicin. Debnath *et al.* (2020) employed *P. aeruginosa* bacteria to synthesize zirconia nanoparticles for wastewater tetracycline bioremediation. Zirconia nanoparticles have a Langmuir isotherm model-calculated maximal tetracycline adsorption capacity of 526.32 mg/g. Zirconia nanoparticles may be used to reduce wastewater tetracycline contamination. Ahmed *et al.* (2023) used green tea leaves to make an iron–copper nanocomposite on alginate limestone. The iron–copper nanocomposite on alginate limestone adsorbs ciprofloxacin and levofloxacin from contaminated media. Kinetic and isotherm models calculated adsorption parameters. This innovative abatement approach removed ciprofloxacin and levofloxacin at 20 ppm, 97.3%, and 10 ppm, 100%, respectively. Abdel-Aziz *et al.* (2019) produced bimetallic nano zero-valent Fe/Cu via green technology using *Ficus benjamina* leaves to remove carbamazepine. Carbamazepine elimination was 95% at 0.4 g/L at pH 5 in 20 min. The Freundlich and Langmuir adsorption isotherms showed that carbamazepine elimination was better than with typical nanomaterials. Misra *et al.* (2018) tested green-synthesized superparamagnetic iron oxide nanoparticles for carbamazepine cleanup. This work generated magnetic nano sorbents by coprecipitation employing *Colocasia esculenta* corms, piper bettle leaves, and *Nelumbo nucifera* stalk extracts to modify nanoparticle surfaces. *N. nucifera* extract-coated nanoparticles had the best removal effectiveness (52% at 5 ppm). Hoslett *et al.* (2021) made tetracycline-removing biochar from food and garden waste, which showed tetracycline removal up to 9.45 mg/g.

12.5.4 Future prospects and challenges of using green nanomaterials for abating pharmaceutical compounds

Nanotechnology is promising in medicine, energy, and the environment. Green nanomaterials for pharmaceutical chemical abatement seem promising. These materials may be cost-effective and efficient. Green nanoparticles are less costly and energy-intensive than typical wastewater treatment processes. Green nanoparticles may also be employed in domestic wastewater treatment and large-scale industrial activities. Green nanoparticles could potentially be used in wastewater treatment systems to eliminate pharmaceutical chemicals. Green nanoparticles may increase pharmaceutical chemical removal in these systems, minimizing their environmental effect.

Green nanomaterials for pharmaceutical compound abatement have some drawbacks. It struggles with raw ingredients, reaction circumstances, product quality, and application. These issues hinder manufacturing and large-scale usage of green nanomaterial (Guan *et al.*, 2022). Researchers found that local plants are suitable for green NP production. These studies imply that native plants may be fully used, although global nanomaterial manufacturing is problematic (Turunc *et al.*, 2017). Some greener synthetic processes need high temperatures and long synthesizing times, which may affect the environment. Although employing ecologically friendly chemicals, the procedure does not necessarily follow sustainable synthesis principles (Muthuvel *et al.*, 2020). A dearth of knowledge of the synthesis pathway makes it hard to find chemical reactions that demonstrate green synthesis (Kora & Rastogi, 2016). Different extracts yield nanoparticles of varied shapes and sizes, and quality assessment is inadequate. Current sources say that particle diameter fluctuates widely, making green technology unsuited for large-scale production and particle size control difficult throughout the manufacturing process (Chahardoli *et al.*, 2018). Green nanomaterials could be harmful. If mishandled, these natural resources may harm the environment and human health. Before mass-producing green nanomaterials, their toxicity must be assessed.

12.6 CONCLUSION

12.6.1 Recommendations for future research

To ensure considerable water quality for drinking while also eliminating micropollutants, novel sophisticated water systems must be adopted. Flexible and diverse water treatment technologies must be employed to improve industrial production processes. One of the primary advantages of nanomaterials over traditional water-based approaches is their capacity to comprise a wide range of characteristics in different systems, such as membranes made of nanocomposite that allow for particle retention as well as pollutant removal. In addition, because of their unique qualities, such as a large surface area, nanoparticles improve the effectiveness of processes. At this moment, it is essential to call attention to a handful of critical disadvantages. Substances enhanced by NPs, for example, integrated into or placed on their surface, have the capacity to be hazardous since NPs may be discharged into the environment and progressively accumulate. Numerous local and global standards and legislation are being developed to mitigate health risks. The main technical constraint of nano-engineered water treatment techniques is that it is barely adaptable to huge-scale operations and is occasionally cost-competitive with traditional treatment methods. Cleaner and more plentiful nano-engineered materials, on the other hand, offer huge promise for progress in the coming decades, especially for decentralized treatment facilities. Before these green-synthesized nanocatalysts and nanomaterials may be deployed at industrial and commercial sizes, additional research into their sustainability and toxicity is necessary. Although the production of these nanomaterials is efficient and sustainable, certain vital and challenging aspects, such as the impact of reactions and instability problems, must be investigated and optimized, as these variables can alter the behavior of nanomaterials, their morphologies, and their efficacy in removing pollutants. More research is required to identify novel nanohybrids and multifunctional nanomaterials that can be used in practical applications. Effectiveness concerns and remedial performance assessments are often created on laboratory scales, replicating the various degrees of genuine exposure settings, but realistic environmental conditions must be explored and assessed.

12.6.2 Final thoughts and implications for practice

Pharmaceutical compounds in wastewater are becoming a growing public health and environmental concern. Nanomaterials have the potential to enhance wastewater treatment. Green nanoparticles are thought to be more environmentally friendly and sustainable than regular nanomaterials. In summary, there is enormous potential in the application of ecologically benign nanomaterials for pharmaceutical chemical abatement in WWTPs. However, further research is needed not only to fully appreciate the processes in use, but also to overcome the challenges associated with their manufacture, capacity, and affordability.

12.6.3 Summary of the study

Using green-synthesized nanomaterials, pharmaceuticals may be effectively eliminated at an affordable cost. Additional studies should concentrate on enhancing the economic feasibility of these nanoparticles and analyzing their interaction dynamics in water treatment systems since low production costs are crucial for the broad usage of these nanomaterials in wastewater treatment. Subsequently, it is also vital to do study on their possible harm to human health and the environment; comprehensive analyses are necessary to ensure that they are safe for use. In addition to eliminating the major drawbacks of existing technologies, nanotechnology-enabled wastewater treatment systems should also offer innovative wastewater treatment solutions that allow non-conventional water resources for water supply to be economically recovered and grown. The potential for the use of green-synthesized nanomaterials for remediation is highly exciting, but serious problems concerning toxicity, biosafety, and the mechanistic characteristics of these materials need to be addressed completely and systematically.

ACKNOWLEDGMENT

We are thankful to Nida Sarfaraz for copyediting and proofreading the manuscript.

REFERENCES

Aarthye P. and Sureshkumar M. (2021). Green synthesis of nanomaterials: an overview. *Materials Today: Proceedings*, **47**, 907–913, https://doi.org/10.1016/j.matpr.2021.04.564

Abdel-Aziz H. M., Farag R. S. and Abdel-Gawad S. A. (2019). Carbamazepine removal from aqueous solution by green synthesis zero-valent iron/Cu nanoparticles with *Ficus benjamina* leaves' extract. *International Journal of Environmental Research*, **13**, 843–852, https://doi.org/10.1007/s41742-019-00220-w

Abouzeid R. E., Khiari R., El-Wakil N. and Dufresne A. (2018). Current state and new trends in the use of cellulose nanomaterials for wastewater treatment. *Biomacromolecules*, **20**(2), 573–597, https://doi.org/10.1021/acs.biomac.8b00839

Aguilar-Pérez K. M., Avilés-Castrillo J. I. and Ruiz-Pulido G. (2020). Nano-sorbent materials for pharmaceutical-based wastewater effluents – an overview. *Case Studies in Chemical and Environmental Engineering*, **2**, 100028, https://doi.org/10.1016/j.cscee.2020.100028

Agunbiade F. O. and Moodley B. (2016). Occurrence and distribution pattern of acidic pharmaceuticals in surface water, wastewater, and sediment of the Msunduzi River, Kwazulu-Natal, South Africa. *Environmental Toxicology and Chemistry*, **35**(1), 36–46, https://doi.org/10.1002/and so on.3144

Ahmad A., Senapati S., Khan M. I., Kumar R. and Sastry M. (2005). Extra-/intracellular biosynthesis of gold nanoparticles by an alkalotolerant fungus, *Trichothecium* sp. *Journal of Biomedical Nanotechnology*, **1**(1), 47–53, https://doi.org/10.1166/jbn.2005.012

Ahmed M. B., Zhou J. L., Ngo H. H., Guo W., Thomaidis N. S. and Xu J. (2017). Progress in the biological and chemical treatment technologies for emerging contaminant removal from wastewater: a critical review. *Journal of Hazardous Materials*, **323**, 274–298, https://doi.org/10.1016/j.jhazmat.2016.04.045

Ahmed S. F., Mofijur M., Ahmed B., Mehnaz T., Mehejabin F., Maliat D., Hoang A. T. and Shafiullah G. M. (2022). Nanomaterials as a sustainable choice for treating wastewater. *Environmental Research*, **214**, 113807, https://doi.org/10.1016/j.envres.2022.113807

Ahmed I. A., Hussein H. S., Alothman Z. A., Alanazi A. G., Alsaiari N. S. and Khalid A. (2023). Green synthesis of Fe–Cu bimetallic supported on alginate–limestone nanocomposite for the removal of drugs from contaminated water. *Polymers*, **15**(5), 1221, https://doi.org/10.3390/polym15051221

Al-Baldawi I. A., Mohammed A. A., Mutar Z. H., Abdullah S. R. S., Jasim S. S. and Almansoory A. F. (2021). Application of phytotechnology in alleviating pharmaceuticals and personal care products (PPCPs) in wastewater: source, impacts, treatment, mechanisms, fate, and SWOT analysis. *Journal of Cleaner Production*, **319**, 128584, https://doi.org/10.1016/j.jclepro.2021.128584

Ali I., Afshinb S., Poureshgh Y., Azari A., Rashtbari Y., Feizizadeh A., Hamzezadeh A. and Fazlzadeh M. (2020). Green preparation of activated carbon from pomegranate peel coated with zero-valent iron nanoparticles (nZVI) and isotherm and kinetic studies of amoxicillin removal in water. *Environmental Science and Pollution Research*, **27**(29), 36732–36743, https://doi.org/10.1007/s11356-020-09310-1

Anastopoulos I., Pashalidis I., Orfanos A. G., Manariotis I. D., Tatarchuk T., Sellaoui L., Bonilla-Petriciolet A., Mittal A. and Núñez-Delgado A. (2020). Removal of caffeine, nicotine and amoxicillin from (waste) waters by various adsorbents. A review. *Journal of Environmental Management*, **261**, 110236, https://doi.org/10.1016/j.jenvman.2020.110236

Azeez L., Adekale I. and Olabode O. A. (2022). Implications of green nanomaterials for environmental remediation. In: Handbook of Green and Sustainable Nanotechnology: Fundamentals, Developments and Applications, U. Shanker, C. M. Hussain and M. Rani (eds), Springer International Publishing, Cham, pp. 1–18. https://doi.org/10.1007/978-3-030-69023-6_18-1

Balarak D., Mostafapour F. K., Akbari H. and Joghtaei A. (2017). Adsorption of amoxicillin antibiotic from pharmaceutical wastewater by activated carbon prepared from *Azolla filiculoides*. Available at *SSRN 3955279*.

Belhachemi M. and Djelaila S. (2017). Removal of amoxicillin antibiotic from aqueous solutions by date pits activated carbons. *Environmental Processes*, **4**, 549–561, https://doi.org/10.1007/s40710-017-0245-8

Beltrame K. K., Cazetta A. L., de Souza P. S., Spessato L., Silva T. L. and Almeida V. C. (2018). Adsorption of caffeine on mesoporous activated carbon fibers prepared from pineapple plant leaves. *Ecotoxicology and Environmental Safety*, **147**, 64–71, https://doi.org/10.1016/j.ecoenv.2017.08.034

Bokare V., Jung J. L., Chang Y. Y. and Chang Y. S. (2013). Reductive dechlorination of octachlorodibenzo-p-dioxin by nanosized zero-valent zinc: modeling of rate kinetics and congener profile. *Journal of Hazardous Materials*, **250**, 397–402, https://doi.org/10.1016/j.jhazmat.2013.02.020

Boukhelkhal A., Benkortbi O., Hamadache M., Ghalem N., Hanini S. and Amrane A. (2016). Adsorptive removal of amoxicillin from wastewater using wheat grains: equilibrium, kinetic, thermodynamic studies and mass transfer. *Desalination and Water Treatment*, **57**(56), 27035–27047, https://doi.org/10.1080/19443994.2016.1166991

Cai W., Weng X. and Chen Z. (2019). Highly efficient removal of antibiotic rifampicin from aqueous solution using green synthesis of recyclable nano-Fe_3O_4. *Environmental Pollution*, **247**, 839–846, https://doi.org/10.1016/j.envpol.2019.01.108

Cerro-Lopez M. and Méndez-Rojas M. A. (2019). Application of nanomaterials for treatment of wastewater containing pharmaceuticals. In: *Ecopharmacovigilance: Multidisciplinary Approaches to Environmental Safety of Medicines*, L. M. Gómez-Oliván (ed.), Springer International Publishing, **66**, 201–219, https://doi.org/10.1007/698_2017_143

Chahardoli A., Karimi N. and Fattahi A. (2018). *Nigella arvensis* leaf extract mediated green synthesis of silver nanoparticles: their characteristic properties and biological efficacy. *Advanced Powder Technology*, **29**(1), 202–210, https://doi.org/10.1016/j.apt.2017.11.003

Chernicharo C. A. L., Van Lier J. B., Noyola A. and Bressani Ribeiro T. (2015). Anaerobic sewage treatment: state of the art, constraints and challenges. *Reviews in Environmental Science and Bio/Technology*, **14**, 649–679, https://doi.org/10.1007/s11157-015-9377-3

Czech B. and Buda W. (2015). Photocatalytic treatment of pharmaceutical wastewater using new multiwall-carbon nanotubes/TiO_2/SiO_2 nanocomposites. *Environmental Research*, **137**, 176–184, https://doi.org/10.1016/j.envres.2014.12.006

Das R., Sarkar S., Chakraborty S., Choi H. and Bhattacharjee C. (2014). Remediation of antiseptic components in wastewater by photocatalysis using TiO_2 nanoparticles. *Industrial & Engineering Chemistry Research*, **53**(8), 3012–3020, https://doi.org/10.1021/ie403817z

De Kwaadsteniet M. I. C. H. E. L. E., Botes M. and Cloete T. E. (2011). Application of nanotechnology in antimicrobial coatings in the water industry. *Nano*, **6**(05), 395–407, https://doi.org/10.1142/S1793292011002779

Debnath B., Majumdar M., Bhowmik M., Bhowmik K. L., Debnath A. and Roy D. N. (2020). The effective adsorption of tetracycline onto zirconia nanoparticles synthesized by novel microbial green technology. *Journal of Environmental Management*, **261**, 110235, https://doi.org/10.1016/j.jenvman.2020.110235

Dehghani M. H., Karri R. R., Alimohammadi M., Nazmara S., Zarei A. and Saeedi Z. (2020). Insights into endocrine-disrupting bisphenol – a adsorption from pharmaceutical effluent by chitosan immobilized nanoscale zero-valent iron nanoparticles. *Journal of Molecular Liquids*, **311**, 113317, https://doi.org/10.1016/j.molliq.2020.113317

Devatha C. P. and Thalla A. K. (2018). Chapter 7 – Green synthesis of nanomaterials. In: Synthesis of Inorganic Nanomaterials, S. Mohan Bhagyaraj, O. S. Oluwafemi, N. Kalarikkal and S. Thomas (eds), Woodhead Publishing, pp. 169–184, https://doi.org/10.1016/B978-0-08-101975-7.00007-5

Ding T., Li W. and Li J. (2019). Influence of multi-walled carbon nanotubes on the toxicity and removal of carbamazepine in diatom *Navicula* sp. *Science of the Total Environment*, **697**, 134104, https://doi.org/10.1016/j.scitotenv.2019.134104

El Messaoudi N., El Mouden A., Fernine Y., El Khomri M., Bouich A., Faska N., Ciğeroğlu Z., Américo-Pinheiro J. H. P., Jada A. and Lacherai A. (2023). Green synthesis of Ag_2O nanoparticles using *Punica granatum* leaf extract for sulfamethoxazole antibiotic adsorption: characterization, experimental study, modeling, and DFT calculation. *Environmental Science and Pollution Research*, **30**(34), 81352–81369, https://doi.org/10.1007/s11356-022-21554-7

Elgarahy A. M., Elwakeel K. Z., Akhdhar A. and Hamza M. F. (2021). Recent advances in greenly synthesized nanoengineered materials for water/wastewater remediation: an overview. *Nanotechnology for Environmental Engineering*, **6**, 1–24, https://doi.org/10.1007/s41204-021-00104-5

Environmental Protection Agency US. (2012). Water: Contaminant Candidate List 3 – CCL.

European Commission. (2013). Directive 2013/39/EU of the European Parliament and of the Council of 12 August 2013 Amending Directives 2000/60/EC and 2008/105/EC as Regards Priority Substances in the Field of Water Policy (2011/0429 (COD).

Fardsadegh B. and Jafarizadeh-Malmiri H. (2019). *Aloe vera* leaf extract mediated green synthesis of selenium nanoparticles and assessment of their in vitro antimicrobial activity against spoilage fungi and pathogenic bacteria strains. *Green Processing and Synthesis*, **8**(1), 399–407, https://doi.org/10.1515/gps-2019-0007

Fernández-López C., Guillén-Navarro J. M., Padilla J. J. and Parsons J. R. (2016). Comparison of the removal efficiencies of selected pharmaceuticals in wastewater treatment plants in the region of Murcia, Spain. *Ecological Engineering*, **95**, 811–816, https://doi.org/10.1016/j.ecoleng.2016.06.093

García-Mateos F. J., Ruiz-Rosas R., Marqués M. D., Cotoruelo L. M., Rodríguez-Mirasol J. and Cordero T. (2015). Removal of paracetamol on biomass-derived activated carbon: modeling the fixed bed breakthrough curves using batch adsorption experiments. *Chemical Engineering Journal*, **279**, 18–30, https://doi.org/10.1016/j.cej.2015.04.144

Gautam P. K., Singh A., Misra K., Sahoo A. K. and Samanta S. K. (2019). Synthesis and applications of biogenic nanomaterials in drinking and wastewater treatment. *Journal of Environmental Management*, **231**, 734–748, https://doi.org/10.1016/j.jenvman.2018.10.104

Gericke M. and Pinches A. (2006). Microbial production of gold nanoparticles. *Gold Bulletin*, **39**(1), 22–28, https://doi.org/10.1007/BF03215529

German Environment Agency–Umweltbundesamt. (2023). https://www.umweltbundesamt.de/en. Accessed on 29/03/2023

Ghazal H., Koumaki E., Hoslett J., Malamis S., Katsou E., Barcelo D. and Jouhara H. (2022). Insights into current physical, chemical and hybrid technologies used for the treatment of wastewater contaminated with pharmaceuticals. *Journal of Cleaner Production*, **361**, 132079, https://doi.org/10.1016/j.jclepro.2022.132079

Gogoi A., Mazumder P., Tyagi V. K., Chaminda G. T., An A. K. and Kumar M. (2018). Occurrence and fate of emerging contaminants in water environment: a review. *Groundwater for Sustainable Development*, **6**, 169–180, https://doi.org/10.1016/j.gsd.2017.12.009

Gopal G., Sankar H., Natarajan C. and Mukherjee A. (2020). Tetracycline removal using green synthesized bimetallic nZVI-Cu and bentonite supported green nZVI-Cu nanocomposite: a comparative study. *Journal of Environmental Management*, **254**, 109812, https://doi.org/10.1016/j.jenvman.2019.109812

Gracia-Lor E., Sancho J. V., Serrano R. and Hernández F. (2012). Occurrence and removal of pharmaceuticals in wastewater treatment plants at the Spanish Mediterranean area of Valencia. *Chemosphere*, **87**(5), 453–462, https://doi.org/10.1016/j.chemosphere.2011.12.025

Guan Z., Ying S., Ofoegbu P. C., Clubb P., Rico C., He F. and Hong J. (2022). Green synthesis of nanoparticles: current developments and limitations. *Environmental Technology & Innovation*, **26**, 102336, https://doi.org/10.1016/j.eti.2022.102336

Gupta R., Sati B. and Gupta A. (2019). Treatment and recycling of wastewater from pharmaceutical industry. In: Advances in Biological Treatment of Industrial Waste Water and Their Recycling for a Sustainable Future, R. L. Singh and R. P. Singh (eds), Springer, pp. 267–302, https://doi.org/10.1007/978-981-13-1468-1_9

GWRC (GlobalWater Research Coalition). (2008). Development of an International Priority List of Pharmaceuticals Relevant for the Water Cycle.

Hamadeen H. M. and Elkhatib E. A. (2022). New nanostructured activated biochar for effective removal of antibiotic ciprofloxacin from wastewater: adsorption dynamics and mechanisms. *Environmental Research*, **210**, 112929, https://doi.org/10.1016/j.envres.2022.112929

Hassaan M. A. and Hosny S. (2018). Green synthesis of Ag and Au nanoparticles from micro and macro algae – review. *International Journal of Atmospheric and Oceanic Sciences*, **2**(1), 10–22, https://doi.org/10.11648/j.ijaos.20180201.12

He F., Cai W., Lin J., Yu B., Owens G. and Chen Z. (2021). Reducing the impact of antibiotics in wastewaters: increased removal of mitoxantrone from wastewater by biosynthesized manganese nanoparticles. *Journal of Cleaner Production*, **293**, 126207, https://doi.org/10.1016/j.jclepro.2021.126207

Homem V., Alves A. and Santos L. (2010). Amoxicillin removal from aqueous matrices by sorption with almond shell ashes. *International Journal of Environmental and Analytical Chemistry*, **90**(14–15), 1063–1084, https://doi.org/10.1080/03067310903410964

Hoslett J., Ghazal H., Katsou E. and Jouhara H. (2021). The removal of tetracycline from water using biochar produced from agricultural discarded material. *Science of the Total Environment*, **751**, 141755, https://doi.org/10.1016/j.scitotenv.2020.141755

Huerta B., Rodriguez-Mozaz S., Nannou C., Nakis L., Ruhí A., Acuña V., Sabater S. and Barcelo D. (2016). Determination of a broad spectrum of pharmaceuticals and endocrine disruptors in biofilm from a waste water treatment plant-impacted river. *Science of the Total Environment*, **540**, 241–249, https://doi.org/10.1016/j.scitotenv.2015.05.049

Husein D. Z., Hassanien R. and Al-Hakkani M. F. (2019). Green-synthesized copper nano-adsorbent for the removal of pharmaceutical pollutants from real wastewater samples. *Heliyon*, **5**(8), e02339, https://doi.org/10.1016/j.heliyon.2019.e02339

Huston M., DeBella M., DiBella M. and Gupta A. (2021). Green synthesis of nanomaterials. *Nanomaterials*, **11**(8), 2130, https://doi.org/10.3390/nano11082130

Ijaz I., Gilani E., Nazir A. and Bukhari A. (2020). Detail review on chemical, physical and green synthesis, classification, characterizations and applications of nanoparticles. *Green Chemistry Letters and Reviews*, **13**(3), 223–245, https://doi.org/10.1080/17518253.2020.1802517

Islas-Flores H., Manuel Gómez-Oliván L., Galar-Martínez M., Michelle Sánchez-Ocampo E., SanJuan-Reyes N., Ortíz-Reynoso M. and Dublán-García O. (2017). Cyto-genotoxicity and oxidative stress in common carp (*Cyprinus carpio*) exposed to a mixture of ibuprofen and diclofenac. *Environmental Toxicology*, **32**(5), 1637–1650, https://doi.org/10.1002/tox.22392

Jafari K., Heidari M. and Rahmanian O. (2018). Wastewater treatment for amoxicillin removal using magnetic adsorbent synthesized by ultrasound process. *Ultrasonics Sonochemistry*, **45**, 248–256, https://doi.org/10.1016/j.ultsonch.2018.03.018

Jain K., Patel A. S., Pardhi V. P. and Flora S. J. S. (2021). Nanotechnology in wastewater management: a new paradigm towards wastewater treatment. *Molecules*, **26**(6), 1797, https://doi.org/10.3390/molecules26061797

Junejo Y., Güner A. and Baykal A. (2014). Synthesis and characterization of amoxicillin derived silver nanoparticles: its catalytic effect on degradation of some pharmaceutical antibiotics. *Applied Surface Science*, **317**, 914–922, https://doi.org/10.1016/j.apsusc.2014.08.133

Kariim I., Abdulkareem A. S. and Abubakre O. K. (2020). Development and characterization of MWCNTs from activated carbon as adsorbent for metronidazole and levofloxacin sorption from pharmaceutical wastewater: kinetics, isotherms and thermodynamic studies. *Scientific African*, **7**, e00242, https://doi.org/10.1016/j.sciaf.2019.e00242

Karunakaran S., Ramanujam S. and Gurunathan B. (2018). Green synthesised iron and iron-based nanoparticle in environmental and biomedical application: a review. *IET Nanobiotechnology*, **12**(8), 1003–1008, https://doi.org/10.1049/iet-nbt.2018.5048

Kermia A. E. B., Fouial-Djebbar D. and Trari M. (2016). Occurrence, fate and removal efficiencies of pharmaceuticals in wastewater treatment plants (WWTPs) discharging in the coastal environment of Algiers. *Comptes Rendus Chimie*, **19**(8), 963–970, https://doi.org/10.1016/j.crci.2016.05.005

Khalil A. M., Memon F. A., Tabish T. A., Fenton B., Salmon D., Zhang S. and Butler D. (2021). Performance evaluation of porous graphene as filter media for the removal of pharmaceutical/emerging contaminants from water and wastewater. *Nanomaterials*, **11**(1), 79, https://doi.org/10.3390/nano11010079

Kim J. E., Bhatia S. K., Song H. J., Yoo E., Jeon H. J., Yoon J.-Y., Yang Y., Gurav R., Yang Y.-H., Kim H. J. and Choi Y.-K. (2020). Adsorptive removal of tetracycline from aqueous solution by maple leaf-derived biochar. *Bioresource Technology*, **306**, 123092, https://doi.org/10.1016/j.biortech.2020.123092

Kora A. J. and Rastogi L. (2016). Catalytic degradation of anthropogenic dye pollutants using palladium nanoparticles synthesized by gum olibanum, a glucuronoarabinogalactan biopolymer. *Industrial Crops and Products*, **81**, 1–10, https://doi.org/10.1016/j.indcrop.2015.11.055

Kumar L., Ragunathan V., Chugh M. and Bharadvaja N. (2021). Nanomaterials for remediation of contaminants: a review. *Environmental Chemistry Letters*, **19**, 3139–3163, https://doi.org/10.1007/s10311-021-01212-z

Kumar R., Qureshi M., Vishwakarma D. K., Al-Ansari N., Kuriqi A., Elbeltagi A. and Saraswat A. (2022). A review on emerging water contaminants and the application of sustainable removal technologies. *Case Studies in Chemical and Environmental Engineering*, **6**, 100219, https://doi.org/10.1016/j.cscee.2022.100219

Letsoalo M. R., Sithole T., Mufamadi S., Mazhandu Z., Sillanpaa M., Kaushik A. and Mashifana T. (2023). Efficient detection and treatment of pharmaceutical contaminants to produce clean water for better health and environmental. *Journal of Cleaner Production*, **387**, 135798, https://doi.org/10.1016/j.jclepro.2022.135798

Liao Z., Zi Y., Zhou C., Zeng W., Luo W., Zeng H., Xia M. and Luo Z. (2022). Recent advances in the synthesis, characterization, and application of carbon nanomaterials for the removal of endocrine-disrupting chemicals: a review. *International Journal of Molecular Sciences*, **23**(21), Article 21, https://doi.org/10.3390/ijms232113148

Lu H., Wang J., Stoller M., Wang T., Bao Y. and Hao H. (2016). An overview of nanomaterials for water and wastewater treatment. *Advances in Materials Science and Engineering*, **2016**, 1–10, https://doi.org/10.1155/2016/4964828

Mahmood N. A. H. J. and Abdulmajeed Y. R. (2017). Adsorption of amoxicillin onto activated carbon from aqueous solution. *International Journal of Current Engineering and Technology*, **7**, 62–67.

Malik S., Dhasmana A., Preetam S., Mishra Y. K., Chaudhary V., Bera S. P., Ranjan A., Bora J., Kaushik A., Minkina T., Jatav H. S., Singh R. K. and Rajput V. D. (2022). Exploring microbial-based green nanobiotechnology for wastewater remediation: a sustainable strategy. *Nanomaterials*, **12**(23), Article 23, https://doi.org/10.3390/nano12234187

Misra T., Mitra S. and Sen S. (2018). Adsorption studies of carbamazepine by green-synthesized magnetic nanosorbents. *Nanotechnology for Environmental Engineering*, **3**, 1–12, https://doi.org/10.1007/s41204-018-0040-4

Mohammadi R., *Massoumi B.* and *Eskandarloo H.* (2015). Preparation and characterization of Sn/Zn/TiO_2 photocatalyst for enhanced amoxicillin trihydrate degradation. *Desalination and Water Treatment*, **53**(7), 1995–2004, https://doi.org/10.1080/19443994.2013.862867

Moulton M. C., Braydich-Stolle L. K., Nadagouda M. N., Kunzelman S., Hussain S. M. and Varma R. S. (2010). Synthesis, characterization and biocompatibility of 'green' synthesized silver nanoparticles using tea polyphenols. *Nanoscale*, **2**(5), 763–770, https://doi.org/10.1039/C0NR00046A

Mukherjee P. (2017). *Stenotrophomonas* and *Microbacterium*: mediated biogenesis of copper, silver and iron nanoparticles–proteomic insights and antibacterial properties versus biofilm formation. *Journal of Cluster Science*, **28**, 331–358, https://doi.org/10.1007/s10876-016-1097-5

Mukherjee P., Ahmad A., Mandal D., Senapati S., Sainkar S. R., Khan M. I., Parishcha R., Ajaykumar P. V., Alam M., Kumar R. and Sastry M. (2001). Fungus-mediated synthesis of silver nanoparticles and their immobilization in the mycelial matrix: a novel biological approach to nanoparticle synthesis. *Nano Letters*, **1**(10), 515–519, https://doi.org/10.1021/nl0155274

Muthuvel A., Jothibas M. and Manoharan C. (2020). Synthesis of copper oxide nanoparticles by chemical and biogenic methods: photocatalytic degradation and in vitro antioxidant activity. *Nanotechnology for Environmental Engineering*, **5**, 1–19, https://doi.org/10.1007/s41204-020-00078-w

Nadim A. H., Al-Ghobashy M. A., Nebsen M. and Shehata M. A. (2015). Optimization of photocatalytic degradation of meloxicam using titanium dioxide nanoparticles: application to pharmaceutical wastewater analysis, treatment, and cleaning validation. *Environmental Science and Pollution Research*, **22**, 15516–15525, https://doi.org/10.1007/s11356-015-4713-2

Nasrollahzadeh M., Sajjadi M., Iravani S. and Varma R. S. (2021). Carbon-based sustainable nanomaterials for water treatment: state-of-art and future perspectives. *Chemosphere*, **263**, 128005, https://doi.org/10.1016/j.chemosphere.2020.128005

Naushad M., Al-Othman Z. A. and Islam M. (2013). Adsorption of cadmium ion using a new composite cation-exchanger polyaniline Sn (IV) silicate: kinetics, thermodynamic and isotherm studies. *International Journal of Environmental Science and Technology*, **10**, 567–578, https://doi.org/10.1007/s13762-013-0189-0

Néstor M. C. and Mariana C. (2019). Impact of pharmaceutical waste on biodiversity. In: Ecopharmacovigilance: Multidisciplinary Approaches to Environmental Safety of Medicines, L. M. Gómez-Oliván (ed.), Springer International Publishing, **66**, 235–253, https://doi.org/10.1007/698_2017_151

Ngeno E. C., Orata F., Lilechi D. B., Shikuku V. O. and Kimosop S. (2016). Adsorption of caffeine and ciprofloxacin onto pyrolitically derived water hyacinth biochar: isothermal, kinetic and thermodynamic studies. *Journal of Chemistry and Chemical Engineering*, **10**, 185–194, https://doi.org/10.17265/1934-7375/2016.04.006

Nino-Martinez N., Martinez-Castanon G. A., Aragon-Pina A., Martinez-Gutierrez F., Martinez-Mendoza J. R. and Ruiz F. (2008). Characterization of silver nanoparticles synthesized on titanium dioxide fine particles. *Nanotechnology*, **19**(6), 065711, https://doi.org/10.1088/0957-4484/19/6/065711

Noirod P., Lamangthong J. and Ninjiaranai P. (2016). Application of chitosan film for the removal of triclosan from aqueous solutions by adsorption. *Key Engineering Materials*. **675–676**, 455–458, https://doi.org/10.4028/www.scientific.net/KEM.675-676.455

Pandit P. and Gayatri T. N. (2020). Introduction to green nanomaterials. In: Green Nanomaterials: Processing, Properties, and Applications, S. Ahmed and W. Ali (eds), Springer, **126**, 1–21, https://doi.org/10.1007/978-981-15-3560-4_1

Paradis-Tanguay L., Camiré A., Renaud M., Chabot B. and Lajeunesse A. (2019). Sorption capacities of chitosan/polyethylene oxide (PEO) electrospun nanofibers used to remove ibuprofen in water. *Journal of Polymer Engineering*, **39**(3), 207–215, https://doi.org/10.1515/polyeng-2018-0224

Patra J. K. and Baek K. H. (2015). Green nanobiotechnology: factors affecting synthesis and characterization techniques. *Journal of Nanomaterials*, **2014**, 219–219.

Pereira A., Silva L., Laranjeiro C., Lino C. and Pena A. (2020). Selected pharmaceuticals in different aquatic compartments: part I–source, fate and occurrence. *Molecules*, **25**(5), 1026, https://doi.org/10.3390/molecules25051026

Ponnusamy, G., Farzaneh, H., Tong, Y., Lawler, J., Liu, Z., & Saththasivam, J. (2021). Enhanced catalytic ozonation of ibuprofen using a 3D structured catalyst with MnO_2 nanosheets on carbon microfibers. *Scientific Reports*, **11**(1), Article 1. https://doi.org/10.1038/s41598-021-85651-2

Prasad T. N., Kambala V. S. R. and Naidu R. (2013). Phyconanotechnology: synthesis of silver nanoparticles using brown marine algae *Cystophora moniliformis* and their characterisation. *Journal of Applied Phycology*, **25**, 177–182, https://doi.org/10.1038/s41598-021-85651-2

Ramírez-Morales D., Masís-Mora M., Montiel-Mora J. R., Cambronero-Heinrichs J. C., Briceño-Guevara S., Rojas-Sánchez C. E., Méndez-Rivera M., Arias-Mora V., Tormo-Budowski R., Brenes-Alfaro L. and Rodríguez-Rodríguez C. E. (2020). Occurrence of pharmaceuticals, hazard assessment and ecotoxicological evaluation of wastewater treatment plants in Costa Rica. *Science of the Total Environment*, **746**, 141200, https://doi.org/10.1016/j.scitotenv.2020.141200

Riaz L., Mahmood T., Coyne M. S., Khalid A., Rashid A., Hayat M. T., Gulzar A. and Amjad M. (2017). Physiological and antioxidant response of wheat (*Triticum aestivum*) seedlings to fluoroquinolone antibiotics. *Chemosphere*, **177**, 250–257, https://doi.org/10.1016/j.chemosphere.2017.03.033

Rout P. R., Zhang T. C., Bhunia P. and Surampalli R. Y. (2021). Treatment technologies for emerging contaminants in wastewater treatment plants: a review. *Science of the Total Environment*, **753**, 141990, https://doi.org/10.1016/j.scitotenv.2020.141990

Sadek A. H., Asker M. S. and Abdelhamid S. A. (2021). Bacteriostatic impact of nanoscale zero-valent iron against pathogenic bacteria in the municipal wastewater. *Biologia*, **76**, 2785–2809, https://doi.org/10.1007/s11756-021-00814-w

Salem D. M., Ismail M. M. and Aly-Eldeen M. A. (2019). Biogenic synthesis and antimicrobial potency of iron oxide (Fe_3O_4) nanoparticles using algae harvested from the Mediterranean Sea, Egypt. *The Egyptian Journal of Aquatic Research*, **45**(3), 197–204, https://doi.org/10.1016/j.ejar.2019.07.002

Seedher N. and Sidhu K. (2007). Studies on the use of tea leaves as pharmaceutical adsorbent. *International Journal of Biological and Chemical Sciences*, **1**, 162–167.

Shukla S., Khan R. and Daverey A. (2021). Synthesis and characterization of magnetic nanoparticles, and their applications in wastewater treatment: a review. *Environmental Technology & Innovation*, **24**, 101924, https://doi.org/10.1016/j.eti.2021.101924

Singaravelu G., Arockiamary J. S., Kumar V. G. and Govindaraju K. (2007). A novel extracellular synthesis of monodisperse gold nanoparticles using marine alga, *Sargassum wightii* Greville. *Colloids and Surfaces B: Biointerfaces*, **57**(1), 97–101, https://doi.org/10.1016/j.colsurfb.2007.01.010

Sinha S., Mehrotra T., Srivastava A., Srivastava A. and Singh R. (2020). Nanobioremediation technologies for potential application in environmental cleanup. *Environmental Biotechnology*, **2**, 53–73, https://doi.org/10.1007/978-3-030-38196-7_3

Stan M., Lung I., Soran M.-L., Leostean C., Popa A., Stefan M., Lazar M. D., Opris O., Silipas T.-D. and Porav A. S. (2017). Removal of antibiotics from aqueous solutions by green synthesized magnetite nanoparticles with selected agro-waste extracts. *Process Safety and Environmental Protection*, **107**, 357–372, https://doi.org/10.1016/j.psep.2017.03.003

Thapa B. S., Pandit S., Patwardhan S. B., Tripathi S., Mathuriya A. S., Gupta P. K., Lal R. B. and Tusher T. R. (2022). Application of microbial fuel cell (MFC) for pharmaceutical wastewater treatment: an overview and future perspectives. *Sustainability*, **14**(14), Article 14, https://doi.org/10.3390/su14148379

Tian P., Tang L., Teng K. S. and Lau S. P. (2018). Graphene quantum dots from chemistry to applications. *Materials Today Chemistry*, **10**, 221–258, https://doi.org/10.1016/j.mtchem.2018.09.007

Tiwari B., Sellamuthu B., Ouarda Y., Drogui P., Tyagi R. D. and Buelna G. (2017). Review on fate and mechanism of removal of pharmaceutical pollutants from wastewater using biological approach. *Bioresource Technology*, **224**, 1–12, https://doi.org/10.1016/j.biortech.2016.11.042

Tripathi R. M., Bhadwal A. S., Gupta R. K., Singh P., Shrivastav A. and Shrivastav B. R. (2014). ZnO nanoflowers: novel biogenic synthesis and enhanced photocatalytic activity. *Journal of Photochemistry and Photobiology B: Biology*, **141**, 288–295, https://doi.org/10.1016/j.jphotobiol.2014.10.001

Turunc E., Binzet R., Gumus I., Binzet G. and Arslan H. (2017). Green synthesis of silver and palladium nanoparticles using *Lithodora hispidula* (Sm.) Griseb.(Boraginaceae) and application to the electrocatalytic reduction of hydrogen peroxide. *Materials Chemistry and Physics*, **202**, 310–319, https://doi.org/10.1016/j.matchemphys.2017.09.032

U.S. Department of Defense (USDoD). (2011). Emerging Chemical and Material Risks. Chemical and Material Risk.

US EPA. (2012). Water: Contaminant Candidate List 3. US Environmental Protection Agency, Washington, DC.

Varma K. S., Tayade R. J., Shah K. J., Joshi P. A., Shukla A. D. and Gandhi V. G. (2020). Photocatalytic degradation of pharmaceutical and pesticide compounds (PPCs) using doped TiO_2 nanomaterials: a review. *Water-Energy Nexus*, **3**, 46–61, https://doi.org/10.1016/j.wen.2020.03.008

Weng X., Ma L., Guo M., Su Y., Dharmarajan R. and Chen Z. (2018). Removal of doxorubicin hydrochloride using Fe_3O_4 nanoparticles synthesized by *Euphorbia cochinchinensis* extract. *Chemical Engineering Journal*, **353**, 482–489, https://doi.org/10.1016/j.cej.2018.07.162

Xing Zha S., Zhou Y., Jin X. and Chen Z. (2013). The removal of amoxicillin from wastewater using organobentonite. *Journal of Environmental Management*, **129**, 569–576, https://doi.org/10.1016/j.jenvman.2013.08.032

Xu S., Zhou S., Xing L., Shi P., Shi W., Zhou Q., Pan Y., Song M.-Y. and Li A. (2019). Fate of organic micropollutants and their biological effects in a drinking water source treated by a field-scale constructed wetland. *Science of the Total Environment*, **682**, 756–764, https://doi.org/10.1016/j.scitotenv.2019.05.151

Yoon S. U., Mahanty B. and Kim C. G. (2017). Adsorptive removal of carbamazepine and diatrizoate in iron oxide nanoparticles amended sand column mimicing managed aquifer recharge. *Water*, **9**(4), 250, https://doi.org/10.3390/w9040250

Zhang K., Zhao Y. and Fent K. (2020a). Cardiovascular drugs and lipid regulating agents in surface waters at global scale: occurrence, ecotoxicity and risk assessment. *Science of the Total Environment*, **729**, 138770, https://doi.org/10.1016/j.scitotenv.2020.138770

Zhang X., Yan S., Chen J., Tyagi R. D. and Li J. (2020b). 3 – Physical, chemical, and biological impact (hazard) of hospital wastewater on environment: presence of pharmaceuticals, pathogens, and antibiotic-resistance genes. In: Current Developments in Biotechnology and Bioengineering, R. D. Tyagi, B. Sellamuthu, B. Tiwari, S. Yan, P. Drogui, X. Zhang and A. Pandey (eds), Elsevier, pp. 79–102, https://doi.org/10.1016/B978-0-12-819722-6.00003-1

Zhang Y., Li Y., Xu W., Cui M., Wang M., Chen B., Sun Y., Chen K., Li L., Du Q., Pi X. and Wang Y. (2022). Filtration and adsorption of tetracycline in aqueous solution by copper alginate-carbon nanotubes membrane which has the muscle-skeleton structure. *Chemical Engineering Research and Design*, **183**, 424–438, https://doi.org/10.1016/j.cherd.2022.05.036

Zhao W., Tian Y., Chu X., Cui L., Zhang H., Li M. and Zhao P. (2021). Preparation and characteristics of a magnetic carbon nanotube adsorbent: its efficient adsorption and recoverable performances. *Separation and Purification Technology*, **257**, 117917, https://doi.org/10.1016/j.seppur.2020.117917

Zinicovscaia I. (2016). Conventional methods of wastewater treatment. In: Cyanobacteria for Bioremediation of Wastewaters, I. Zinicovscaia and L. Cepoi (eds), Springer International Publishing, pp. 17–25, Cham, https://doi.org/10.1007/978-3-319-26751-7_3

IWA Publishing's authorised EU representative for General Product Safety Regulations is Diane D'Arras, 15 rue Duret, 75116 Paris, France, e-mail: safety@iwap.co.uk.

Printed and bound by CPI Group (UK) Ltd, Croydon, CR0 4YY

12/05/2026

02108877-0005